Palgrave Stu

Series Editor
Thomas Teo
Department of Psychology
York University
Toronto, ON, Canada

Palgrave Studies in the Theory and History of Psychology publishes scholarly books that draw on critical histories and theoretical concepts and methods, from a variety of approaches in the psychological humanities, to examine the discipline, profession, and practice of psychology.

This series publishes scholarly books that use historical and theoretical methods to critically examine the historical development and contemporary status of psychological concepts, methods, research, theories, and interventions. Books in this series are characterised by one, or a combination of, the following: (a) an emphasis on the concrete particulars of psychologists' scientific and professional practices, together with a critical examination of the assumptions that attend their use; (b) expanding the horizon of the discipline to include more interdisciplinary and transdisciplinary work performed by researchers and practitioners inside and outside of the discipline, increasing the knowledge created by the psychological humanities; (c) "doing justice" to the persons, communities, marginalized and oppressed people, or to academic ideas such as science or objectivity, or to critical concepts such social justice, resistance, agency, power, and democratic research. These examinations are anchored in clear, accessible descriptions of what psychologists do and believe about their activities. All the books in the series share the aim of advancing the scientific and professional practices of psychology and psychologists, even as they offer probing and detailed questioning and critical reconstructions of these practices. The series welcomes proposals for edited and authored works, in the form of full-length or short monographs; contact beth.farrow@palgrave.com for further information.

Gregg Henriques

A New Synthesis for Solving the Problem of Psychology

Addressing the Enlightenment Gap

palgrave
macmillan

Gregg Henriques
James Madison University
Harrisonburg, VA, USA

Palgrave Studies in the Theory and History of Psychology
ISBN 978-3-031-18495-6 ISBN 978-3-031-18493-2 (eBook)
https://doi.org/10.1007/978-3-031-18493-2

This Palgrave Macmillan imprint is published by the registered company Springer Nature Switzerland AG.
The registered company address is: Gewerbestrasse 11, 6330 Cham, Switzerland

Praise for *A New Synthesis for Solving the Problem of Psychology*

"This integrative masterpiece exemplifies the best of cognitive science. It pursues a synoptic integration of many different literatures to generate a new, highly plausible framework for understanding the mind. Henriques is a modern Aristotle whose overarching vision is not only cogently and rigorously argued, but it also affords a conceptual vocabulary and theoretical grammar for the scientifically grounded practice of psychotherapy and the existential understanding of the human condition needed to address the current meaning crisis. This book should be required reading for any undergraduate psychology program, and it should be studied and discussed in depth within graduate programs. The field of psychology needs to be transformed in a way that can address both its internal problems and the external problems of mind and mental health facing the world today. Henriques' book does this masterfully."

—John Vervaeke, Ph.D., Associate Professor of Cognitive Science at the University of Toronto, Canada, and author of *Awakening from the Meaning Crisis*

"We live in an age that requires a new and better vision of what psychology is and can be, which is to say that today we need a new *metapsychology*. Gregg Henriques offers one of the most comprehensive and academically rigorous metapsychological visions available today. Moving across 'Big History,' epistemology, and nearly every subfield of psychology, Henriques offers an impressive metamodern synthesis, and provocative cultural play. If this book has the reception it deserves, the field of psychology will never be the same."

—Dr. Zak Stein, Co-Founder Civilizational Research Institute & Consilience Project, Co-President; Center for World Philosophy and Religion, USA. Author of *Education in a Time Between Worlds: Essays on the Future of Schools, Technology, and Society*

"*A New Synthesis for Solving the Problem of Psychology* is a dazzlingly ambitious attempt to close the gap between lawful objective and contextual interpersonal science; a gap left wanting since the advent of the Enlightenment. Through the promising voice of metamodernism – along with synthetic elegance – Henriques moves us toward a fuller, richer, and wiser psychological worldview."

—Kirk J. Schneider, Ph.D., author of *Existential-Integrative Psychotherapy, The Spirituality of Awe*, and the forthcoming *Life-Enhancing Anxiety: Key to a Sane World*

Contents

List of Figures

List of Tables

CHAPTER 1

Preface: A New Solution to the Problem of Psychology

In reflecting on how to teach undergraduate students about the philosophy of mind, Renee Smith (2016) noted how difficult the concepts are for students to grasp. She explained that the philosophy of mind is "written by philosophers of mind for philosophers of mind" and that the highly technical field is often experienced as having "no obvious practical value" (p. 177). She then proceeded to list several major issues and questions that are at the center of the field, such as the mind-body problem (framed via the question "What is the nature of the mind?"), the hard problem of consciousness ("What is the nature of phenomenal consciousness?"), the problem of mental causation ("What is the causal relation between the mind and the body and between mental states themselves?"), and the problem of the self ("What is the nature of the self?"). Although this book is not situated within the philosophy of mind literature, it nonetheless affords a new way forward on these exact questions. It shifts the context from the convoluted, highly technical philosophy of mind literature into the science of psychology and lays out a new vision and a new vocabulary for talking about behavior, mind, consciousness, and the self.

I am a clinical psychologist, and the present work starts by diagnosing the problems that led to the current situation. The inability to answer the major questions in the philosophy of mind literature stems from a deep and profound gap in our knowledge systems that emerged in the context of the scientific Enlightenment. I call this the *Enlightenment Gap*, and it

G. Henriques, *A New Synthesis for Solving the Problem of Psychology*, Palgrave Studies in the Theory and History of Psychology, https://doi.org/10.1007/978-3-031-18493-2_1

can be seen as a combination of two great philosophical problems. The first is the mind-body problem, which, as Smith notes, is so fraught with complications that one needs to be a graduate student to understand how philosophers talk about these issues.

The second and related problem is how to understand the proper relationship between scientific knowledge and subjectively experienced and socially constructed knowledge. At its core, this problem relates to whether science can reveal objective, transcendent truths about the universe or whether scientific knowledge must be placed in the subjective or social epistemological context of the human knower. The reader may recognize this conundrum as the fundamental dispute between the modernist and postmodernist sensibility. Modernists tend to believe that science does reveal hard-won transcendent truth claims, whereas postmodernists see them as inevitably contextual and social.

This book is written from a new, emerging perspective called metamodernism, which is the sensibility that comes after postmodernism. The intellectual center of a metamodern sensibility can be located via those who seek an effective synthesis between the modernist thesis about science's capacity to generate truth claims that transcend culturally contextualized knowledge (e.g., a claim like the idea that the Periodic Table of the Elements is a fundamentally more accurate representation of the material dimension of complexity than the Greeks' notions of the fundamental elements of earth, air, fire, and water) and the postmodern antithetical claims that scientific knowledge systems are systems of justification built by human knowers in particular contexts inevitably framed by power relations. A metamodern sensibility sees both claims as at least partially true and works to place them in proper relation and effectively synthesize them.

The present work contributes to the metamodern project by outlining a system of understanding that can bridge and resolve the Enlightenment Gap. It posits that one of the most well documented but also most ignored and overlooked problems in the academy is located at the epicenter of the Enlightenment Gap. This is the problem that gives the present work its title. Originally called "the crisis of psychology," the problem of psychology can be identified in the literature as early as 1899. It is most clearly spelled out in 1927 by the great Russian psychologist Lev Vygotsky. The problem refers to the fact that psychology has no clearly identified subject matter. The core nature of the problem can be seen if we return to Smith's concerns about the philosophy of mind. If philosophers cannot define the concept of mind, how can psychologists do scientific research on it?

Psychology is constituted historically by different schools of thought attempting to frame "mind-and-behavior." The behaviorists framed it as the experimental analysis of behavioral responses, the psychoanalysts largely focused on unconscious thought, the structuralists on human conscious experiences, and the functionalists on adaptive patterns and "mental life" in both animals and humans. The crisis that Lev Vygotsky and others pointed out was that these are different entities in the world and there was no way to stitch them together into a coherent whole that was consistent with a natural science view of the universe. Consequently, the schools of thought in psychology proceeded to carve up the domain of the mental in radically different ways, but no one ever figured out how to put Humpty Dumpty back together.

What did emerge in mainstream academic psychology is a kind of unification centered on the methods of science. That is, mainstream psychology is characterized as a science because it applies behavioral scientific methodology to its subject matter. Psychology is most generally defined as the science of behavior and mental processes, and the division between behavior and mental processes is based on the epistemology of science. That is, the enterprise of science is characterized by observing, measuring, and experimenting on behaviors from a third-person epistemological stance. It is this epistemological position that defines "behavior" in mainstream psychology. Consequently, the modern psychological scientist observes behavior and then hypothesizes about possible causes, generally by inferring some "unobservable" set of mental processes, and then proceeding to gather data to see if the findings are supportive of the proposal. The technical term for this approach is "methodological behaviorism," and it is now deeply ingrained in the institution and dominates American psychology. The only real exception is the 5% or so of psychologists who follow Skinner's epistemology and philosophy of behavior (i.e., radical behaviorism).

The book introduces a new vision for scientific psychology called *mental behaviorism*. Mental behaviorism specifies the nature of the mental with metaphysical, ontological, epistemological, and metatheoretical clarity. That is, it operates from a worldview that tells us how to clearly specify the nature of the mind, understand what consciousness is, understand mental causation, and be specific regarding what is meant by the self.

Mental behaviorism is grounded in a metapsychological framework called the Unified Theory of Knowledge (UTOK). The UTOK is a metamodern framework for filling in the Enlightenment Gap and obtaining the proper relations between matter and mind and scientific and social and subjective knowledge in a way that solves the problem of psychology and yields a coherent scientific ontology of the mental.

Central to UTOK is a new, descriptive metaphysical framework called the Tree of Knowledge (ToK) System. The ToK System provides us a new map of "Big History," which is the perspective that traces the evolution of complexification from the Big Bang to the present. In addition to mapping cosmic evolution on the dimensions of time and complexification, the ToK System shows how to divide the natural world into four different planes of existence called Matter, Life, Mind, and Culture. Moreover, it allows us to align those dimensions with the behavior of four kinds of entities (atoms, cells, animals, and persons) and four broad classes of science (physical, biological, psychological, and social). This alignment between the planes of existence and behavior patterns in nature and domains of science affords us a new tool for defining psychology's subject matter. Mind with a capital "M" is the third dimension of behavioral complexification. It can be defined as the set of mental behaviors and can be seen and talked about as clearly and objectively as the living and material worlds. And, according to mental behaviorism, it is the proper subject matter of a scientific psychology.

This coherence in clearly defining the field's subject matter is in marked contrast to mainstream psychology's errant conception of behavior and mental processes defined via the epistemological methods of science. The reason that the mainstream approach is flawed is that behavior and mental processes can mean many different things. For example, behaviors can mean both movements that are observable and the functional activities of animals operating on affordances in the environment. In contrast, UTOK shows us how to organize behavior across the spectrum from quarks to culture (Volk, 2017), and how to understand that science is about mapping behaviors in nature in general. This means that psychology is a science that is interested in a specific class of behaviors, and we can use the ToK System to specify why these behaviors should be characterized by the adjective "mental."

The UTOK system further highlights that both scientists and philosophers use the concept of mental processes to refer to radically different things in the world. Specifically, some use the concept to refer to neurocognitive functional processes, whereas others use it to refer to felt experiences of being in the world, whereas still others use mental to refer to human capacities for self-reflection and reasoning. As this book will show, UTOK maps these interrelated meanings with precision. Not only that, but it also gives us metatheoretical formulations for both the Mind-Animal and Culture-Person planes of existence. Specifically, UTOK's Behavioral Investment Theory is a metatheory of the Mind plane. It enables us to see how the nervous system evolved as an information processing and investment value system that coordinates animal action toward paths of investment based on energy expenditure, cost, and risk framed by broad principles pertaining to evolution, development, learning, and computational control. This framework integrates insights from bioenergetics, neurobiology, ethology, and sociobiology, and the behavioral, cognitive, and developmental systems perspectives.

Via Justification Systems Theory (JUST), UTOK also specifies in rich detail the transition our ancestors made from being primates to becoming persons who operate on the Culture-Person plane of existence. JUST posits that the emergence of symbolic syntactical language was a major shift in the development of complexification because propositions gave rise to the problem of justification. In particular, the emergence of propositional claims quickly gave rise to the questioning of such claims, forcing the dynamic of justification, and ultimately leading to the development of systems of justification that coordinate and legitimize action.

There is a reason the philosophy of mind is a convoluted field that has no obvious utility. The emergence of physics resulted in an Enlightenment Gap, and the philosophy of mind tradition fell into that gap and has been groping around in the dark ever since. The Enlightenment Gap also engulfed scientific psychology, resulting in its long-standing crisis of identity that it has coped with via a combination of denial and rationalization that it can be defined via the methods of science. The UTOK is a game changer when it comes to both the philosophy of mind and the science of psychology. It not only specifies the ontological domains, but it also clarifies the epistemological differences between the objective behavioral and subjective phenomenological vantage points and provides metatheories that knit the dynamic picture together in a clear and consilient way. By following this lighted path, we can clarify the confusions that have entangled both mainstream empirical psychology and the philosophy of mind such that we can transcend the Enlightenment Gap and move toward a second enlightenment in the twenty-first century.

References

Smith, R. (2016). How to teach philosophy of mind. *Teaching Philosophy, 39*(2), 177–207. https://doi.org/10.5840/teachphil201651649

Volk, T. (2017). *Quarks to culture: How we came to be*. Columbia University Press.

Vygotsky, L. (1927/1987). *The historical meaning of the crisis in psychology: A methodological investigation*. The collected works of Lev Vygotsky. Plenum Press.

PART I

The Problem of Psychology

CHAPTER 2

Psychology, We Have a Problem

Psychology is a fascinating field that has much to offer, but it has always had a fundamental problem. Despite its attempts to be a science for approximately 150 years, psychologists still lack agreement about the meaning of its most basic concepts. There is no agreement on what terms like behavior, mind, cognition, and consciousness mean, not to mention the word psychology itself. For a contrast, consider chemistry. Chemistry is about atoms and how they link up to form molecules, and the kind of energy patterns these interactions produce. The Periodic Table of the Elements and the atomic theory of matter provide chemists with a shared language from which to operate. Sociologists have long noted that a core of consensual understanding is what constitutes the essence of scientific knowledge, and no one questions whether chemistry is a science. In contrast, psychology lacks anything even remotely like a Periodic Table of the Elements. There is no shared foundational set of definitions and concepts that experts agree define the essence of the field or its subject matter. Given this, it is hardly surprising that psychology's status as a "true science" remains a point of contention.

Part of the confusion is perhaps located in the fact that the root of the word psychology is the term "psyche," which historically corresponded to the word soul and carries associations for many with the supernatural realm of existence. Yet even when modern scientific terms like "behavior" or "the mind" are used as the primary referents for what psychology is

G. Henriques, *A New Synthesis for Solving the Problem of Psychology*, Palgrave Studies in the Theory and History of Psychology, https://doi.org/10.1007/978-3-031-18493-2_2

about, we do not achieve consensual clarity. The fact is that even after over 150 years there is no shared lexicon and no clear agreement about what it is that the science of psychology references in the world. Instead, there is a massive, fragmented plurality of views and a multitude of different language games that compete for attention and validity. We can thus consider the field of psychology to both be "multi-paradigmatic" in that it consists of many different models, schemes, and schools of thought, and "pre-paradigmatic" in that it has never had a shared meta-paradigm that was up to the task of effectively defining the science in its entirety.

Scholars in the field have long been aware of this confusion. The Russian psychologist Lev Vygotsky identified this as "the crisis" of psychology in 1927. He lamented that theorists, researchers, practitioners, and even lay persons had pointed out that the field seemed to refer to fundamentally different things in the world, depending on the school of thought (Vygotsky, 1927/1987). For example, behavioral psychologists like John Watson framed the field as being about behavioral responses analyzed in the context of experiments. Other psychologists followed Freud's vision and were concerned with unconscious mental forces and the role they played in neurotic struggles. And yet, the father of psychology, Wilhelm Wundt, emphasized that the primary subject matter was subjective conscious experience accessed by introspection. Still others, like the father of American psychology William James, emphasized the functional adaptive properties of mental life in both animals and humans.

The domains of behaviors, unconscious mental forces, subjective conscious experiences, and functional adjustment and adaptation refer to fundamentally different things in the world, and none of them effectively captures the whole of what is meant by the terms "mind and behavior." In 1927, Vygotsky described the situation as follows:

> Lately more and more voices are heard proclaiming that the problem of general psychology is a problem of the first order. What is most remarkable is that this opinion does not come from philosophers who have made generalisation their professional habit, nor even from theoretical psychologists, but from the psychological practitioners who elaborate the special areas of applied psychology: psychiatrists and industrial psychologists; the representatives of the most exact and concrete part of our science. The various psychological disciplines have obviously reached a turning point in the development of their investigations, the gathering of factual material, the systematisation of knowledge, and the statement of basic positions and laws. Further advance along a straight line, the simple continuation of the same work, the gradual accumulation of material, are proving fruitless or even impossible. In order to go further we must choose a path....

> To take the mind, the unconscious, or behaviour as the primary concept implies not only to gather three different categories of facts, but also to offer three different ways of explaining these facts...To what extent are the psychological facts elicited and known at the moment, and what changes in the structure of the science do they require in order to make possible the further acquisition of knowledge on the basis of what is already known? ... The history of the science is important for us insofar as it determines the degree to which psychological facts are cognised.

The state of confusion that Vygotsky was remarking upon almost 100 years ago has remained. I have rechristened Vygotsky's "crisis" as *the problem of psychology* (Henriques, 2008). Because it is the normative state of the field, it is no longer appropriate to consider it a crisis. In addition, as will become clearer as my argument advances, the problem framing gives more precision to exactly what is the difficulty.

The crisis or problem is never solved, but instead the field habituated to it and proceeded to gloss it over as a problem that could not be solved, and thus could be safely ignored. It has long been my contention that the field's denial has been misguided. The problem of psychology should be known to every scholar and practitioner in the field, and it should be broadly understood as one of the great problems in science. That is, it should be analogous to the problem of "quantum gravity" in physics. Everyone who is familiar with physics knows that the classic Newtonian "matter in motion" paradigm was overturned in the beginning of the twentieth century by the "twin pillars" of quantum mechanics and general relativity. Yet these two theories offer different pictures for how the universe behaves at its most fundamental levels. The inconsistency results in one of the greatest problems in all of science, which can be stated as follows: *Can quantum mechanics and general relativity be coherently merged into a theory of quantum gravity?* The solution to this question frames the much-vaunted "theory of everything" in physics, and it has so captured the popular imagination that hundreds of books, courses, and even movies have been made about it.

Much like the problem of quantum gravity, the problem of psychology can be stated simply: *Why has it been impossible to develop consensus among the experts on how we define the science of psychology and its subject matter?* Another way of stating this problem is that the physical and biological sciences have a shared subject matter and consensually agreed upon paradigmatic theories that frame the science. Specifically, quantum mechanics and general relativity frame physics, which is defined as the science of the behavior of matter and energy across scales. Evolutionary theory, genetics, and cell theory provide the foundational frames for biology, which is consensually defined as the science of the domain of life or the behavior of organisms.

In contrast, psychologists have foundational disputes about psychology's subject matter, its organizing theories, and even its institutional identity. There are core disagreements about whether psychology is: (a) fundamentally about behavior or about mental operations or about subjective conscious experiences; (b) concerned with animals in general, with humans being a specific case or primarily focused on humans alone; (c) predominantly classified as a natural science or should be thought of as a social science endeavor; or (d) primarily a basic science concerned with describing and explaining behavior and mental processes or whether it is an applied health service profession concerned with enhancing human well-being. As with quantum gravity, the problem of psychology is well known, at least to theoretical psychologists. Indeed, if there is one thing upon which theoretical psychologists agree it is that there is no consensus about how to define what psychology references in the world. However, in stark contrast to quantum gravity, the problem of psychology is not generally well known by the public. Indeed, even many researchers and practitioners in the field are only vaguely aware of it. The academic institution has largely sidestepped the issue and proceeds to advertise to students that the field is unified by its commitment to applying the methods of science. However, as will become clear, this is a wholly inadequate solution to the problem.

What Do We Mean by Mind?

To deepen our understanding of the problem of psychology, let us consider the term *mind*. To what, exactly, does this term refer? Scholars do not agree. And it is not that they quibble over minor differences, debating what might be the edges of the concept, akin to when biologists debate the edges of what is considered alive and argue about how to classify things like viruses. In contrast, psychologists disagree about the essence of what the term mind refers to in the world. For example, one common meaning of mind refers to the higher domains of human thought, such as self-conscious reflection and reasoning. This meaning most clearly has its roots in René Descartes, who thought of the mind as consisting of the stuff of reason and the province of humans alone. A second definition of mind is much broader and more inclusive. It is more recent and grows out of psychology and the cognitive and neuroscience revolutions in the middle and latter parts of twentieth century. It refers to the processing of information done by the nervous system, which plays a key role in coordinating the functional behavior of animals in situational contexts. We can call this the "neurocognitive functional" view of the mind or mental processes.

In *The Conscious Mind,* Zoltan Torey (2014) complains about how vague the term "mind" can be. He insists that there is "no justification for the confusion" (p. 81) and asserts that it refers to human self-consciousness and language-based reasoning. He then launches into a critique of the neurocognitive meaning as follows:

> Nor is brain science innocent of misusing the term [mind]. Take, for example, David Oakley's (1985) claim that 'the emergence of neural modeling corresponds to the emergence of mind'—a wild generalization, for, if all neuronal representations, from simple sensory alertness to our reflective consciousness, are regarded as instances of mind, the term loses all specificity and is rendered useless...The mind is an exclusively human neural system, first instantiated when, empowered by the motor-wiring of the speech-areas, the brain gained access to itself...This is clearly a good deal less than the total range of our brain's functioning. Therefore, to equate the total range of our brain's functioning with the mind, that is, a system of the brain that only has limited access to it, is quite wrong. (Torey, 2014, p. 82)

As will be made clear as this book progresses, I think it is Torey who is wrong, and that we should embrace the neurocognitive functional definition and consider self-consciousness to be a specific kind of mental process. But that is not the point I am making here. *Rather the point is that neurocognitive processes and self-conscious reasoning are two very different referents in the world.* If one scientist uses the concept of mind to refer to the former and another uses it to refer to the latter, then we have completely incommensurate language systems. This situation is akin to chemists having multiple meanings for the word "molecule," with some using it to refer to atoms and others using it to refer to proteins.

There is yet another common referent for the domain of mind that is used by both scientists and lay people alike. For many, the mind primarily corresponds to the domain of subjective conscious experiences, such as our felt experience of perceptions, emotions, or imaginal wonderings.

Such mental processes are often referred to as inner experience, qualia, sensory consciousness (as opposed to self-consciousness), or phenomenology. Given that the domain of subjective conscious experience represents yet another primary referent for the mind, we now have *three* different possible meanings of what mind or mental processes might refer to in the world (i.e., neurocognition, subjective conscious experience, and self-conscious reasoning). It is this confusion about core terms and the absence of a shared vocabulary that is at the center of the problem of psychology.

The confusion worsens when we consider that psychology is not only about the mind or set of mental processes, but is also about "behavior." Indeed, there is a long tradition of behavioral psychology that denies the claim that the mind or mental processes are viable scientific constructs. The philosophical division here is the long-standing debate in psychology between the mentalists and the behaviorists. To understand this division, take a moment and reflect on your understanding of the mind and its relationship to behavior. When you think of the mind, do you think of something that is inside of people's heads that (somehow) causes or at least influences people to act the way that they do? Put differently, when you see someone act, do you think they are acting that way at least partly because of their minds? If so, then you are a mentalist or a cognitivist. For a cognitivist, the mind is separate from overt action and functions to play a causal role in observable behaviors. Cognitive psychology traditionally takes this perspective. It is the science of figuring out how the mind works by studying overt behavior and inferring and modeling how such actions are caused by underlying mental processes.

A behaviorist comes at these issues from a completely different perspective. For the behaviorist, "the mind" is not really a thing or force or entity. It is most definitely *not* some separate, hidden entity that *causes* behavior. To the extent that the mind is anything at all, it is a kind of behavior. As you read this book, you are both exhibiting overt actions and having covert or internal experiences. Both are "behaviors" to the behaviorist. And the behaviorist is interested in changes in the environment that cause changes in the frequency of emitted behaviors, both overt and covert. From this perspective, to say behaviors are caused by the mind is about as meaningful as saying they were caused by God or some supernatural force.

Although it is a minority perspective, there remains a vocal group of radical behaviorists who argue that the key to solving the problem of psychology is a firm rejection of any kind of mentalism. A few years ago in the

APA Monitor, Ed Wasserman (2018) passionately issued yet another call for psychologists to adopt this view:

> Will psychology ever 'lose its mind'? From its inception, psychology has concerned itself with private experience, particularly with the ever-alluring yet obscure notion of consciousness. But natural science must study observables, whether they are the overt behaviors of organisms or the biological activities within organisms. We must at long last 'lose our mind' and embrace the same basic paradigm that has proven so effective in physics, chemistry and biology. Only then can we escape from the obscurantism of mentalism and develop a truly scientific psychology. The tools of behavior analysis and neuroscience have matured to the point where this aim is attainable. Let's get going!

As this quotation makes clear, the hallmark of the behavioral tradition in psychology is to ground the field in the traditional "matter in motion" model of the universe that guides physics. It is because of this view of reality that many behaviorists reject mind as an unworkable concept. As such, traditional behaviorists are "anti-mentalistic" in that they are explicitly defined against the idea that there is a conscious mind that is separate from behavior. This, of course, is the opposite of cognitivism.

Learning to See the Problem of Psychology

When you learn how to look for the confusion in psychology, you can see it everywhere. Consider the cognitive behavioral therapy (CBT) paradigm, which represents a main class of psychotherapy that blends cognitive and behavioral approaches to intervention. From a practice standpoint, this blending makes good sense. I am well versed in CBT, and I use the tools and principles regularly in my therapeutic work supervising doctoral students. However, a second look informed by a theoretical and philosophical lens reveals a deep *conceptual* problem with CBT. This is because cognitivism and traditional forms of behaviorism represent two fundamentally different paradigms for thinking about mind and behavior. Thus, at its philosophical foundations, CBT can be considered a "mentalistic anti-mentalistic" approach to psychological treatment. This oxymoron is just one example of the problem of psychology lying in plain sight.

Another way to see the problem is to ask the question: *What, exactly, is psychology?* In doing so, we can follow Richards (2002) and make a distinction between psychology and Psychology. The former refers to the field's subject matter (i.e., the entities in the world we are interested in understanding, such as behavior, mind, or consciousness) and the latter, which I will capitalize as Richards does, refers to the institution. Thus, psychology is about things like habits, perceptions, memory processes, and learning, whereas Psychology is a university major and a field of study that confers degrees and regulates professional practice. On the surface, there are elements of consensus about both psychology and Psychology. For example, almost all in Psychology agree that psychology is a field of study that attempts to describe and explain why people behave the way they do via the methods of science, and, via psychotherapy and other applications, it offers ideas for how to help people who are suffering with psychological problems.

Given that we are concerned with analyzing both Psychology and psychology, it is useful to be clear about the intellectual space within which this book is located. This book is written from a "metapsychological" vantage point (Henriques, 2019). A meta-view is one that shifts the focus from being inside the stream of thought or activity to a zoomed-out position where the activity becomes the object of analysis. Framed this way, metapsychology can be thought of as the space between psychology and philosophy. As will become clear, this book adopts a meta-view on natural science in general, and on psychology specifically. Thus, rather than being "inside" of psychology or even inside modern empirical natural science writ large, the perspective we are operating from is a philosophical stance that takes both of these as entities we are considering. Science, psychology, and Psychology all are objects of analysis in the present work.

When we take a zoomed-out view of the field, we can note that there is some consensus regarding what we might call the "vertical placement" of psychology's subject matter relative to that of biology and the social sciences. This is found in the "biopsychosocial" model, which is prominent and frequently referenced in the field. Indeed, the American Psychological Association (APA) structures the knowledge requirements for accredited programs to train doctoral students in the biological, psychological, and social bases of behavior. This framing suggests that the subject matter of psychology can be vertically located somewhere "above" biology and "below" sociology. Even more concretely, there is a readily identifiable institution of Psychology and college degrees that appear to give the discipline its structure and identity. From this vantage point, we can say that

Psychology is an institutional structure devoted to whatever it is that psychologists study.

Although this captures the current situation, it is not intellectually satisfying. The problem of psychology becomes manifest when one talks with different psychologists who have different emphases and orientations and reside in different enclaves within the institution. One of the best ways to see this is to attend the APA's annual conference and visit with the more than 50 divisions. Vastly different conceptions of both psychology's subject matter and its institutional identity emerge when one listens to the behavior analysts (Division 25) compared to the psychoanalysts (Division 39). The humanistic psychologists (Division 32) see psychology in a remarkably different light than those who are trained in the domain of experimental psychology and cognitive science (Division 3). Behavioral neuroscientists (Division 6) have a completely different vocabulary than those who work in Division 44 (Society for the study of LGBT issues and concerns). The remarkably different divisions represent the varied interests, subject matters, paradigms, and lenses that have been taken in American psychology. The incredible diversity only grows when we consider other approaches outside the Western tradition, such as Eastern spiritual traditions (e.g., the yogic sciences) or the multitude of indigenous perspectives.

The Problem of Psychology and the Fragmentation in Psychotherapy

I began to realize there was a serious problem in the field when I was a graduate student in the early 1990s taking my second class in psychotherapy. The course was taught from the vantage point of the psychotherapy integration movement, which turns out to be a wonderful way to get a grip on the problem of psychology. As seasoned practitioners know, the field of psychotherapy is a jungle of different approaches, anchored to different views of the human condition and different perspectives on how to best help folks who are suffering with mental health and relationship problems. Consider, for example, that Lambert (2013) estimated that there were over 400 different name brand psychotherapies, with each approach offering a different framework for helping patients with psychological distress.

The psychotherapy integration movement is explicitly concerned with understanding this diversity and deliberately reflecting on what it means and what might be done to generate more integrative approaches. My professor was both a serious academic and a gifted clinician who deeply appreciated the key insights that each of the major models of psychotherapy offered. Unlike highly partisan individuals who proclaim that their model is the one true model, my professor recognized the depth and complexity of each of the major approaches and taught us how to respect what they had to offer. This instilled in me a desire to incorporate the best of the best from each paradigm into how I approached my therapeutic work.

Learning about the psychotherapy integration movement, I began to shift my frame from assuming what might be called a "horse race" mentality (i.e., Which is the best school of thought to bet on and invest in as a scientist or practitioner?) to embarking on a quest to "find the elephant." That is, I began to sense that the single schools' relationship to the truth might be analogous to the six blind men who happen upon an elephant in the famous parable by John Godfrey Saxe. In the parable, each man grabs a hold of a piece of the elephant and makes strong proclamations about its true nature. One, holding onto its trunk, claims it is like a snake; a second pats down its leg and proclaims it is like a tree trunk; a third feeling its tail proclaims it is like a rope; a fourth touching the point of the tusk says it is like a spear; a fifth grabbing its ear says it is like a fan; and the sixth pushing up against its side says it is like a wall. And, in justifying the validity of his perspective, each man dismisses the others as being blind to the truth.

I came to believe there should be a way to integrate the key insights from the various approaches and see the whole elephant. In delving into psychotherapy's problems with competing paradigms, I began to shift my focus on integration from the field of psychotherapy to the science of psychology. The reason for the shift was that it made sense to me that the various psychotherapies should all be conceptually grounded in the science of human psychology. The field of medicine offers a clear analogy for how I was thinking about the problem. Modern Western medicine is reasonably well unified, at least compared to the field of psychotherapy. That is, the ideas of mainstream medicine are not fundamentally fragmented the way they are in psychotherapy. (We can acknowledge that the issues become complicated when considered in the context of Eastern or "alternative" medicine traditions, but we can sidestep that issue for now.) Modern medicine rests on the science of biology, especially the branches of physiology, anatomy, and pathophysiology. Physiology is the science of cells, organs, and organ systems; anatomy is the science of the structure of organisms; and pathophysiology is how the systems malfunction and break down. By taking a step back and thinking about the organization of modern medicine, we can see a basic harmony and correspondence between the domains of physiology and anatomy and the medical specialties.

A real-life example can help clarify. In 2014 my son Jon had an unusual dysfunction in his hip. He played soccer all the time, and one time he kicked the ball awkwardly and felt an unusual pull and snap in his hip. He rested for a week or so and then was back playing again. But the event happened again a couple of months later. And then it happened again. We took him to see his regular doctor, who conducted an examination, did an X-ray, and determined that he had fractured a bone in his hip. (Technically, he had an avulsion fracture.) Upon that diagnosis, we were referred to a pediatric orthopedic specialist. An orthopedist is an individual who specializes in disorders of the muscular-skeletal structure. She recommended a path of rest and physical therapy for rehabilitation. Unfortunately, this did not fix his problem and every time he went out and played hard, he would snap off the piece of bone again. Finally, we had to go to yet another specialist, this time an orthopedic surgeon, who developed a plan for operating on his hip by placing a screw into the bone to reinforce the weak area. Thankfully, it turned out to be a successful strategy. Indeed, because it was such an unusual problem and solution, the treatment was written up and added to the scientific knowledge of the medical community (Carr et al., 2017).

The point here is that orthopedics is a subspecialty of medicine that is organized and focused on a specific organ system (i.e., the muscular-skeletal system), which is a clear domain within human physiology and anatomy. If Jon had a problem with his heart, he would have seen a cardiologist; if it had been with his bladder, he would have seen a urologist; if had been with his eyes, he would have seen an ophthalmologist, and so on. All of these are different specialties that connect to a picture of human physiology and anatomy, which are well placed in the larger science of human biology.

Now imagine if, instead of being connected to that larger picture, each subspecialty claimed that it had *the* key insights for biomedical health and functioning *overall*. Imagine also that it had different language systems, different training philosophies, and the leading experts were proponents of one system who were politically and culturally defined against experts of the other systems. In such a world, we can imagine that as they complete their training, physicians going into their residency would write essays justifying why they were committed to believing that their organ system was the key for promoting human health. This, after all, is what modern-day doctoral students of professional psychology must do when applying for their doctoral internships. That is, they must write an essay explaining their specific theoretical orientation in psychotherapy in relationship to the various schools of thought (e.g., why they adopt CBT versus a client-centered humanistic approach). In such a thought experiment, we could envision exchanges between a cardiologist, an orthopedist, and an endocrinologist unfolding as follows:

> 'The key to health is a well-functioning heart and circulatory system,' proclaims the cardiologist.
>
> 'You are wrong,' the orthopedist says. 'The key to health is a strong body, with strong muscles and good bone density.'
>
> 'You are both way off base,' says the exasperated endocrinologist. 'Hormones are where we need to be focusing.'

This conversation sounds silly because we live in an era in which modern medicine is anchored to a relatively clear and unified understanding of biology, physiology, and anatomy. We recognize that the medical specialists are focused on organ systems that are part of a larger whole that is networked together as a system. With that recognition, we can easily see why the idea that all our biological health could be reduced to just one organ system is nonsensical. No one would suggest that there should be a horse race between cardiologists and endocrinologists in terms of who offers the better pathway to health *in general.* Instead, the biological map of the physiology and anatomy of the whole organism makes clear the subsystems that go into it. It is from this shared understanding that we have generalist physicians who can diagnose and treat common ailments, and specialists who treat more complicated, specific disorders of organ systems, such as the pediatric orthopedists that treated my son's hip.

The analogy with medicine points to a potential solution for the problem of psychotherapy's fragmentation and the competition between the warring paradigms: the field of psychotherapy should be, first and foremost, conceptually anchored to the science of human psychology. This was why I turned my attention from psychotherapy to the science of psychology. The idea was that if there was a way to unify the science of psychology, then we could move toward the effective integration of psychotherapy. I was also incentivized by the fact that as my career progressed, I saw that there were obvious problematic consequences associated with the field of psychotherapy being organized by competing and incomplete paradigms. In making this analogy, I do not want to imply that professional psychologists should wear white coats and view psychological therapies as being directly akin to medical treatments. Indeed, that does not reflect my approach to psychotherapy at all. Rather, my point is that there should be a more unified understanding and language system that organizes psychology, and that would function to better align the concepts, theory, research, and practice done by therapists.

The Problem of Psychology Has Real-World, Practical Consequences

To understand the practical limitations associated with the conventional, single-school approach in mainstream psychotherapy, I will share some observations stemming from my work in Aaron T. Beck's cognitive psychotherapy lab. I joined Beck at the University of Pennsylvania's Psychopathology Research Unit in 1999 as a postdoctoral fellow, and shortly afterward I was formally assigned the role of Project Director for a multi-million-dollar study on the impact of providing a brief cognitive therapy intervention for individuals who had recently made a suicide attempt. The individuals we treated in this study had an enormous number of serious life problems. Consider that the modal number of psychiatric diagnoses each participant was given was three. Two-thirds had a serious problem with substance abuse, the average scores on a depression inventory were in the severe range, most were unemployed, a quarter were homeless, histories of abuse were the norm, and over 70% made less than 15,000 dollars a year. And to get entry into the study, they had to have recently made a genuine suicide attempt.

Ultimately, the study was a success in that we conducted a sophisticated randomized controlled trial that demonstrated that individuals who were assigned to the treatment condition faired notably better on several important outcome measures than those who received a slightly enhanced form of treatment-as-usual. This finding was published in the *Journal of the American Medical Association* (Brown et al., 2005). On its surface this appears to be an exemplar of an empirically based, clinical psychological science advancing our knowledge. However, as I detail in Henriques

(2011, p. 200–205), the back story reveals a significantly different picture than the one described in the professional write-up. The original study design failed. Our patients would not show up for regular clinic appointments, and the therapy was almost impossible to deliver in the standard format. We also lost track of them frequently. The problems were so great that after a year of work and getting half of our study participants enrolled, it was determined that the study was failing, and we needed a whole new approach. This resulted in what we would call the shift from "Study 1" into "Study 2."

Importantly, virtually all the changes we made to Study 2 were outside the scope of cognitive therapy. We changed the protocol for tracking participants and assigned an undergraduate research assistant to each participant to maintain contact to facilitate follow-up. We also changed the way we paid subjects, changed the way we interviewed them, and especially ensured that we maintained contact as they were transitioned out of the hospital. Perhaps most importantly, we changed the basic structure by which the therapy was provided and who provided it. Instead of referring patients to the Center for Cognitive Therapy, the Psychopathology Research Unit became responsible for delivering the therapy. I personally provided therapy to many of the patients and directly supervised many others. The reason for this was that I had much more flexibility in my schedule and could assume much more responsibility for getting the participants to their sessions. I could drive them to and from the sessions, or even conduct therapy at their home when convenient.

There was quite a dramatic shift in the results from Study 1 to Study 2. In the second study, we successfully tracked most of the patients, and 75% of those in the cognitive therapy condition completed four or more sessions, as opposed to only 33% during the first phase. And, as mentioned, the intervention was found to have a significant and substantial impact on patient functioning. And here comes the interesting point. We had good evidence that what we were doing in the first phase of the study had no notable impact, but what we did in the second phase did. Crucially, the cognitive therapy intervention was constant across both phases. What had changed was how it was delivered and who had provided it. In short, we had strong data suggesting that the novel developments we implemented in delivering the intervention were central to its efficacy and effectiveness.

From a purely empirical perspective, this is an important finding. But we were not operating from a purely empirical perspective, because, as we will see, there is no such thing. Rather, data are always interpreted in relationship to some model, schema, or theory, which in turn is embedded in a larger paradigm or shared understanding of the way the world works. "Facts are theory laden" is a common phrase that captures this point. The paradigm we were embedded in was the Beckian cognitive model. Because of this framing, the real goal of the research was to explore and advance the *cognitive psychotherapy brand* (see, e.g., Robert Woolfolk's (2015) critique of research in psychotherapy as being brand promotion). The changes made from Study 1 to Study 2 were consequently interpreted not as key aspects of the intervention, but as necessary "tweaks" that allowed us to deliver it to this population.

Now consider if, instead of the Beckian paradigm, our focus had been on social work and the structure of treatment delivery for underserved and marginalized populations. If this had been the case, then everything we did would have been seen in a different light. The title of that study might well have been "Methods for Effectively Engaging Patients Who Have Attempted Suicide." This point highlights the fact that empirical research is framed by the knowledge, language, motives, and general paradigm that it is grounded in. This underscores something well known by philosophers of science, which is that scientific research is about *both* empirical findings and the conceptual frameworks in which they are embedded. This means that if the conceptual frame is not up to the task, then the value of the empirical data will be greatly limited. This fact about the relationship between empirical findings and conceptual framework drives a central impetus for the present work. At its core, the Unified Theory is about developing the broadest, most coherent, and most practical frame possible that effectively organizes the field's key concepts (such as behavior, mind, cognition, and consciousness) and assimilates and integrates the key insights and interventions from the major paradigms, and the reliable and valid findings from empirical research.

The Problem of Psychology and Its Connection to Culture and Human Values

Part of the reason psychology and psychotherapy lack a coherent conceptual frame is that the field is vast, and the conceptual issues are deep and knotty. The landscape covered by psychology's subject matter potentially ranges from neuronal networks and animal reflexes to human consciousness and even human culture and the dramatic effect socialization has on shaping the human mind (e.g., consider cultural psychology and indigenous psychological perspectives). In addition, the applied professional side deals directly with issues of human suffering, human functioning, well-being, and fostering human betterment. As such, it connects deeply with core human values. The fact that the field deals directly with humans, cultural beliefs about reality, and human values makes the science especially complicated.

To understand this point, we will continue examining the suicide attempter project, only this time from the vantage point of a social justice lens. The central axiom of a Beckian cognitive approach is that psychopathology stems not from situations in and of themselves, but rather how people interpret and explain the events around them. That is, people's mental health problems with depression, anxiety, or anger stem from rigid or extreme or otherwise maladaptive beliefs that emphasize loss, defeat, threat, or hostility in others. Although I believe this perspective has much value, it is nevertheless clear that this is only one piece of the puzzle of human psychopathology. The analysis above showed how the model fails to address many of the basic structural treatment delivery issues that would line up with the perspective of social work. We can also show that it fails to consider some foundational human values, especially when placed in social contexts that involve substantial inequity in the social structure.

I am a straight, cis gendered white male who was born into an intact, upper middle-class family. Both of my parents have their doctoral degrees, with my mother's degree being in early childhood education. The study participants were largely minorities (mostly Black) who had limited educational backgrounds, many had histories of sexual and physical abuse, and many lived at or below the poverty line. Now consider the basic cognitive psychotherapy maxim that suffering comes not from the world per se, but from how one interprets the world. This quickly leads to a potentially embarrassing situation. Consider the setup for the start of therapy. Here I am, a white man from a privileged background working with a suffering person from a disadvantaged background from a frame that starts with the assertion that their suffering does not really stem from the world per se, but rather stems from how they are seeing the world. It does not take much education from a social justice perspective to see this frame as being seriously problematic. Of course, a good cognitive therapist would not deny issues of privilege or power or social justice. And yet, despite this caveat, it is nevertheless the case that the Beckian system lacks a clear framework for attending to it.

I am not singling out cognitive therapy because I think it is a weak paradigm. Quite the contrary, as I have always seen it as representing an important set of insights that are on par with those from psychodynamic, humanistic, behavioral, and interpersonal or family systems approaches. Indeed, it has developed as one of the most important frameworks in the field of psychotherapy precisely because it offers many advantages. My view is that all the major approaches have both significant value and serious blind spots and limitations. Although I just offered an example of how the cognitive approach can be seen to be limited from a sociocultural perspective emphasizing issues of social justice, we can easily flip this around.

Over the past decade, many social justice perspectives have become too narrowly focused on vulnerability, identity politics, and excessive concern about microaggressions and trigger warnings, such that the meta-messages can result in endorsing a problematic culture of victimhood. The social psychologist Jonathan Haidt has offered powerful analyses and critiques that suggest that a pervasive philosophy of "safteyism" is likely contributing to the unhealthy levels of anxiety, reactivity, and grievance in adolescents and young adults (see, e.g., Lukianoff & Haidt, 2019). He cogently argues that to counteract this trend, we need to embrace principles of Stoicism. How does he recommend we go about this? By learning Beckian cognitive principles, which can be very helpful in toning down hypersensitivity, as it down-regulates neurotic, reactive thinking and the excessive perception of threats that results in histrionic reactivity or a disempowered, defeatist attitude.

The fundamental point here is that *every* mainstream paradigm in the field (i.e., psychodynamic, behavioral, cognitive, evolutionary, sociocultural) can be readily shown to be not up to the task of solving the problem of psychology. This critique includes empirically based research programs that appear to be a success on the surface. For example, consider Attention Deficit/Hyperactivity Disorder (ADHD). Keith Conners was a psychologist and an early advocate for the identification and treatment of ADHD. When he started his investigations in the 1970s, he was looking at a relatively small subset of children, perhaps 1–2% of the population at most, who had unique challenges and temperaments that suggested some potential irregularities in their neuro-physiological makeup. His research was instrumental in raising awareness about the condition. Given all the attention that ADHD now receives, one might think that in the twilight of his distinguished career Connors would have felt triumphant. Not so. He was deeply disturbed by the enormous increase in numbers of diagnoses (now upwards of 15% of high school children) since the 1980s. He called it "a national disaster of dangerous proportions" (Schwarz, 2013).

The reason Conners was so concerned was that he did not think the last 30 years had seen an epidemic rise in ADHD rates, a conclusion he called "preposterous." Rather, he believed the dramatic rise was the result of a "concoction to *justify* the giving out of medication at unprecedented and unjustifiable levels." Justify is in italics because it is a central concept in the current approach, and it is highlighted to forecast this fact. What had happened was that after the label of ADHD emerged, it was found that some stimulants could help control kids who were simply more active and distractible than average. This set up the situation whereby these drugs were sold as "medicine" by the psychopharmacology industry to parents and teachers who wanted their kids to conform to the highly novel and artificial modern educational setting, all to the tune of billions of dollars.

The point here is that the concept of ADHD that Conners investigated in the lab transformed significantly as it moved into the broader culture. Tracing how that happened provides a powerful example of the complicated interplay between scientific constructs that presumably describe human behaviors, interventions designed to manage those behaviors, and how those constructs and practices change how humans think about how they ought to be and the institutions and policies that are then instantiated on this set of beliefs. An important aspect of this pattern is called *concept creep*, where the threshold of a condition gets lower and lower, with enormous implications for how the society thinks about the issue. As Haslam (2016) details it, there are many examples of concept creep. The modern definitions of child abuse, depression, anxiety, ADHD, bullying, and many other problematic behaviors and syndromes have expanded, such that there is a much lower threshold now than when these concepts were originally discussed in the literature and in the policies that regulate them.

The general point is that psychology and Psychology are tangled up in immensely complicated dynamics. Consider the issues we have raised in the chapter, including: (1) the different definitions associated with mind and behavior; (2) the vast potential scope of psychology from neurons to cultures; (3) disputes about the core identity of the discipline; (4) the many different paradigms with different language systems and assumptions; (5) the nature of the applied work of psychotherapy; (6) the complicated feedback loop between how psychological knowledge is produced and how it proceeds to impact the culture; and (7) its close ties to core human values. This list makes it easy to see why the field has lacked a coherent framework for understanding. Looked at this way, it is little wonder that the elephant is hard to see and raises questions as to whether it even exists. And yet, despite the magnitude of the issues involved, this book argues that there is another way. There is an elephant that warrants the label psychology that can be seen through the fog of confusion when one puts on the right metapsychological lens to perceive it.

Conclusion

One point of consensus among theoretical and philosophical psychologists is that there is no consensus about what the term psychology refers to in the world. In a related vein, psychologists lack clear definitions of the field's key concepts, like behavior, mind, and consciousness. Mind, for example, can refer to neurocognitive processes that regulate overt activity or subject conscious experiences or human self-conscious reflections and justifications for acting and being. It can also refer to the behavior of the animal as a whole. This problem of agreement about psychology and its subject matter becomes obvious when we look out across the landscape of the field and consider the multitude of different approaches to psychotherapy and how the paradigms are focused on particular slices of mental experience and adaptation, rather than the whole of the human condition. The field is both multi- and pre-paradigmatic and none of the specific schools of thought are up to the task of defining the field's core subject matter as the science of behavior and mental processes.

Rather than tackling the problem head on, Psychology's solution has been to define the field around the methods of modern scientific inquiry. As such, most scientific psychologists see their task as empirically defining their constructs and then testing hypotheses in various kinds of research programs and designs to determine their validity. Unfortunately, as we will see in the next chapter, the empirical solution that defines the field based on its focus on scientific methodology is not up to the task. Rather, it is like clapping with one hand.

References

Brown, G. K., Ten Have, T., Henriques, G. R., Xie, S. X., Hollander, J. E., & Beck, A. T. (2005). Cognitive therapy for the prevention of suicide attempts: A randomized controlled trial. *Journal of the American Medical Association, 294*, 563–570.

Carr, J. B., Conte, E., Rajadhyaksha, E. A., Laroche, K. A., Gwathmey, F. W., & Carson, E. W. (2017). Operative fixation of an anterior inferior iliac spine apophyseal avulsion fracture nonunion in an adolescent soccer player: A case report. *JBJS Case Connector, 7*(2), e29.

Haslam, N. (2016). Concept creep: Psychology's expanding concepts of harm and pathology. *Psychological Inquiry, 27*(1), 1–17.

Henriques, G. (2008). The problem of psychology and the integration of human knowledge: Contrasting Wilson's consilience with the tree of knowledge system. *Theory and Psychology, 18*, 731–755.

Henriques, G. R. (2011). *A new unified theory of psychology*. Springer.

Henriques, G. (2019). Toward a metaphysical empirical psychology. In T. Teo (Ed.), *Re-envisioning theoretical psychology: Diverging ideas and practices* (pp. 209–237). Palgrave Macmillan.

Lambert, M. (2013). *Bergin and Garfield's handbook of psychotherapy and behavior change* (6th ed.). Wiley.

Lukianoff, G., & Haidt, J. (2019). *The coddling of the American mind: How good intentions and bad ideas are setting up a generation for failure.* Penguin.

Richards, G. (2002). The psychology of psychology: A historically grounded sketch. *Theory & Psychology, 12*(1), 7–36. https://doi.org/10.1177/0959354302121002

Schwarz, A. (2013, December 14). The selling of attention deficit disorder. *The New York Times.* https://www.nytimes.com/2013/12/15/health/the-selling-of-attention-deficit-disorder.html

Torey, Z. (2014). *The conscious mind.* MIT Press.

Vygotsky, L. (1927/1987). *The historical meaning of the crisis in psychology: A methodological investigation.* The collected works of Lev Vygotsky. Plenum Press.

Wasserman, E. (2018, July/August). What's next? We asked 33 influential psychologists to identify the critical questions the discipline must answer. *APA Monitor, 49*(7), 40–49. https://www.apa.org/monitor/2018/07-08/cover-next#wasserman

Woolfolk, R. L. (2015). *The value of psychotherapy: The talking cure in an age of clinical science.* Guilford Publications.

CHAPTER 3

Modern Empirical Psychology and Its Inadequacies

Given that psychology is not currently unified by any conceptual framework that clearly defines its subject matter, we can ask: *What gives the field its identity?* As an institution, mainstream academic Psychology is unified by the empirical approach that it takes to its subject matter. Specifically, the field adopts a modern scientific epistemology and the methods of behavioral science to delineate whatever the paradigm or researcher identifies as the domain of "behavior and mental processes" of interest. Our task in this chapter is to understand what is meant by a scientific empirical epistemology anchored to what is technically termed "methodological behaviorism" (Moore, 2012), why it emerged as the current frame for mainstream psychology, and why it is inadequate.

Science Is Grounded in a Third-Person Empirical Epistemology

We can start our review with a basic question: *What does it mean when we say something is scientific?* In *A Scientific Method in Brief*, Gauch (2012) offered what might be called the "standard model" view of modern science. He argued that science is defined by a set of presuppositions that include scientific objectivity and the correspondence theory of truth. This refers to the idea that one can develop true or accurate statements about the state of affairs that correspond to reality. For example, one can posit

G. Henriques, *A New Synthesis for Solving the Problem of Psychology*, Palgrave Studies in the Theory and History of Psychology,
https://doi.org/10.1007/978-3-031-18493-2_3

that the Earth is either shaped more like a sphere or more like a flat disc. Gauch also argued that science is based on the legitimacy of reason, and a scientific realism, which is the idea that there is a "real" world out there that exists (relatively) independently of our beliefs about it (i.e., the shape of the Earth is either round like a sphere or flat like a disc regardless of human beliefs about it). Science also generally assumes that the universe at large is a closed system (meaning no outside supernatural entities) and that it follows cause-effect processes that are reliable and discoverable and comprehendible by the human mind.

The modern scientific mindset also includes the following characteristics: (a) attitudes of openness to possible naturalistic explanations and a skepticism about social convention, subjective revelations, faith, and traditional social authority; (b) an emphasis on objectivity and measurement and systematic and preferably experimental methods that control for the biases or values of a particular observer or group; (c) an emphasis on logical coherence and quantitative or mathematical analyses; and (d) the belief that humans can build systems of knowledge that more or less accurately correspond to the way the world actually works. With these frames in hand, scientists use empirical methods to gather data and test models of the domain of reality they are concerned with. Scientific theories about that reality are never final, but scientific epistemology and its empirical methodology allow scientists increasing degrees of confidence that the models are accurate, and that the better and more established theories make truth claims that are more likely to accurately model, correspond to, or map key aspects of reality under consideration.

The standard picture of the scientific method starts with a general observation and intuition, which is followed by a more systematic description and some preliminary ideas about underlying features or causes. These notions lead to hypotheses that can be examined. This process results in developing methods of measurement and then, in the exemplars that make up the scientific method, experimentation to test the validity of the proposed model relative to other possible explanations. These results should be "objective" in the sense that they are based on models and data that can be seen from an exterior, third-person point of view and are at least partially quantified and reliably reproduced. The scientific process at the institutional level also involves peer-reviewed journal publications and conferences that give rise to a public fund of knowledge. These elements constitute the core of science's epistemology, meaning the way scientists generate knowledge claims and determine the extent to which they are justified and generalizable.

A central argument in the present work is that an empirical epistemology is necessary, but not sufficient for authentic scientific understanding. For a system of knowledge to be truly scientific, it also needs to be grounded in a shared conceptual system that is up to the task of organizing the field of empirical findings into a coherent picture of reality. This second aspect enables scientists to clap with both hands in producing genuine cumulative knowledge. However, before we make that argument, we must dive a bit deeper into scientific epistemology, specifically focusing on what is meant by empiricism.

Two Meanings of the Word Empirical and the Epistemological Gap

Crucial to scientific epistemology is the concept of empiricism. The word empirical refers to data gathered via observation, and the nature of scientific knowledge is that such claims must be based on empirical data. The common aphorism from empirically oriented scientists is "show me the data." However, it is essential to understand how the word empirical has evolved and is now used in science. As we will see going forward, the distinction has deep relevance for the problem of psychology.

There are two primary meanings of empirical that are often conflated. One meaning is a first-person point of view grounded in subjective conscious experience. For example, I can empirically observe that my computer mouse is red. This redness is experienced by me in the subjective, first-person sense. We can call this the "interior" meaning of empirical, because it refers to an epistemic position of the individual and it is situated "from the inside" of that perspective. However, empirical also has a second meaning. The second meaning refers to data that are gathered based on systematic observation, measurement, and experimentation and can be seen from the third-person vantage point. We can call this the "exterior" meaning, because it is an epistemic position taken from the outside in. It is different from the first-person meaning, and it is most often the meaning that scientists have in mind.

To see the difference between the interior and exterior meanings of the word empirical, consider that we can we say that empirical research has verified the fact that antibiotics help to reduce bacterial infections. This is, of course, the second, scientific definition of the word. It is not a fact that can be directly apprehended via interior observation, such as my experience of my red mouse. Rather it is a fact that has been observed by trained experts who can apply the research methods that involve measurement and quantification, experimentation on independent variables that result in dependent consequences, and other advanced technologies (e.g., microscopes). The data that ground the experiments and subsequent conclusions can be observed by any of the scientists who are trained in how to perform these procedures and interpret them.

Crucially, then, the empirical findings that ground science are not dependent upon an idiosyncratic subjective, interior point of view. Rather, the methods of science are in large part about factoring out the idiosyncratic subjective interior empirical perspective and replacing it with a model of investigation that yields data that can be observed via the exterior position by anyone with the relevant training. Of course, different scientists might interpret the findings in different ways. Nevertheless, if they are deemed scientific, the data themselves must be available via a third-person, exterior empirical perspective.

Prior to the emergence of modern science in the sixteenth and seventeenth centuries, there was not such a strong and systematic distinction between the interior and exterior empirical vantage points. However, it became a crucial distinction in modern scientific justification systems. Empiricist philosophers like John Locke and David Hume divided the world into "primary" and "secondary" qualities. Primary qualities were things that could generally be quantified and were presumed to exist independently of one's perception of them. For example, the mass of a stone would be a primary quality. These quantifiable properties could be accessed from a third-person empirical perspective. Secondary qualities were interior. Modern philosophers refer to them as "qualia." These are the perceptual representations available from a first-person, interior perspective. That is, they exist "in the mind" of the person. The color of a stone would be a secondary quality. The experience of color is different than the electromagnetic waves that can be measured in the outside or exterior world.

It is crucial to be clear about the difference between the interior and exterior meanings of empiricism when considering the problem of psychology. The divide between what can be seen from the unique first-person subjective interior and the generalized third-person objective exterior viewpoints represents what I call the *epistemological gap*. It is the gap that is well framed by Ken Wilber's (1995) interior versus exterior quadrants, which I discuss in more detail later in the chapter. The two positions represent two fundamentally different vantage points from which to view the world. Moreover, as a justification system, modern science was defined in large part by the third-person empirical vantage point. That is, it is a system of knowledge that views the world from an exterior empirical epistemological perspective.

One of the elements in science that grounds the shift to the exterior empirical position is the process of measurement. It is generally not considered scientific to say that a body of water is "cold." Cold is a subjective, first-person experience—it is a secondary quality. However, when we say that the water is five degrees Celsius, we now have moved to a more scientific description, one that is anchored to a measured quantity. Notice that, via the translation to a quantitative scale, anyone with a working thermometer can observe that the water is five degrees. Moreover, by translating temperature into numbers, many important advantages emerge. Perhaps most notably, via quantification, mathematics can be utilized to describe events, variables, and objects in a much more standardized and logical way. This also allows for much greater specification, confidence in thinking, and precision regarding predictions. It was not accidental that what is arguably the largest single contribution to science was Newton's theories about matter in motion, which were closely paired to his development of a new mathematical representation of change processes in the form of calculus.

To summarize, there are two fundamentally different meanings of the word empirical: one that is based on the first-person, interior experience, the other based on systematic observation that can be readily seen from the perspective of a general or trained observer. There is a profound difference between these two meanings, and there is an epistemological gap between them, most notably that the interior perspective can not be directly seen by the exterior perspective. In addition, part of what emerged during the Enlightenment was a realization of these two different domains of the empirical, and modern science became anchored to a third-person, exterior empiricism. Indeed, this is a core feature of science that becomes the ground on which the edifice of modern psychology is built.

Modern Psychology Is Grounded in a Scientific Empirical Epistemology

To the extent that there is consensus in the science of psychology, it is found in the institution's commitment to a scientific empirical epistemology. It is a connection that is featured prominently in virtually every mainstream introductory psychology text. For example, Lilienfeld et al.'s (2017) *Psychology: From Inquiry to Understanding* equated the field of psychology with thinking empirically about psychological topics, and their book is organized by six key principles of scientific thinking. In *How to Think Straight About Psychology*, Keith Stanovich (2013) explained that psychologists ask questions that can be researched via empirical methods, and he goes on to articulate how the heart of psychology is learning to think like a behavioral scientist. In his *Introduction to Psychology* (Stangor & Walinga, 2014), Charles Stangor acknowledged that the field is lacking conceptual coherence. He noted (p. xiii):

> When I first started teaching Introduction to Psychology, I found it difficult—much harder than teaching classes in statistics or research methods. I was able to give a lecture on the sympathetic nervous system, a lecture on Piaget, and a lecture on social cognition, but how could I link these topics together for the student? I felt a bit like I was presenting a *laundry list of research findings rather than an integrated set of principles and knowledge* [italics added]. Of course, what was difficult for me was harder still for my students. How could they be expected to remember and understand all the many phenomena of psychology? How could they tell what was most important?

Stangor proceeded to explain that a focus on behavior and an empirical scientific attitude allowed him to organize the field's findings: "The idea of empirical research testing falsifiable hypotheses and explaining much (but never all) behavior—the idea of psychology as a science—was critical, and it helped me differentiate psychology from other disciplines" (p. xiii).

These texts embrace a scientific empirical epistemological version of psychology, which I will refer to here as "Empirical Psychology" (Henriques, 2019). Empirical Psychology is grounded in what Moore (2012) calls methodological behaviorism. In methodological behaviorism, behavior refers to the domain of psychology that can be accessed by science. That is, behaviors are the observable, dependent variables that scientists quantify, whereas mental processes are inferred to be the cause of such events. Consider, for example, how George Mandler explained that every cognitive psychologist is a "behaviorist" in this way:

> [N]o cognitive psychologist worth his salt today thinks of subjective experience as a datum. It's a construct....Your private experience is a theoretical construct to me. I have no direct access to your private experience. I do have direct access to your behavior. In that sense, I'm a behaviorist. In that sense, everybody is a behaviorist today. (Mandler in Baars, 1986, p. 256)

Mandler is pointing out that many (if not all, depending on one's definition) mental processes can only be inferred by the scientist, but overt action can be observed and measured from an exterior epistemological position. Moore (2012) rightfully argued that methodological behaviorism is the central position in mainstream scientific psychology. He summarized the situation as follows (p. 146):

> [M]ethodological behaviorism currently underlies mainstream research programs in psychology as well as professional socialization in that discipline. It underlies courses in research methods, experimental design, and statistics in most psychology departments at colleges and universities. It underlies such standardized tests in the discipline as the Graduate Record Examination. Research and psychological explanations that are not consistent with these features are given less weight, if any weight at all, in the scientific community, for example, as reflected in the editorial practices of journals and research support from granting agencies.

As Moore notes, the only other substantive position in American Psychology is B. F. Skinner's radical behaviorism. As briefly discussed in the previous chapter, radical behaviorism eschews mental processes as causal explanations for behavior and instead focuses on empirical observation of behavior change, both overt and covert, and the environmental mechanisms involved in the controlling of behavior. However, radical behaviorists currently make up only a small minority of psychologists. The rest of scientific psychology is grounded in methodological behaviorism.

That behavior affords scientific access to the mind is well illustrated by Spielman (2017) in her introductory text *Psychology*. She defines the field as follows (p. 6):

> The word psychology was coined at a time when the concepts of soul and mind were not as clearly distinguished.... The root 'ology' denotes scientific

> study of, and psychology refers to the scientific study of the mind. *Since science studies only observable phenomena and the mind is not directly observable, we expand this definition to the scientific study of mind and behavior.*

In this passage, Spielman lays out in plain sight the basic epistemological structure of mainstream Empirical Psychology. Behavior can be independently observed, and thus is accessible to the lens of science. This results in the field dividing its subject matter into behavior and mind (or mental processes). Scientific psychologists are trained as "behavioral scientists" who apply the methods of science to accessible behaviors and infer mental processes and share those findings with the public. The largely implicit logic underlying this arrangement is that mental processes are presumed to reside across the epistemological gap that is framed by the interior versus exterior distinction.

This summary captures how the defining feature of mainstream Empirical Psychology is its focus on an exterior empirical epistemology grounded in a methodological behaviorism. And yet, despite this proclaimed emphasis on science, debate still rages as to whether psychology should be characterized as genuine scientific discipline, on par with obvious "hard" sciences like physics, chemistry, and biology. A brief overview of this debate will enable us to begin to see the conceptual problems lurking beneath the surface.

The "Is Psychology a Science?" Debate

Several years ago, an opinion piece by the microbiologist Alex Berezow (2012) in the *Los Angeles Times* definitively declared that psychology was not a science. This was followed by several opinion pieces in *Psychology Today* and *Scientific American* by psychologists and others declaring definitively that psychology is, in fact, a science. Somewhat more recently, Jan Smedslund (2016), a long-time scholar of the field, authored the paper *Why Psychology Cannot Be an Empirical Science*, and again the debate was engaged. Smedslund pointed out that the field struggles to adequately define its core concepts, that many supposed "hypotheses" are actually based on definitions rather than predictions, that observed correlations between variables tend to be too weak to be meaningful in everyday life, and that the influence of the social world and feedback of knowledge is so great that he concluded that "an objective, accumulative, empirical and theoretical science of psychology is an impossible project." Smedslund's analysis is just one of many different voices that have argued that psychology should not be thought of as a straightforward, empirically grounded scientific discipline.

Empirical Psychology was also dealt a significant blow during the last decade via its own methods when the now well-known *replication crisis* emerged. Brian Nosek and colleagues (Open Science Collaboration, 2015) admirably launched a major research endeavor to determine if key findings reported in the field could be reliably replicated. The original project involved repeating 100 studies that were chosen based on being represented in flagship journals and a few other considerations to obtain a relatively representative sample. These studies were then distributed to a network of collaborating researchers who proceeded to attempt to replicate the research and determine if the findings held. The executive summary is that about one third of the studies replicated well, one third were mixed, and one third largely failed, with quite a few studies finding explicitly contrary results. The Empirical Psychology community was shaken by these outcomes. However, most of the dialogue has remained within the framework of a scientific empirical epistemological approach, and there have been many calls for better designs, more data, more replication studies, and more humility in making broad pronouncements and generalized conclusions rather than deep questions about the fundamental frame that psychologists adopt regarding how their field is defined.

These concerns and the more general debate about the status of psychology as a science leave us with the following question: *If academic psychologists have a scientific mindset and adopt empirical methodologies, why are there still so many skeptics proclaiming that psychology is not a "real" science?* According to the perspective taken here, such reasons are not found in the methods or the mindsets of psychologists, both of which are "scientific," at least in the assumptions most psychologists make and the empirical methods they use. Nor can we say this conclusion is because psychology is a young science. This is a myth. With the field approaching the age of 150 (i.e., its official "birthday" is considered the opening of Willem Wundt's lab in 1879), there are many "real" sciences like molecular biology and quantum field theory that are much younger than psychology.

The reason many are rightfully skeptical about its status as a science pertains to the body of scientific knowledge produced by Psychology. As we reviewed in the second chapter, the fact of the matter is the field of psychology remains pre-paradigmatic. That is, in contrast to the disciplines of physics, chemistry, and biology, psychology has no consensually agreed upon core corpus of knowledge that represents the center of the field. This is seen in the lack of agreement about the core subject matter and the failure to develop a shared lexicon regarding its foundational terms like behavior, mind, and consciousness. One can pick up any psychology textbook or skim any basic introduction to the field and what is offered is a summary of major, competing schools of thought that spell out a vast but vague territory between biological and social dimensions of analysis.

The major approaches that are usually included are behaviorism, cognitivism, humanism, psychoanalysis, evolutionary, and sociocultural perspectives. Each of these approaches has merit, but none of them have clear definitions of behavior and mental processes that can sort out the philosophical issues needed for coherent understanding. The result is a shaky foundation, devoid of a central, consensual core of knowledge. This preparadigmatic state is fundamentally what makes psychology a "soft" science and separate from the "hard" sciences. Here we can interpret hard as referencing the fact that there is a core of consensual agreement that forms a firm foundation upon which the science rests. This is in direct contrast to the softness that is at the heart of scientific psychology. It is that softness that makes the debate about whether psychology is a science so long-standing.

The Sandcastle Problem and the Need for Conceptual Grounding

Although mainstream psychology emphasizes the need for empirical methodology in developing scientific formulations, the critique I offered of the work we produced grounded in Beck's cognitive psychotherapy paradigm makes it clear that researchers do not operate on empirical data alone. Rather, data are interpreted by conceptual tools and schemes. We can see the logic of this fact in the work of Immanuel Kant, and his synthesis of rationalist and empiricist philosophies. Whereas the empirical philosophers like Locke and Hume emphasized the notion that all knowledge comes from the senses, Kant cogently argued that knowledge required concepts and categories that organized the empirical experiences. Kant was famous for saying that "Concepts without percepts are empty; percepts without concepts are blind."

Concepts are a different kind of thing than empirical facts. To be clear about the difference, imagine you are walking down the street and you observe a car crash. You run over to see if everyone is ok. A police officer arrives and queries you about what you saw and heard. Cars, injuries, and police officers are concepts and categories that you have available to make sense out of what you saw. These concepts can be separated from the empirical facts of the events. Indeed, concepts can be considered in many ways as being "pre-empirical." Consider that you can imagine an endless number of possible factual scenarios involving different kinds of cars and crashes, different degrees of injury, and different questions from a variety of different police officers. These are empirical variations, but they are all still framed by the concepts. *Conceptual analyses are every bit as important to the scientific enterprise as empirical data gathering and the testing of hypotheses.* Yet, as Machado et al. (2000) point out, mainstream psychology is obsessed "with a narrow and mechanical view of the scientific method and [has] a misguided aversion to conceptual inquiries" (p. 1).

To be sure, mainstream research psychologists do often concern themselves with definitions of their concepts. However, most focus on defining terms *operationally*. Consistent with an empirical epistemological approach, an operational definition is tied to describing the world in a way that allows for, and is grounded in, measurement that results in quantification. Thus, one might operationalize romantic love by defining it as a set of feelings and attraction that is measured in terms of the urge to be near the loved person, the sexual energy directed at that individual, the amount of time the person spends thinking about the loved individual, and the strength of self-reported feelings and thoughts. This process of quantifying variables and then studying them empirically is a crucial element of science. However, operationalizing definitions is not enough for cumulative knowledge. This is because, absent a shared language system that effectively frames the field of inquiry, what emerges is what Arthur Staats (1983) called "the problem of proliferation." The problem of proliferation, when combined with a soft conceptual foundation, gives rise to what I call the sandcastle problem.

The Sandcastle Problem

Basing a science on the methods of empirical epistemology produces an explosion of operational definitions and overlapping but inconsistent research programs. This problem was well seen by one of clinical psychology's greatest minds, Paul Meehl (1978/1992), who wrote:

> It is simply a sad fact that in soft psychology theories rise and decline, come and go, more as a function of baffled boredom than anything else; and the enterprise shows a disturbing absence of that *cumulative* [italics in original] character that is so impressive in disciplines like astronomy, molecular biology and genetics. (p. 524)

We can call this coming and going of findings in a manner that lacks cumulative understanding *the sandcastle problem*. The problem is that to the extent that each program of research is grounded in its own operational definitional systems, the collection of research projects will amount to something akin to building sandcastles on the beach. Although the specific programs are interesting (as are elaborate sandcastles), they are constructed out of an ephemeral ground of understanding. Like castles in the sand, each new tide of definitions will sweep the old findings out and clear the field for a new batch of researchers to build new programs of research in ways that are not cumulatively connected to research in the past.

Consider the recently published article *The Tendency for Interpersonal Victimhood: The Personality Construct and Its Consequences* (Gabay et al., 2020). True to much research in personality and social psychology, the article describes four studies that together yield the conclusion that "the tendency for interpersonal victimhood" is a "stable and meaningful personality tendency." To be sure, the article does make some useful points. For example, the way the researchers tie together the features of the (a) need for recognition; (b) moral elitism; (c) lack of empathy; and (d) rumination is clarifying. However, because the field is lacking a coherent consensus about the nature of personality, the role of development, the social environment and its construction of roles and identities, and the nature of human change, the idea that these authors have discovered a new personality type rather than invented a new tool for labeling people is not at all clear.

This goes back to how mainstream psychology is organized. Empirical Psychology is grounded in a methodological behaviorism that says: If you do a series of studies of an "X, Y, Z" variety, and find confirmation, then you can post a new personality construct. But a massive problem emerges when we ask the basic question: *Where does this finding fit in relationship to personality theory more generally?* There simply is no adequate theory of the person or personality that allows for personality, social, and clinical psychologists—not to mention the general public—to make sense of these findings. Rather, there is a morass of findings and each person or psychologist or community can then decide what is meant by "the tendency for interpersonal victimhood" and how that fits into our culture, how we make attributions about what people do, and how state-like or trait-like such behavior patterns are. This process of knowledge generation produces an explosion of operational definitions and overlapping but inconsistent research programs. The reason is that, absent a shared generalizable frame of reference, lines of research develop interesting findings, but they are not connected to a larger network of understanding.

This is what Paul Meehl meant when he wrote about the sad fact that soft psychological theories come and go. The sandcastle problem is a crucial and unfortunate consequence of the field's commitment to operational definitions and empirical research absent a unifying language system. The empirical projects of psychologists are too often akin to building castles in the sand; the information gained might be impressive, but it is temporary, contextual, and socially dependent, and ends up being washed away when new programmatic research tides come in.

Given the critical stance I am taking, it is perhaps useful to reiterate here that I am a scientist, and the criticisms I have been leveling here emerged only after I spent many years enamored with the mainstream empirical approach. Moreover, it is perhaps worthwhile to note that there are many stellar examples of mainstream psychology that continue to impress. To provide just one example, Daniel Kahneman's (2011) work on cognitive heuristics and decision making, narrated in his stimulating and accessible book *Thinking, Fast and Slow*, represents significant advances in our understanding of human cognition. I highlight this to be clear that I am not claiming psychologists are making *no* progress in our scientific investigations of psychological topics. Rather, my critique is found in the sandcastle problem. I am contending that we are creating massive amounts of information that fails to cohere into a general picture, and we can do better in generating cumulative knowledge about mind and behavior. Specifically, with the right frame of understanding, we can move the state of our knowledge from the current fragmented pluralism into a more organized, coherent, integrated pluralism.

Many mainstream psychologists readily acknowledge the cumulative knowledge problem, and that one of psychology's big problems is that new paradigms simply throw the babies out with the bathwater. Far fewer offer much in the way of solutions. From the perspective taken here, *the fundamental problem centers on the fact that we lack a coherent conceptual system that effectively defines what the terms behavior and mental processes refer to, such that no amount of observation or data gathering will clear up that confusion and the foundation of the field will be forever muddled.* To address the problem of psychology at its roots, we need a zoomed out, "metapsychological" vantage point to view the field.

Science, Paradigms, and the Missing Meta-Paradigmatic View

As was made clear in my recounting of my experiences in Beck's research lab exploring a cognitive therapy intervention for suicidal behavior, scientists do not operate from data alone. Rather they operate from schools of thought or paradigms that serve to frame the theories which set the stage for the research questions that drive most studies. Using this formulation, we can create a hierarchical continuum that ranges from paradigms to data (Fig. 3.1).

Applying this continuum to psychology raises some interesting points. The history of the field is marked by the emergence of schools of thought or (partial or mini) paradigms, whereby advocates compete for control

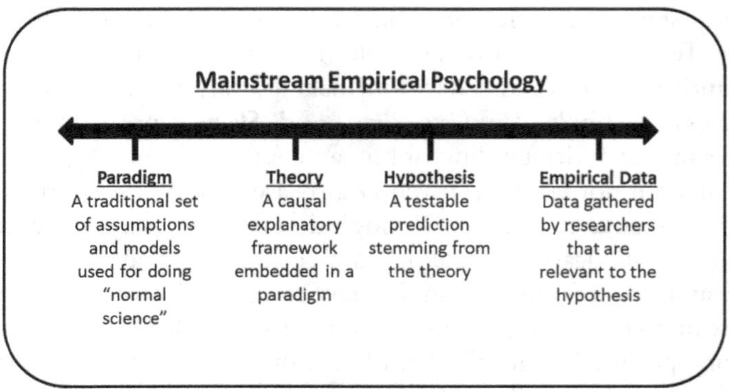

Fig. 3.1 Mainstream Empirical Psychology ranges from paradigms to empirical data

and depict the other frames as inadequate or misguided. The current work represents a fundamental shift in how the field is approached. Rather than grounding the science in empirical epistemology and methods, the current solution shifts the focus to metaphysics, ontology, and metatheory. That is, it steps outside of the specific schools of thought and adopts a metapsychological perspective that considers how the key insights from the various schools of thought and foundational concepts (i.e., mind and behavior) are defined and defined in relationship to one another.

Before we see how this might be accomplished, we need a better understanding of the ways in which Empirical Psychology is framed and why it fails to deliver. And one of the clearest ways to do that is to pay close attention to how the field's core concepts are defined and to see how often confusions and inconsistencies emerge. Doing so will enable us to clarify the nature of the problem of psychology's subject matter and the kind of solution that is required going forward.

Conceptual Problems with Defining Behavior and Mental Processes

David Myers is the author of several best-selling introductory psychology texts, and thus his work serves as a good representative for the mainstream view. In several popular academic texts, he defines psychology as "the scientific study of behavior and mental processes," which, as we have noted, is the standard definition in Empirical Psychology (see, e.g., Myers & DeWall, 2017). Myers defines behavior as "anything an organism does—any action we can observe and record," and mental processes as "the internal, subjective experiences we infer from behavior—sensations, perceptions, dreams, thoughts, beliefs and feelings." We can note how these key definitions are consistent with a methodological behavioral approach. The behaviors are the things that an exterior observer can track, whereas the mental processes are available from the inside. We can thus see that the standard definition of behavior and mental processes in Empirical Psychology consists of both (a) a referent in the world and (b) how we know about that referent from the vantage point of an observer.

To put this in more concrete terms, imagine you and I are having a conversation. As you talk, you are waving your arms and gesturing to make a point. According to Myers' definitions, we can ask: Are those arm movements behaviors or mental processes? If we take my perspective, I observe your arm movements as behaviors, and I infer that covert or hidden mental processes are operating inside of you that play a role in how and why your arms were moving the way they were. So far so good, as I am taking an exterior empirical perspective relative to you.

Now consider what happens when we shift the perspective to your point of view. Did you observe your arm movements as overt behavior and then infer you had some hidden mental processes guiding them? Not really. In fact, putting it this way does not make much sense. From your interior point of view, you are simply being-in-the-world. This means you had the private, subjective conscious experience—both visually and bodily—of your arms moving in harmony with the message you were trying to convey. Thus, according to Myers' definition, your subjective experience of observing your arms moving should be categorized as a mental process.

This brief example demonstrates that by defining the terms of behavior and mental processes based on who is doing the observing, the point of view of the observer ends up changing the referent. What had been referenced as behaviors from my point of view become mental processes from your point of view, and vice versa. Thus, in the definition of behavior and mental processes offered by mainstream psychology, the epistemological perspective one takes potentially determines what gets counted as a mental process relative to what gets counted as behavior. This fact brings us to one of the great errors of Empirical Psychology, which is that it confounds and conflates epistemology (which is how we know about something) with ontology (which is one's theory about reality). Because the relationship between epistemology and ontology is a central theme of the current work, we will need to deepen our analyses of these concepts.

Earlier I briefly mentioned Ken Wilber's (1995) epistemological quadrants, which he developed in his magnum opus *Sex, Ecology and Spirituality*. It is a four-quadrant model of epistemology, framed by two axes. The first axis is the interior–exterior axis, which we have discussed in some detail and refers to whether one is considering knowledge from a first-person, subjective perspective (i.e., interior) or whether one is considering knowledge from a third-person, objective perspective (i.e., exterior). The second axis is whether one is situated at the level of the individual or at the level of the social or collective. The combination of the two axes gives rise to the four quadrants: interior-individual; exterior-individual; interior-collective; and exterior-collective. These quadrants are often labeled as phenomenology, behavior, culture, and systems.

Consistent with the argument we have been making, Wilber (1995) argues that the exterior epistemological vantage point is the traditional natural science position. The interior subjective view is contained within the individual and cannot be analyzed directly via exterior means. In saying this, we should note that even as we can allow for the fact that we can *imagine* ourselves from different points of view, we still ultimately remain anchored to our interior perspective. It is well represented in Wilber's interior-exterior distinction and a crucial point that we have labeled the epistemological gap between the interior and exterior meanings of the word empiricism.

For greater clarity about what I mean by ontological reference, let me introduce the semiotic triangle of reference (Fig. 3.2). The triangle of reference refers to the linkages between words or symbols, thoughts, and the entity in the world. For example, there are real dogs in the world (referent), there is the idea or image of a dog (thought/reference), and the word "dog" (symbol). Thus, when a young child points to a dog and says "dog," the semiotic analysis is that the dog in the world is the referent, the image the child has is the thought or reference, and the word is the symbol. We can now clarify what is meant by the ontological referent. It is the thing in the world that one's thoughts and symbols are referring to.

A central feature of the problem of psychology is that the field's key concepts have confusing relations between words, concepts, and referents, and they vary significantly across the schools of thought. Consider the confusion that surrounds the term behavior. If we start with the idea that behavior refers to anything an organism does that we can observe, we can ask: If a cat falls out of a tree, is that something that it "does"? We can say that falling is a "behavior," at least in some senses of the term. It is, for example, movement that we can observe and measure. Yet, it does not seem to be the meaning intended. Indeed, although both a dead cat and a living cat might "behave" as falling objects, the latter also behaves very

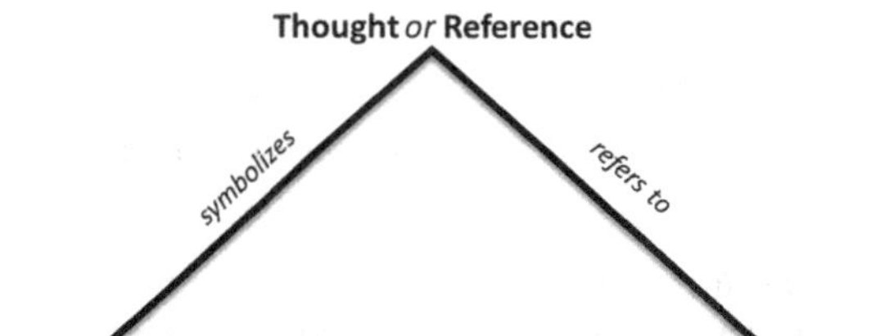

Fig. 3.2 The semiotic triangle

differently. Whereas a dead cat might bounce lifelessly on the grass, a live cat lands on its feet and takes off. This means that some of the movements we observe are behaviors in the psychological sense, but some are not. *As such, we need to somehow differentiate behaving in a physical, movement sense from behaving in terms of doing or engaging in a functional animal activity.*

The point here is that there is the potential for confusion regarding the exact ontological referent for what the concept of behavior in psychology is specifying. The same confusion exists regarding mental processes. In the second chapter, we reviewed that there were at least three possible references for what one is referring to when one uses the referent "mental processes" (i.e., neurocognition, subjective conscious experiences, and self-conscious justification). Myers defines mental processes primarily in terms of subjective experiences. But this definition leaves open the question of whether nonconscious neuronal information processing would count as mental processes. For example, much research has demonstrated that people are not conscious of many of the processes that guide how they make decisions. Given his focus on subjective experience, it is unclear if these nonconscious processes that shape decision making are included in Myers' definition of mental processes.

A central focus of the current work is on obtaining a clear ontology of both behavior and mental processes. That is, we will be delineating what exactly those terms refer to in the world and why. The general point here is that, if a system of knowledge is going to be both deep and broad, it needs to effectively align its ontological referents of the things in the world with its epistemological position and methods and justifications for knowing about those things. This brief analysis foreshadows serious problems with Empirical Psychology's core definitions. However, the definition of psychology as the science of behavior and mental processes became mainstream for some good reasons, and prior to deepening our critique, we need to understand those reasons as clearly as possible.

Behavior and Mental Processes from the Vantage Point of Empirical Psychology

To be sure, there is some utility in dividing the activities of animals and people into the dual categories of behavior and mental processes. Think about the people you know. You can see their overt behavior and monitor their responses. From this, you can infer those individuals are experiencing sensations, feelings, and thoughts, and that these are playing a role in their actions. Indeed, you likely often find yourself trying to read the minds of others, at least the aspects of which are hidden from your point of view.

The utility of this point becomes extended when we consider the fact that scientific knowledge is not based on the perspective of any single person, but rather on what we might call an "epistemologically generalizable" third-person view. As we have discussed, such a view is obtained via some form of systematic observation or measurement that is not wholly dependent on one subjective vantage point. This is why many describe science as operating from an objective point of view. For our purposes and consistent with Wilber's quadrants, we can consider the general exterior view as being represented by whatever could be recorded by a camera. When we use this view, it aligns with Myers' definitions, and it helps clarify the lens used by mainstream Empirical Psychology. Specifically, by using the camera as a stand-in for the general exterior epistemological perspective, we can say that anything it records represents the domain of behavior. This point also holds when we consider what a video camera cannot capture. A video camera can never directly capture our phenomenology (i.e., the first-person view from inside of our heads). To put it concretely, if we look inside a person's head with a video camera, what we find is the person's brain, not their experience of being.

It is worth noting that the division between behavior and mental processes also works reasonably well dealing with animals. For example, I just walked over to my fish tank, and as I turned the light on and grabbed the fish food, I observed the fish becoming more energized, as they swam to the top of the tank, presumably predicting and seeking the food that normally follows these events. Indeed, this is how much cognitive and behavioral neuroscience works in understanding animal behavior. That is, scientists observe and measure the frequency of animal responses and then alter environmental contingences and physiological variables and then measure changes in the rate of responses and develop models of learning or decision making based on the structure of the nervous system and the flow of information through it. These analyses help us to see that the mainstream definition has some utility.

Three Major Problems with the Mainstream Definition

Despite the utility of the mainstream definition at least three significant problems remain. First, there is the debate between the radical behaviorists and the mentalists. According to radical behaviorism, if I am to take a scientific view of animal actions (including humans), it is pointless to try to infer what is going on in the "black box" that some call the mind. Rather, behaviorism is about observing how the fish alter in their responses as a function of changes in their environment. This is why radical behaviorists define psychology as simply the science of behavior. The stuff that happens inside the animal is the domain of neurophysiology, and it is different than the domain of the behavior of the animal as a whole. Of course, the cognitive psychologists disagree. They argue that neuro-information processing mediates the functional responses that can be observed via a camera and must be included to understand how the behavior of the animal as a whole unfolds. If we return to the example of the fish in my tank, cognitive psychologists frame the fishes' overt behavior as a function of the fishes' nervous systems taking in sensory inputs and processing those inputs based on a hierarchy of rules in ways that regulate their behavioral outputs. Broadly defined, this is the neuro-computational theory of the mind. For many cognitivists, the primary referent for psychological science is understanding the neuro-information processes involved in such activities.

The metapsychology offered by the Unified Theory provides a novel solution to bridging this divide between radical behaviorists and mentalists. In numerous previous publications, I have shown we can describe the kind of behaviors my fish exhibit when they come to the top of the tank for food as "mental behaviors." As we will see, the concept of mental behavior gives us a clear ontological referent for the kinds of things psychological scientists are interested in understanding, which is the property of "mindedness." For now, we can note that Myers is a cognitivist and proceed from that point of view and ignore the radical behavioral objections. Unfortunately, however, another problem emerges rather quickly. This is the important difference between neurocognition (defined broadly in terms of neuro-information processing) and subjective conscious experience. Recall that Myers defined mental processes primarily as subjective experiences, which refer to things like the felt pain from a pinprick. However, as we have noted, the subjective experience of a pinprick is a very different reference point than neurocognitive processing more generally.

The difference becomes obvious when we think about the fish in my tank and ask: *What, exactly, are their subjective experiences?* Do fish feel pain the same way humans do? Do they see or taste in the same way? Can we say that the fish enjoyed the taste of the food and that is why they swim to the top of the tank? These questions are hotly debated in the literature and are difficult to answer. The epistemological gap between the interior and exterior points of view means that we can never directly perceive the interior world of another sentient being from the exterior perspective. We do not have a camera that allows us to scientifically observe what it is like to be a fish from the fish's subjective experience. In contrast, it is scientifically feasible to track how the fish's nervous system takes in information and processes it and functions to coordinate and regulate overt actions. Thus, we can develop a neurocognitive functionalist analysis that maps the fishes' overt actions in responses to changes in their environment, all from an exterior epistemological stance that grounds scientific knowing.

This analysis gives rise to the crucial difference between the neurocomputational and the phenomenological meaning of mind. This difference was well seen by the cognitive psychologist Ray Jackendoff (1987), who called it "the mind-mind problem." He described the problem as follows (p. 20):

> The upshot is that psychology has not two domains to worry about, the brain and the mind, but three: the brain, the computational mind and the phenomenological mind...The 'phenomenological mind-body problem'... is, how can a brain have an experience? The 'computational mind-body problem' is how can the brain [process information]? In addition, we have the mind-mind problem, namely, What is the relationship between computational states and experience?

Although the question of whether fish have an inner experience is an interesting one that we will return to in this book, there is no doubt that humans have an inner experience. Clearly at some point in the evolution of mental processes in the animal kingdom, the problem of phenomenology emerges as separate from neurocognitive functionalism. The standard definition of mental processes in Empirical Psychology does not differentiate nonconscious neurocognitive processes that take place in the brain from subjective conscious experiences. Using Jackendoff's analysis, it completely sidesteps the "mind-mind" problem.

Although this is a major problem, our earlier analysis of the various meanings of mental processes suggests that we can go further. Doing so brings us to the third major problem hidden in the standard definition of behavior and mental processes. In addition to the difference between neurocognitive and subjective conscious experiences, there is yet another "mind" problem that emerges when we consider persons and their mental processes. This is the domain of self-conscious justification that flows through human language. As will become clear, human language is a game changer when it comes to these issues. Consider, for example, that both neurocognitive processes and subjective conscious experiences can be said to take place *within* the confines of the nervous system. As such, they can be effectively and reliably differentiated from overt actions, which take place *between* the animal and the environment. Thus, following the language of mainstream cognitive psychology, it is reasonable to characterize neurocognitive and phenomenological domains as mental processes that exist *within* the animal and overt actions as behaviors that exist *between* the animal and environment and consider them reliably different in important ways, both ontologically and epistemologically.

Language-based thoughts, however, are different. This is because propositions move through the nervous system and the boundary of the skin and out into the world and back again without losing their informational form. This is one of the reasons that radical behaviorists often argue that the boundary of the skin is an illusion, especially regarding verbal behaviors. Our capacity to share in language means that a direct highway of mental information is opened between us. Indeed, this has been happening the entire time you have been reading this book. *Although I cannot directly share with you my subjective conscious experience of redness, I can directly share my self-conscious, language-based description of redness and any other language-based thoughts I have.* This means that the interior versus exterior divide that we have been emphasizing is radically different when it comes to language-based mental processes. With this final point about the nature of mental processes involving human language made, we can now effectively specify the conceptual problems that we are confronting in attempting to effectively define Psychology's subject matter.

The Behavior-Mind-Mind-Mind or BM^3 Problem

Our analyses have shown that there are many hidden layers to developing an effective science of behavior and mental processes. We started by showing that the standard definition of behavior and mental processes conflated the ontological referents with the epistemological vantage points taken. Although we attempted to salvage that by emphasizing the fact that science takes a generalizable third-person vantage point, we quickly stumbled into the fact that there was no clear differentiation between neurocognitive and subjective conscious experiences. This point was helpfully framed by Jackendoff as the "mind-mind" problem, which, as he noted, also must be defined in relationship to the brain.

Our analysis of the tensions between behavioral and mentalistic approaches and confusions regarding the definition of behavior allows us to extend the mind-mind to the mind-mind-behavior problem. That is, we need to consider mind as behavior and its relation to the overall domain of the mental. Finally, when we move to humans, another layer of complexity is added. Self-conscious justification mediated by symbolic language represents a separate set of mental processes that has a different ontological referent and different epistemological considerations. This analysis makes clear the fact that psychology has massive problems with clarifying its ontological referents relative to the epistemological position of the observer. We can characterize this set of difficulties as the behavior-mind-mind-mind problem which can be shortened to the BM^3 problem. We can note that there is also the problem of the brain's role in the BM^3 problem, such that I am tempted to call it the B^2M^3 problem. Indeed, we can also note that we need to consider physical bodies and living bodies as well. Thus, the problems could be extended even further to be framed as the B^4M^3 problem or the "physical body-living body-animal behavior-brain-mind-mind-mind" problem. We will leave it at the BM^3, and hopefully the point is clear. The BM^3 problem is central to the problem of psychology, and the central task of the current work is to lay out a framework that can afford an effective solution.

The BM^3 problem is obvious when you know to look for it. However, neither it nor any variation of it is alluded to in most introductory psychology textbooks. Instead, consistent with the fact that psychology is anchored to a methodological behaviorism, in virtually all introductory textbooks the conceptual issues are bypassed quickly, and the description of psychology moves to the central grounding of Empirical Psychology, which is its scientific epistemology. For example, in Myers' case, he tells students that "The key word in psychology's definition is *science*" (italics in original). Psychology, he proclaims, "is less a set of findings than a way of asking and answering questions," by which he means that psychologists approach their subject matter through empirical methods. He goes on to state that his hope for what introductory psychology students learn is "how psychologists play their game," which consists of how psychological scientists ask data-based questions and conduct empirical research.

This brings us to the second key focus of the present work, which is an analysis of exactly what science is. Much as is the case with behavior and mental processes, science is a complicated term with many different meanings and aspects. Moreover, just as was the case for behavior and mental processes, it will be crucial that we separate out the ontological from epistemological aspects in science. As we will see, the Unified Theory comes with new ways to frame natural scientific knowledge and define it in relationship to social knowledge more generally, setting up ways to transcend the impasse between modernist and postmodern epistemological positions regarding the nature of truth and the social and subjective construction of knowledge.

Conclusion

The problem and thus the stage for the current work has been set up in these two chapters. Psychology has long faced a previously unresolvable conceptual problem, which it has subsequently glossed over and dealt with via a shift to a methods-based approach to science. However, it is clear that its solution in the form of methodological behaviorism is inadequate. To solve the problem of psychology, we must tackle both the concept of science and the BM^3 problem and generate a framework that effectively untangles the ontological from the epistemological considerations. This is precisely what the Unified Theory of Knowledge does. It provides a conceptual framework for understanding natural science in a way that allows for a clear ontology of behavior and mental processes. It is a synthesis that Enlightenment scientists and philosophers failed to generate, and the consequences have been the unsolved problem of psychology. The result is a new vision for psychology being a science of mental behavior, grounded in a clear descriptive metaphysical system, a broad naturalistic ontology, and organized by metatheoretical frameworks that can assimilate and integrate the key findings from the empirical literature and insights from the major schools of thought into a coherent whole.

References

Baars, B. J. (1986). *The cognitive revolution in psychology.* Guilford.

Berezow, A. (2012, July 13). Why psychology isn't science. *Los Angeles Times.* https://www.latimes.com/opinion/la-xpm-2012-jul-13-la-ol-blowback-pscyhology-science-20120713-story.html

Gabay, R., Hameiri, B., Rubel-Lifschitz, T., & Nadler, A. (2020). The tendency for interpersonal victimhood: The personality construct and its consequences. *Personality and Individual Differences, 165*, 110–134.

Gauch, H. G., Jr. (2012). *Scientific method in brief*. Cambridge University Press.

Henriques, G. (2019). Toward a metaphysical empirical psychology. In T. Teo (Ed.), *Re-envisioning theoretical psychology: Diverging ideas and practices* (pp. 209–237). Palgrave Macmillan.

Jackendoff, R. (1987). *Consciousness and the computational mind*. The MIT Press.

Kahneman, D. (2011). *Thinking, fast and slow*. Farrar, Straus and Giroux.

Lilienfeld, S. O., Namy, L. L., & Lynn, S. J. (2017). *Psychology: From inquiry to understanding* (4th ed.). Pearson.

Machado, A., Lourenço, O., & Silva, F. J. (2000). Facts, concepts, and theories: The shape of psychology's epistemic triangle. *Behavior and Philosophy, 28*(1/4), 1–40.

Meehl, P. E. (1978). Theoretical risks and tabular asterisks: Sir Karl., Sir Ronald, and the slow progress of soft psychology. *Journal of Consulting and Clinical Psychology, 46*, 806–834.

Moore, J. (2012). Methodological behaviorism as a radical behaviorist views it. *Behavior and Philosophy, 39*, 145–202.

Myers, D., & DeWall, N. (2017). *Exploring psychology* (10th ed.). Worth Publishers.

Open Science Collaboration. (2015). Estimating the reproducibility of psychological science. *Science, 349*(6251), aac4716. https://doi.org/10.1126/science.aac4716

Smedslund, J. (2016). Why psychology cannot be an empirical science. *Integrative Psychological & Behavioral Science, 50*(2), 85–95. https://doi.org/10.1007/s12124-015-9339-x.s

Spielman, R. (2017). *Psychology*. Rice University Press.

Staats, A. (1983). *Psychology's crisis of disunity: Philosophy and method for a unified science*. Praeger.

Stangor, C., & Walinga, J. (2014). *Introduction to psychology*. BCcampus.

Stanovich, K. E. (2013). *How to think straight about psychology* (10th ed.). Pearson.

Wilber, K. (1995). *Sex, ecology, spirituality: The spirit of evolution*. Shambhala Publications.

PART II

The Unified Theory of Knowledge and Its First Two Key Ideas

CHAPTER 4

The Unified Theory of Knowledge: A New Metapsychology for the Twenty-First Century

With the inadequacies associated with Empirical Psychology clearly spelled out, we are now in a place to introduce the Unified Theory of Knowledge (UTOK) as a new system of understanding. This chapter provides an overview of the system, which I will variously refer to as the Unified Theory, the UTOK when starting a sentence, or, most simply, UTOK. This book makes the argument that UTOK frames the nature of natural scientific knowledge and does so in a way that allows us to clearly disentangle the meanings of mind as (animal) behavior, mind as neurocognition, mind as subjective conscious experience, and mind as self-conscious reflection, and reweave these meanings back together into a coherent whole. By providing a wholistic vision that can box in the ontology of the mental and resolve the BM^3 problem, UTOK sets the stage for a new vision for psychology. Instead of Empirical Psychology's methodological behaviorism anchored to and defined by an empirical epistemology, UTOK gives rise to a scientific psychology called "mental behaviorism." It is an approach anchored to a descriptive metaphysical system that provides a clear, scientifically grounded ontological referent for the mental that is tied together by metatheoretical frameworks that assimilate and integrate key insights from the major schools of thought. If successful, the implications for psychology and philosophy are difficult to overstate.

The core of UTOK consists of eight key ideas that are divided into three major projects. The first four key ideas make up the first major

G. Henriques, *A New Synthesis for Solving the Problem of Psychology*, Palgrave Studies in the Theory and History of Psychology,
https://doi.org/10.1007/978-3-031-18493-2_4

project, which is framed as the "unified theory of psychology." The second set of four key ideas make up the second major project, which is the development of the "unified approach to psychotherapy." The third major project is concerned with developing a scientific humanistic philosophy or "metaphysical empirical metapsychology" that effectively and coherently combines a scientific, naturalistic worldview with social scientific inquiry, humanistic values, sociocultural epistemology, and the subjective experience of being into a unified theory of knowledge that allows for the cultivation of wisdom in the twenty-first century and beyond.

Coherence, or logical harmony and synthesis, is a long-standing tradition in science and philosophy. The idea is that, despite the incredible complexity and diversity found in the world that can be obtained by many perspectives, there are nonetheless unifying principles that undergird the plurality of modern lines of knowledge and inquiry. The word unified in UTOK owes its lineage to how E. O. Wilson (1999) framed coherence in *Consilience: The Unity of Knowledge.* According to Wilson, consilience refers to the idea that, with the right understanding, there is a "jumping together" of facts and concepts, such that an integrative, holistic, comprehensive, and coherent account of the universe, human experience, and human knowledge emerges. As used here, unified is synonymous with consilience, and, at its broadest contours, UTOK is offered as a consilient theory of knowledge into wisdom for the twenty-first century.

Aligning UTOK with Wilson's vision of consilience allows us to locate the kind of work being promoted. As a naturalistic approach, UTOK is ultimately anchored to the core theories of physics (i.e., quantum mechanics and general relativity) and it rejects substance dualism (i.e., the belief that matter and mind reside in two completely separate worlds) as metaphysically incoherent. It models nature as an unfolding wave of behavior framed in terms of energy and information, in which new properties, levels of stratification and complexification, and planes of existence emanate and emerge. More specifically, UTOK explicitly depicts this process in a way that affords a coherent picture of the cosmic evolutionary emergence from energy into matter in motion into living processes into the animal-mindedness domain, and, finally, into self-conscious humans constructing systems of justification that function to propositionally map this history based on experience, logical analysis, and intelligibility. As this extended description suggests, UTOK is not reductive. Rather it affords what might be called an "extended naturalism," a view that is grounded in natural science, but extends into the transcendent and seeks to revitalize the human soul and spirit in the twenty-first century.

In some ways, UTOK's picture of knowledge into wisdom closely aligns with several ancient views of the cosmos. For example, in *Thinking Being*, Perl (2014) summarizes Aristotle's view of metaphysics in relationship to Plato as follows (p. 105–106):

> The greatest difference between Plato and Aristotle, then, does not lie in their answers to the question 'What is being?' For Aristotle no less than for Plato, being, τὸ ὄν, or reality, οὐσία, is form, εἶδος, and as such is divine. The difference, rather, is that whereas Plato exalts the good beyond form, being, and intellect as the very source of reality, of intelligibility, Aristotle looks no further than pure form, pure being, pure intellect. For Aristotle, the hierarchical ordering of the different kinds of beings is based on the extent to which form predominates over matter in each. Non-living things have the lowest degree of form, of unifying selfhood, of activity that proceeds from themselves. Although they have some form, some nature, some behaviors of their own, without which they would be nothing at all, they come closer than all other things to being purely material, purely passive. A living thing, characterized by organic unity and the ability to nourish, maintain, and reproduce itself, is far more one, more active, exhibits a far higher degree of formal identity. A sentient living thing, an animal, exercises not only these life-functions but also consciousness, which, as the capacity to receive forms without matter, is a still higher degree of formality, of immateriality. A human being, in turn, has not only life and sense but the capacity for the wholly immaterial activity of intellection, which has as its content, and thus is one with, purely immaterial ideas. And at the peak of the ascending ranks of beings we find the divine, which is nothing but form, nothing but intellection, and as such is life itself and being itself.

In this description, we see a metaphysical worldview that has deep parallels with UTOK, such that intelligibility and the epistemic forms by which the knower maps the known are central and necessary concepts for a comprehensive, holistic understanding of reality and our knowledge of it. Perl describes a four-stage grouping that moves from the inanimate to the animate to the animal to the justifying person, which, as we will see, is deeply congruent with the worldview espoused by UTOK. In addition, there is an appreciation for a wisdom orientation, which can be broadly or metaphorically framed as being oriented toward the "divine." However, despite this deep connection to prominent historical frames, it is also the case that

UTOK gives a substantially deeper view of natural science, scientific psychology, the subjective experience of being in the world, metaphysics, ontology, and wisdom than has been previously articulated. Indeed, ultimately UTOK gives rise to a novel vision of human knowledge writ large.

A New Unified Theory of Psychology (Henriques, 2011) demonstrated how the first four key ideas of UTOK, specifically, the Tree of Knowledge System, Justification Systems Theory, Behavioral Investment Theory, and the Influence Matrix, interlock to generate the outline of an effective metatheory for the science of psychology. This book builds on that argument. However, the focus in the present work shifts the emphasis. Whereas my first book focused on metatheory, the current work emphasizes metaphysics and ontology. The reason that such a shift is necessary is that the ultimate root of the problem of psychology is found in a set of epistemological and ontological confusions and conundrums that emerged with the modern scientific enterprise and the way the Enlightenment thinkers attempted to map the human mind in relationship to the physical world framed by Newtonian mechanics. The UTOK allows for a new perspective on the perennial philosophical issues that emerged in the wake of the Enlightenment, and it sets the stage for new ways to solve, resolve, and/or dissolve these problems.

The Problem of Psychology Arises from the Enlightenment Gap

It might seem odd that a system that claims to solve the problem of psychology would be framed as a unified theory of human knowledge. The primary reason for this has to do with the systems of knowledge that came out of the Enlightenment. To get straight to the point, we can frame these problems in terms of a set of confusions that arose between the science of Isaac Newton and the philosophy of Immanuel Kant. The former gave us the "matter-in-motion" model of the physical universe that would be the exemplar of objectivist, realist science. The latter bridged empiricist and rationalist philosophical approaches and gave us the idea that the human mind has a priori epistemological categories that allow us to perceive the world and make causal inferences. Yet, according to Kant, we only have access to our "phenomenology" rather than the noumena or the "thing-in-itself." As such, Kant's philosophical picture of the world was more idealist and epistemological rather than materialist and ontological in its framing and emphasis.

The Enlightenment never produced a clear, consilient philosophy that effectively joined Newton's matter in motion materialistic ontology with a proper understanding of epistemology grounded in human phenomenology. Hegel famously attempted such a synthesis. But American Psychology did not follow Hegel's line of thought, and the modern version of the discipline can be framed by the broad views of a Newtonian materialistic ontology and Kantian epistemology. The failure to generate a coherent synthetic philosophy that effectively mapped the proper relation between material behavior and human phenomenology ultimately and inevitably gave rise to the problem of psychology. Indeed, we can reconceptualize the problem of psychology as being the downstream consequence of fundamental weaknesses in the knowledge systems that were generated during the Enlightenment.

I call this absence of coherence and clear intelligibility the *Enlightenment Gap*. It specifically refers to the failure of modernist systems of science and philosophy to effectively produce a consilient system for understanding the proper relationships between (a) matter and mind and (b) scientific knowledge relative to social and subjective forms of knowing. These two separate but related problems combine to make up the Enlightenment Gap. That these are two central problems that stem from Enlightenment thinking does not need much justification, as they represent some of the most salient problems in philosophical discourse. Consider, for example, that the "mind-body problem," reframed here as the BM^3 problem and referenced in the Preface in the context of the philosophy of mind literature, is ubiquitous. The many confusing dualities that emerge (e.g., subjective versus objective, consciousness versus matter, mental versus physical, free will versus determinism, self versus no self) and the endless debates about the proper framing are evidence of this fact. Similarly, the disputes between modernist and postmodernist sensibilities regarding the nature of truth and the position of scientific knowledge relative to the social construction of the knowledge are also pervasive in modern discourse. In short, when one looks across the landscape of our current chaotic, fragmented knowledge systems, the Enlightenment Gap can be considered a given.

When the Enlightenment Gap is brought into focus, we can then place the problem of psychology as an inevitable downstream consequence of it. Indeed, given Psychology's identity as attempting to be a modern, empirical, natural scientific approach to the domain of the mental, its core mission can be seen as residing at the epicenter of the long shadow of confusion cast by the Enlightenment Gap. The connection between the Enlightenment Gap and the problem of psychology becomes even more clear when we consider the place of psychology in the modern academy. Indeed, a core feature of psychology is that it resides at a central nexus in human knowledge and the academy. That is, arguably more than any other discipline, psychology's borders connect to the natural sciences, the social sciences, and the humanities. In *A New Unified Theory of Psychology*, I characterized the issue as follows (Henriques, 2011, pp. 41–42):

> The problem of psychology is the joint observation that the field cannot be coherently defined and yet it connects more deeply than any other discipline to the three great branches of learning. Taken together, these observations suggest that the problem of psychology is a profound problem in academia at large. This conclusion is bolstered by the fact that as psychology has lumbered along acquiring findings but not foundational clarity, the fragmentation of human knowledge has grown exponentially. All of this suggests that the question, 'What is psychology?' is profoundly important, one of the central questions in all of philosophy. Asking the right questions is often the most important step in getting the right answer. My interest in psychotherapy integration ultimately led me to ask the question, 'What is psychology?'. Although I had no idea at the time, it turns out that this is the right question. And, as startling as it sounds, because psychology connects to so many different domains, the correct answer to it opens up a whole new vision for integrating human knowledge.

The combination of psychology's connection with many different fields and its lack of coherence highlights why it is such an important philosophical problem. When placed in the shadow of the Enlightenment Gap, the nature of the difficulty both becomes clearer and grows substantially. The deep confusion that arises from the Enlightenment Gap and entangles both psychology and the philosophy of mind is the Gordian knot that the Unified Theory of Knowledge seeks to untangle. The present work shows how UTOK affords scholars a new synthetic philosophy that can address the Enlightenment Gap and resolve the problem of psychology. Because so many issues we are dealing with are philosophical in nature, it is necessary to clarify key terms such as metaphysics, ontology, epistemology, and metatheoretical analyses in greater detail.

Metaphysical and Metatheoretical Analyses

In the previous chapter, I introduced a continuum that placed empirical findings in relationship to hypotheses grounded in theories that in turn were grounded in "mid-level" paradigms in psychology (aka schools of thought). The example of my work with Beck highlighted the fact that if we were using a different paradigm (e.g., a social work model that focused on treatment access for underprivileged populations), the consequence would have been very different interpretations of the empirical findings than what we ended up sharing with the larger science and practice communities.

In their work on metatheory, Witherington et al. (2018) explain how we can zoom out beyond the mid-level paradigms and consider the conceptual analyses that ground our theories. As these authors note, normal science practices "are designed to be operationalized, expected to yield testable, observable predictions, and, as such, are subject to adjudication through empirical activity, in keeping with instrumentalist tradition" (p. 182). Metatheory is different; it operates at the pre-empirical level of definitions and foundational frames, presuppositions, and assumptions. More specifically, (pp. 182–183):

> Metatheories involve a set of *background concepts*—various philosophical beliefs and assumptions that we, as humans and as scientists, hold concerning the nature of reality (ontology) and how we come to know that reality (epistemology) [Overton, 2015]. They establish what does and does not make sense to even consider or investigate in the observations that we make and the theories that we construct. All of our scientific work, therefore, necessarily presupposes, and is preconditioned by, the background concepts of metatheory [Overton, 2015]. Critically, this means that, unlike their theoretical counterparts, background concepts are not amenable to empirical investigation and adjudication.

As a metatheoretical system, the Unified Theory orients us to pay attention to the paradigms and their interrelations. In my first book, I argued that each major school of thought offers key insights that are useful and cannot be dismissed, and that none are clearly superior to the others, nor can the constellation be reduced to any single paradigm. I also pointed out that none dominate in terms of application to mental health problems and treatments, which also suggests the need for an integrative effort in psychotherapy. Finally, none of the schools of thought are up to the task of coherently solving the deep ontological and epistemological problems that characterize the problem of psychology. This gives rise to many questions from a metatheoretical perspective, such as: *What should the relationship between the paradigms be? Is there a coherent higher level of abstraction that exists beyond the paradigms? Could there be a truly unifying meta-paradigm for the field?*

A New Unified Theory of Psychology focused on how the system works as an effective metatheoretical framework for psychology. It does this both vertically and horizontally. By vertically, I mean that the system integrates the physical, the biological, the psychological, and the social dimensions of existence into their proper alignment. This is accomplished via the Tree of Knowledge System, which is the central idea that grounds UTOK in a coherent theory of reality and our scientific knowledge about it. The book also summarized the three other key ideas—Behavioral Investment Theory, the Influence Matrix, and Justification Systems Theory—that work together to generate a metatheoretical solution that assimilates and integrates the key insights from behavioral, cognitive, psychodynamic, and humanistic paradigms into a coherent whole. This allows for the horizontal integration of ideas across the dimensions of human mental behavior (i.e., the behavioral and mental processes of human individuals), and the work showed how the key insights from cognitive, behavioral, humanistic, and psychodynamic approaches could be readily aligned. The domains that were integrated across the vertical and horizontal dimensions are depicted in Fig. 4.1.

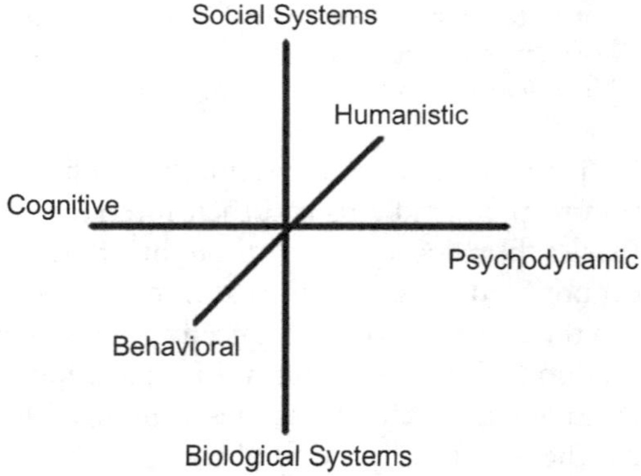

Fig. 4.1 A biopsychosocial view with the four major schools of thought

The current work deepens and extends the argument. *It does so by shifting the primary focus from the metatheoretical to the metaphysical and ontological levels of analysis.* That is, rather than focusing on the key insights from the paradigms, the primary focus here is on delineating the ontology of mental behavior and elucidating the core concepts and categories that define scientific psychology. This means diving deeply into what is meant by science in general, and, more explicitly, a science of "behavior and mental processes." Figure 4.2 makes clear the focus of the current work, relative to the first book. It extends the continuum offered in the previous chapter and includes the metatheoretical and metaphysical points for consideration. This book explores the metaphysical concepts and categories that define psychology, and then uses these to develop a coherent naturalistic ontology of the mental.

Because of the Enlightenment Gap, academics in general and psychologists in particular have lacked a shared system of understanding that effectively defines mind in relationship to matter and clarifies the relationship between scientific and social and subjective knowledge. As such, no system of thought emerged that could effectively frame a genuine meta-paradigm for psychological science. The current proposal is that UTOK is structured in such a way that allows us to clarify what is meant by modern science and use that to effectively frame the domain of "mental behaviors." That is, it provides us with a naturalistic scientific ontology that explicitly disentangles the BM[3] problem and defines the various meanings of

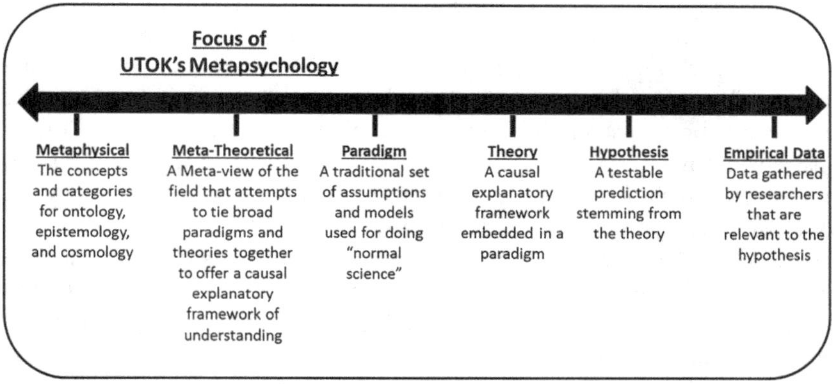

Fig. 4.2 The solution provided by UTOK involves the left side of the continuum

behavior and mental processes and their relations in the natural world and how that world is understood by human knowers. In doing so, it affords psychologists a coherent metapsychology that points the way toward a shared lexicon that solves the problem of scientific psychology's subject matter that has long eluded the field. The hope is that UTOK's solution to the BM^3 problem will provide future psychologists a way of "speaking psychology" that is comprehensive, coherent, and consilient, both within the field and between it and other fields of inquiry, and in society writ large.

On Metaphysics, Empiricism, and the Metaphysical–Empirical Dialectic

To ground this argument, we need to turn to metaphysics and understand how it relates to science. In contrast to the virtual omnipresence of the word empirical, the word metaphysics rarely shows up in discussions regarding science in general and scientific psychology in particular. Outside of academic settings, the word sometimes carries a meaning associated with New Age, mystical, or "alternative" ways of thinking. In scientific circles it is commonly used in a pejorative sense to communicate things that are not very serious or things that are unknowable. For example, if a scientist were to say, "Now you are just talking metaphysics," she would likely mean the other person was just talking nonsense or was engaged in pure conjecture.

Using metaphysics in this way stems in large part from the emergence of modern scientific ways of thinking, which emphasized the importance of empirical investigations over philosophical inquiry (or worse, pure speculation). Although understandable, it is unfortunate that metaphysics came to be ignored by so many in psychology because, in its formal sense and as it is being used here, it refers to the concepts and categories one is using to frame reality and how one knows about it. Given the confusions Empirical Psychology has with the BM^3 problem, there should be no doubt that metaphysical considerations are essential.

Merriam-Webster's Dictionary defines *metaphysics* as: (1) a division of philosophy that is concerned with the fundamental nature of reality and being that includes ontology, cosmology, and epistemology; and (2) abstract philosophical studies, including what is outside of objective experience. In this work, I am going to be emphasizing the related notions of *descriptive metaphysics* and *metaphysical systems*. Both refer to foundational concepts and categories one is using to describe reality (i.e., ontology) and how we obtain knowledge of it (i.e., epistemology). Descriptive metaphysics is a term introduced by Peter Strawson (1959) that refers to understanding the core concepts and categories humans use to carve up the world. This includes considerations such as what constitutes an object or an individual. This book is concerned with the metaphysical descriptions of concepts like matter, mind, consciousness, and behavior.

A metaphysical system refers to the process by which concepts and categories are interrelated to generate a cosmology or a system for mapping or understanding reality. This meaning is consistent with the description of metaphysics given by Koons and Pickavance (2014). In their introduction to the topic, they state that metaphysics is about understanding:

> the fundamental structure of reality as a whole. How do things fit together in the world? Plato describes this task of philosophy as 'carving nature at the joints,' comparing metaphysics to a skillful and knowledgeable act of dissection. Here are four relations that seem to be among the fundamental relations of this worldly structure: the relation between things and their properties, between wholes and parts, between causes and effects, and things related to each other in space and in time. (p. 34)

This book makes the argument that UTOK affords a new descriptive metaphysical system that frames both reality and our scientific knowledge of it in a way that addresses the Enlightenment Gap and resolves the problem of psychology.

If we return to the metaphysical–empirical continuum, we can now contrast descriptive and systemic metaphysics with metatheory. Descriptive metaphysics and the generation of metaphysical systems in this context is about delineating the proper definitions of concepts like science, mind, and behavior, and showing their systemic interrelations. It also entails being clear about the difference between reality and our claims about that reality. This brings in issues of ontology and epistemology, which will be clarified below. Metatheory extends from descriptive metaphysics and involves more consideration about the underlying causal-explanatory structure that functions to explain why events unfold in the way that they do.

With metaphysics being concerned with the system of concepts and categories one uses to describe both reality and one's knowledge of it, we can ask: *What is the relationship between the metaphysical and empirical elements of knowledge systems?* Consistent with the continuum depicted previously, these two domains of knowledge are, in many ways, defined in contrast to one another. A descriptive metaphysical system refers to one's conceptual map of the world and the labeled categories within it, whereas empiricism is about one's direct sensory experience of the world or the systematic data gathered via the scientific process. Recall from the previous chapter, the way that we differentiated the concepts of cars and police officers from the empirical specifics of actual accidents. The concepts of cars and police correspond to descriptive metaphysical concepts, which would then be networked together to afford a metaphysical system of understanding the infinite variety of empirical events in the category of car accidents.

The second definition of metaphysics refers to that which is outside of direct experience. This contrasts with direct observational experience, which constitutes empiricism. An idea that refers to something completely disconnected from sense experience or measurement can be referred to as a *purely* metaphysical idea. For example, if one has the belief that there is a God who exists in a separate realm of reality, but one could never prove or disprove that claim via any empirical evidence, then we can call that a *purely* metaphysical conception of God. The potential for metaphysics to be disconnected from the empirical is, in part, why metaphysics became associated with negative connotations by scientists. A basic principle of modern science is that its claims must be grounded in empirical evidence. This book is not about pure metaphysics, divorced from empirical knowledge. Rather it is about developing a descriptive metaphysical system for scientific knowledge grounded in the empirical process. Indeed, the goal of the present work is to develop a descriptive metaphysical system that maps ontology in a way that is consistent with both a first-person empirical experience of being in the world and a generalizable third-person scientific understanding.

A central claim of UTOK is that metaphysical and empirical considerations provide complementary components to knowledge systems. The metaphysical aspects refer to the concepts and categories that one is using to describe reality and how we know about it. The empirical refers to information acquired via direct observation in specific individuals (i.e., first-person, interior, or subjective empiricism) or systematic measurement and experimentation in science (i.e., third-person, exterior, or objective empiricism). Although pure metaphysical concepts are not scientific, scientific concepts are metaphysical in the sense that they are tied to ideas about what is objectively real, how that reality came to be, and how we know about it. In this way, metaphysics is the conceptual yin to the empirical yang, and together they make up the key ingredients of scientific knowledge. Thus, in direct contrast to mainstream Empirical Psychology, according to UTOK, we need a "descriptive metaphysical empirical systems" approach to psychology, and science more generally. In a previous essay, I have called this a Metaphysical-Empirical Psychology (Henriques, 2019).

Modern Empirical Natural Science and Its Rejection of Metaphysics

Central to addressing the Enlightenment Gap is the argument that modern science was developed as a system of justification that became defined against metaphysics in an unproductive way. Metaphysics was historically concerned with the intrinsic nature of things, and it is here that we encounter a key feature of the tension that emerged between metaphysical considerations and the modern, empirical, natural scientific enterprise. The primary set of understandings that informed Europe prior to the modern science revolution was Scholasticism, which was a blend of Christian and Greek worldviews. Much intellectual energy in the Scholastic tradition was focused on metaphysics, which was conceived of in terms of getting to the essence of existence. From the vantage point of UTOK, many of those intellectual efforts were exercises in "pure metaphysics," in that they were generally divorced from empirical description and contingency.

Pure metaphysics has led countless scholars down many unproductive rabbit holes filled with circular arguments that fail to generate cumulative knowledge. This was happening to the Scholastic academics of the 1600s. In his analysis of how the work of Galileo and Descartes gave birth to modern science, Ricardo Nirenberg (1996) described the situation as follows: "these [metaphysical speculations about the nature of matter] had been thoroughly abused by the Scholastics, and much fun was poked at them in the 1600s and later; writers of comedies had ridiculous doctors solemnly proclaim that opium made one sleep because of its 'dormitive virtue', as if that explained anything."

Directly aligned with the present argument, Nirenberg argued that the central difference between the Scholastics and modern science was a question of metaphysics. Whereas the Scholastics embraced such questions to the point of abuse, Galileo—who is generally considered to be the father of modern science—largely rejected metaphysics. His theories of matter in motion were defined by empirical facts (i.e., what he and others could see or measure and repeat) mapped by mathematics. Nirenberg's analysis of Galileo's empirical approach to matter in motion relative to metaphysics is worth quoting:

> Galileo's principle defies metaphysics, in the sense that motion turns out not to be an intrinsic quality of things. The consequence of this was enormous, once scientists and thinkers started reducing all phenomena in nature to the one phenomenon of motion: the consequence was that physics didn't care for metaphysics. But when metaphysics is expelled, it will come back through the back door: one asks, what is it that moves? Answer: planets, particles, atoms, etc. Why does a planet (say) move? Because it has mass and an initial impulse (as Newton would state some years later). And what is mass? An inner, active principle in things…
>
> But Galileo didn't deal with [the problem of metaphysics], nor did he care for it….All he cared for was motion (which was, needless to say, a lot). So, Galileo's decision, his way out of the crisis, was this: he sharply separated physics from metaphysics—had he used today's academic jargon he would have said: 'They are two separate, untranslatable discourses, and I talk only physics.' This makes him the first professional scientist.

Nirenberg (1996) makes several relevant points for our understanding of the modern scientific system of justification. First, it was Galileo's mathematical and empirical analysis of matter in motion that makes him "the father of modern science." Second, it shows how modern science was grounded in empiricism, defined as being anchored to quantifiable data and observation tied to logical analyses. Framed this way, a defining feature of modern science is a rejection of metaphysical speculation. Third, although rejecting *pure* metaphysical speculation is reasonable for science, Nirenberg makes clear that even physics does not escape metaphysics entirely. Deep and profound questions endure regarding exactly what motion is, what constitutes change versus stasis, what objects are, what gravity is, what space and time are, and how these concepts and categories relate to reality relative to how the human mind perceives reality.

The UTOK posits that these issues must be addressed if one is going to generate a clear and consilient theory of scientific knowledge. The revolutions of quantum mechanics and general relativity drive home this point. Indeed, a strong argument can be made that modern-day physics is fundamentally defined by (or at least entangled with) the *metaphysical* revolutions wrought by the way quantum mechanics and general relativity undermined the classical Newtonian matter-in-motion view of the universe (i.e., issues of how we think about time, space, and cause–effect relations). The descriptive metaphysical issues are even more pressing for a scientific psychology that attempts to study mind and behavior and does so in a way that blatantly conflates and confounds ontological and epistemological considerations. All of this points to the need for a new descriptive metaphysical system that can organize both our interior, subjective, qualitative experience of being-in-the-world and our exterior, quantitative empirical data-based descriptions of behavioral patterns derived via scientific methodology.

The Unified Theory of Knowledge

The Unified Theory of Knowledge (UTOK) is so named because it allows for novel solutions to the unresolved difficulties Enlightenment thinkers had in resolving the proper conceptual relationships between both matter and mind and social and scientific systems of knowledge. The UTOK can usefully be considered a synthetic scientific humanistic philosophy or a metapsychology. The term metapsychology dates most clearly back to Freud. Although other metapsychologies have been offered, it has been used most regularly to characterize his body of psychoanalytic work. The time is ripe for a new metapsychology to emerge, as it is becoming clearer and clearer that the old ways of knowing are not up to the task of effective sense- and meaning-making in the twenty-first century. The prefix "meta" refers to that which is beyond or above or that which is more comprehensive or operates at a higher level of abstraction. Metacognition, for example, is cognition about cognition. Metapsychology can thus refer to a frame that reflects on the field of psychology and its subject matter (i.e., mind and behavior), or it can refer to a comprehensive approach that is used to specify the structure of the institution. The UTOK does both.

Metapsychology is not properly considered to be located within the science of psychology or the profession, but rather represents the space in between psychology and philosophy and the bridge between them. As a metapsychology, UTOK is ultimately structured to deal with the foundational metaphysical, ontological, epistemological, and ethical elements of knowledge, with a particular focus on the philosophy of mind, science, and the conduct of human action and the base of values for assessing such action. Framed this way, many of the great questions of philosophy and works of scholars like Aristotle, who reflected on the nature of the mind, of knowledge, and questions of what we should value, can be considered metapsychological in nature. Switching to more recent times, we can see the centrality of these issues in the way the philosopher Nick Maxwell (2019, May) frames the proper task of philosophy in general:

> The proper task of philosophy, even more important today, perhaps, than ever before, is to keep alive rational—that is, imaginative and critical—thinking about our most urgent and fundamental problems of thought and life. It is, above all, to keep alive such thinking about our most fundamental problem of all, which can be put like this: how can our human world, the world as it appears to us, the world we live in and see, touch, hear, and smell, the world of living things, people, consciousness, free will, meaning and value—how can all of this exist and best flourish embedded as it is in the [natural] Universe?

Maxwell's central question for philosophy applies equally well as the central question for UTOK. Subsumed under this grand philosophical project are two separable but related projects pertaining to developing a unified theory for the science of psychology and a unified approach for the practice of psychotherapy.

The Unified Theory is a holistic system of knowledge that consists of several different ideas that interlock to generate the whole. At the center of UTOK are four key ideas that function to make up the unified theory of psychology, and four other key ideas that make up the unified approach to psychotherapy. It is through the combination of a framework for scientific psychology that can track human mental behavior via generalizable causal processes and a humanistic approach to psychotherapy that UTOK affords a coherent synthesis between the academy's two cultures of the sciences and humanities.

The Four Key Ideas That Make up the Unified Theory of Psychology

As laid out in *A New Unified Theory of Psychology*, the unified theory of psychology consists of the following four key ideas: (1) The Tree of Knowledge System; (2) Justification Systems Theory; (3) Behavioral Investment Theory; and (4) the Influence Matrix. These ideas are briefly described here and will be unpacked in more detail over the course of this book. The Tree of Knowledge (ToK) System is the first key idea, and it is the centerpiece of the UTOK metapsychology. The most novel feature of the ToK System is that it consists of a new map of both cosmic evolution and our scientific knowledge of it. More specifically, it provides a new way to conceptualize the emergence of behavioral complexification over time as consisting of four different dimensions or planes of existence. These are labeled: (1) Matter-Object, (2) Life-Organism, (3) Mind-Animal, and (4) Culture-Person. In addition, the ToK System posits that, consistent with modern physics, the Matter-Object dimension emerged from an Energy-Information implicate order that resides as the base of the universe and serves as a kind of "ultimate common denominator" (Henriques, 2021).

The ToK System depicts four metatheoretical "joint points" that connect these dimensions of complexification. Joint points are the complexity building feedback loops that give rise to the next plane of existence. They represent central nodes of understanding that allow us to integrate foundational insights into a causal-explanatory schematic. Two of these joint points are well known. The Big Bang, along with insights from general relativity and quantum field theory, can be thought of as the joint point between the Energy and Matter planes of existence. The modern evolutionary synthesis, which merged Darwin's theory of natural selection with genetics, forms the outline of the Matter-to-Life joint point. As we will see, UTOK comes with ideas that fill in the Life-to-Mind and Mind-to-Culture joint points.

In addition to mapping the various planes of existence and the broad frames of explanation that can account for the emergence of these planes, the ToK System also maps the evolution of scientific knowledge as a specific kind of justification system that emerges from Culture in a way that loops back and maps the four dimensions of reality. This creates a new way to conceive of the relationship between reality and how science functions to map it. Chapter 6 provides an overview of how the ToK System emerged in a flash of insight, and the subsequent chapters delineate how the system provides a new way to frame scientific ontology writ large.

Justification Systems Theory (JUST; formerly named the Justification Hypothesis) is the second key idea, and its development and structure are detailed in the next chapter. JUST is a metatheory that ties together the evolution of human language, self-consciousness, the structure and function of propositional thought, and reason-giving in the lived sociocultural context. It is positioned as the joint point between Mind and Culture the ToK System, and it clarifies the fact that human culture and the behavior of human persons are framed and shaped by justification systems. This fact resulted in the shift in perspective that gave rise to the ToK System's novel depiction of reality and science. In essence, JUST is a theory of how our hominid primate ancestors evolved into modern human persons. It includes the hypothesis that the emergence of propositional language created a significant tipping point approximately 100,000–200,000 years ago that resulted in the acceleration of human cultural evolution. Coupled with the ToK System, it frames the Culture-Person plane of existence as being structurally and functionally organized via systems of justification at various aggregate scales (i.e., ranging from micro systems between dyads to large-scale systems that coordinate nations or global endeavors). In addition, as this book will make clear, together JUST and the ToK System provide a new approach to scientific onto-epistemology relative to the social construction of knowledge in a specific historical and cultural context.

Behavioral Investment Theory (BIT) is the third key idea in UTOK and is positioned as the joint point between Life and Mind in the ToK System. It functions as a metatheory for the mind, brain, and animal behavior sciences. Placed in the context of the descriptive metaphysics provided by the ToK System, BIT posits that Mind is the set of "mental behavior," which have the property of mindedness. The mental behaviors of animals with brains and complex active bodies operate on the third dimension of behavioral complexity, a complex adaptive plane of existence that is separate from both the Life-Organism plane beneath it and the Culture-Person plane above it. BIT is a framework that synthesizes B. F. Skinner's theory of behavioral selection with modern models of neurocognition, and thus it serves to bridge mainstream mentalistic approaches with Skinner's central insights regarding the selection of animal behavior via its consequences. Specifically, BIT posits that mental behaviors can be framed in terms of neurocognitive functional processes that guide the animal toward paths of behavioral investment that are anticipated to afford the best outcomes. Neurocognitively, this process unfolds via what the cognitive scientist John Vervaeke and colleagues call recursive relevance realization (see, e.g., Vervaeke et al., 2009). In addition, as we will see, BIT also readily sets the stage for understanding the evolutionary emergence of subjective conscious experience in the animal world.

The Influence Matrix is the fourth key idea. It is an extension of BIT, and it works to map the human relationship system. Specifically, the Matrix represents the human self-in-relation-to-other mental architecture that guides and drives human perception, motivation, and emotion in the social world. It provides a map that merges John Bowlby's attachment theory and Timothy Leary's Interpersonal Circumplex and functions to specify the social influence dynamics that guide human social interaction and their felt sense of being in the relational world. The Influence Matrix highlights four key relational dimensions that are tracked by the human preverbal social cognitive processing architecture. The primary axis on the Matrix is relational value and social influence, which refers to the extent to which the individual has influence and feels known and valued by important others. There are also three relational process dimensions called power, love, and freedom that place and track the individual in social space. The Influence Matrix serves as an important bridge between BIT and JUST. It aligns with both the basic behavioral investment tendencies mapped by BIT and how JUST maps how humans tend to justify their actions on the social stage to maintain their relationships and status. Together, these three ideas provide a map of human mental behavior, framed in terms of processes and dynamics of investment, influence, and justification.

A New Unified Theory of Psychology showed how the ToK System, BIT, JUST, and the Influence Matrix interlock to generate the outline of an effective metatheory for the science of psychology. Specifically, the ToK System affords a big picture zoomed-out view that allows for the vertical integration of psychology with the physical and biological sciences from below and the social sciences from above. In addition, the four key ideas function to assimilate and integrate the central insights from the behavioral, cognitive, humanistic, and psychodynamic traditions, which are the dominant individual-level paradigms in the field. This book shifts the focus from the capacity of the unified theory of psychology to assimilate and integrate the major paradigms to UTOK as a metapsychology that addresses deep philosophical problems that have prevented consilience from being achieved.

The Four Key Ideas That Make up the Unified Approach to Psychotherapy

The Unified Approach to psychotherapy returns to the original problem I encountered in my graduate training in clinical psychology, which was the applied problem of psychotherapy integration. This is the problem of how we bridge research and theory in a way that assimilates and integrates the key insights from the major perspectives to develop the most comprehensive and effective systems of understanding that are up to the task of grounding the work of licensed health service psychologists. These ideas were alluded to in *A New Unified Theory of Psychology*; however, they were only being initially formulated at that time. It is important to note that the Unified Approach relates directly to my primary professional role as a clinician and my educational responsibilities as a professor who trains "psychological doctors." A brief summary of this aspect of UTOK is important to understand the origin and scope of the theory; however, the Unified Approach is the subject of a separate forthcoming book and will not be elaborated on in this present work.

In 2003, I joined the "Combined-Integrated" Doctoral Training Program in clinical, counseling, and school psychology at James Madison University (JMU). I became training director for the program in 2005 and remained in that role until 2017. Consequently, I was intimately involved in both learning the model of Combined-Integrated training and then developing and implementing it. When I arrived at the program, the founders were already on the leading edge of the argument that the future of professional psychology needed to transcend the traditional practice boundaries of clinical, counseling, and school psychology, and could do so in a way that offered a more effective, integrated identity. I joined the program as it hosted the Consensus Conference on Combined-Integrated Training (Henriques & Sternberg, 2004), which explained why the innovative model affords a way of training that effectively merges the traditional professional specialty areas of clinical, counseling, and school psychology into a generalist approach.

The JMU Combined-Integrated program has always advocated for a theoretically integrative vision of professional psychology. The idea of a more "unified professional psychology" (Henriques & Sternberg, 2004) consists of the amalgamation of three different movements in psychology toward: (a) a more unified science of psychology; (b) an independent professional psychology; and (c) generalist training models that cut across the traditional practice areas of clinical, counseling, and school psychology. There have been several developments in professional psychology that accord with this view. The most notable has been the development of the Health Service Psychology as an omnibus category that cuts across the practice areas. As of January 1, 2017, Health Service Psychology became the official term used by the American Psychological Office of Accreditation to denote professional psychologists across the practice areas. The concept of health service psychology stemmed in part from the vision of the combined-integrated programs articulated in the early 2000s.

I briefly allude to this background because, although it will not be a focus in the current work, the Unified Approach is a central aspect of the overall UTOK. Indeed, the Unified Approach is central to my role as the current President of the Society for the Exploration of Psychotherapy Integration, where my presidential theme is focused on elucidating the "common core" of psychotherapy (Henriques, 2022, March). In addition to grounding the key elements regarding the practice of psychology, the Unified Approach more generally highlights the fact that UTOK is not just a set of abstract ideas about knowledge and theory. Rather it is ultimately a concrete project about the philosophy of living the cultivation of an ecology of practices that are centered on the alleviation of suffering, living a rich and fulfilling life, and enhancing human flourishing in a way that emphasizes healthy relational communities. It is this linkage that moves the system from being focused primarily on analytic philosophy and science (i.e., what Kant called "pure reason") to one that is deeply concerned with values, ethics, and morality, and living the good life (i.e., "practical reason"). The ultimate vision is that UTOK is positioned as a theory of knowledge that helps humanity move toward wisdom in the twenty-first century.

Like the Unified Theory, the Unified Approach also consists of four key ideas. Character Adaptation Systems Theory (CAST; Henriques, 2017) is the first key idea in the Unified Approach and offers a taxonomy of five different systems of character adaptation: (1) the habit system; (2) the experiential system; (3) the relationship system; (4) the defensive system; and (5) the justification system. This "new big five" offers a way to bridge modern personality theory with the major paradigms in individual psychotherapy. The habit system corresponds to the behavioral paradigm in psychotherapy, the experiential system corresponds to the neo-humanistic emotion-focused paradigm, the relationship and defensive systems correspond to the modern psychodynamic paradigm, and the justification system corresponds to the cognitive, existential, and narrative paradigms.

The Wheel of Development is the next key idea. Centrally grounded in CAST, which is placed as the axis on which the wheel of development turns, it identifies five key domains of personality development across the lifespan. The domain of identity refers to the self-concept and meaning-making system for one's place in relation to the world and other people. The domain of traits refers to the dispositional setpoints that emerge based on genetic dispositions, but must be thought of as tracking into the epigenetic development of the various patterned tendencies of adaptation. The domain of abilities and talents refers to the specific set of skills and procedural knowledge that individuals accumulate. The developmental domain of values and virtues refers to moral, ethical, and spiritual development. Finally, challenges and pathologies refer to systematic domains of difficulty and entrenched maladaptive patterns, such as those categorized by the Diagnostic and Statistical Manual of Mental Disorders (APA, 2013). Together, CAST and the Wheel of Development provide a comprehensive map of character adaptation and functioning and the major domains of personality development.

The Nested Model of Well-Being (Henriques et al., 2014) functions as a descriptive metaphysical system for defining human well-being. It identifies four nested layers or domains, as follows. First, the subjective domain refers to the first-person, conscious experience of being and includes both the positive and negative emotional and mood-based valences, as well as the reflective evaluations of how an individual assesses their levels of satisfaction and fulfillment. Second is the health and functioning domain, which refers to both the biophysiological functioning and neurocognitive and personality functioning, including things such as character structure and cognitive capacities. The third nested level is the environmental domain, which consists of the affordances and stressors present in both the material and social worlds. Fourth, there is the evaluative domain, which consists of the ideology and values of the evaluator. The final important point is that all these domains are put on a developmental timeline to contextualize them. The concluding summary of the Nested Model can be thought of as validating Kant's assertion that well-being can be effectively conceptualized as "happiness with the worthiness to be happy."

Finally, there is the CALM-MO approach to psychological mindfulness (Henriques, 2018, May/June). It identifies the key elements associated with fostering adaptive, reflective, and attuned responses to social and emotional conflicts and adverse experiences. Specifically, it guides individuals to engender a "metacognitive observer" (an MO) that cultivates an attitude that is *c*urious, *a*ccepting, *l*oving/compassionate, and *mo*tivated to grow toward valued states of being in the short and long term. The focus is to bring a CALM-MO observer to the "vertical organization" of one's psychological experience, starting with the base into the body (position, feelings of pleasure, pain, basic safety, appetites), into the heart (felt sense of attachment and belonging), into the head (the self-conscious rational and narrating portion of the human mind), and finally into the spirit (i.e., one's larger or trans-egoic sense of meaning and purpose). The CALM-MO frame is a heuristic for the key principles that enable processes to be enacted that allow for the metabolizing of difficult events, reduce unnecessary suffering, and generate good faith efforts toward adaptive growth grounded in the ethical sense of the good.

The ideas that make up the Unified Approach and how they set the stage for practical reasoning and psychological fulfillment at the individual and relational levels of analysis are a topic for a different book. Such a work would focus on: (1) the emerging unified psychotherapy movement; (2) the primary ways UTOK conceptualizes personality and psychopathology; (3) the key principles and processes that guide the Unified Approach to psychological assessment and intervention; and (4) the delineation of a unified health service psychologist identity and approach to training and education. In addition, it would include the kinds of adaptive and developmental trajectories that tend to lead to optimal functioning, in contrast to the reverse. Although these ideas will not be featured prominently in this work, it is crucial that they are mentioned here because they serve as key elements that make up UTOK. That is, the Unified Approach is central to the architecture that enables the UTOK metapsychology to guide practical reason toward the cultivation of wisdom in the real world. Moreover, no solution to the problem of psychology would be complete if it did not directly address the profession and the practice of intervention, assessment, and consultation in various systems.

Ultimately, the UTOK metapsychology is structured in a way that forms a new bridge between scientific and professional psychology that can knit together the currently chaotic, fragmented pluralistic field and afford the vision for moving toward a coherent integrated pluralism. And in so doing, it gives rise to a web understanding that enables one to obtain a new metaphysical, ontological, and metatheoretical grounding of coherence that was previously absent, because it had been covered by the long shadow cast by the Enlightenment Gap. This new metaphysical, ontological perspective afforded by UTOK will hopefully seed new understandings and new ways of being in the twenty-first century.

Toward a Coherent Metaphysical, Ontological, and Metatheoretical Frame for the Science of Psychology

The central task of the current work is to delineate how UTOK affords us a descriptive metaphysical system that can coherently frame the science of behavior and mental processes. This means that we need to (a) further clarify what is meant by science and a naturalistic scientific ontology and do so in such a way that (b) generates a coherent frame for (redefining) behavior and mental processes. The next two chapters deepen the description of the two foundational ideas that make up UTOK, JUST, and the ToK System. Together, they provide the insights that allow UTOK to develop a much more coherent frame regarding the relationship between both matter and mind and scientific and social and subjective knowledge than has been offered to date.

Section III of this book consists of two chapters that advance the argument that the UTOK affords us a coherent naturalistic scientific ontology via the ToK System. The second chapter in the section corresponds the ToK System to the Big History movement and explains how Big History frames the standard natural scientific ontology but does so in a way that fails to generate a picture that is both complete and coherent. This is because it neglects the problem of psychology and thus does not provide an adequate descriptive metaphysics for the ontology of the mental. The subsequent chapter in the section demonstrates how the ToK System is aligned with philosophical perspectives in systemic metaphysics that justify the framing that there are different orders of nature that move from the energetic to the material to the living to the animal mental and finally the human culture. This affords convergent validation for the ToK System's central claim that there are four ontologically separable dimensions of existence. This aspect of the argument sets the stage for clarifying the ontology of the animal-mental dimension and how it can be crisply defined as arising out of the living dimension from below, and can be differentiated from human cultural processes mediated by propositional language from above.

Section IV builds from the map of science and reality generated given by the ToK System and bridges this analysis to the concept of behavior and its relationship to both psychology and science writ large. The first chapter in this section recounts the history of behaviorism in psychology and tracks how and why the concept of behavior migrates from psychology down into the biological and physical scientific lexicon. The chapter argues that the ultimate reason for this migration is that behavior can be framed as the central concept in science writ large. The chapter shows how and why behavior maps onto the metaphysical, epistemological, and ontological aspects of modern empirical natural science. This is a fact that has been overlooked in the philosophy of science, but it has major implications especially for psychology and how we think about what behavior refers to in the world.

The second chapter in Section IV elaborates on this claim. Specifically, the map of science and reality provided by the ToK System posits that science is about mapping patterns of behavioral frequencies in nature across aggregates, levels and dimensions of existence. Given the basic logic of the ToK System, these ontological emergent patterns should be divided into two separate axes, one of which represents the emergence of behavioral frequencies *within* a dimension (i.e., complexity emerges within the Matter plane) and the other of which represents the shifts *between* the dimensions (i.e., Life is a novel plane of complexification that arises out of Matter).

This insight gives rise to the Periodic Table of Behavior (PTB). It posits that each dimension of Matter, Life, Mind, and Culture has a primary entity of analysis, which are, respectively, atoms in the Matter dimension, cells in the Life dimension, animals in the Mind dimension, and persons in the Culture dimension. These primary entities can be analyzed in terms of parts, wholes, or groups across aggregate scales. The PTB is a table that places the dimensions as columns and the levels as rows and shows how they come together to map the primary stratified layers in nature in a way that is commensurate with the organization of the sciences. Together, these two chapters show that by separating and then realigning the metaphysical, epistemological, and ontological aspects of behavior, a coherent picture emerges of the arrangement of the sciences across the various levels and dimensions of complexification in nature.

With key aspects of science and reality clarified by the concept of behavior, Section V moves to frame the metaphysics, ontology, and metatheoretical analyses of mental behavioral processes and works to address the BM^3 problem specifically. The UTOK metapsychology enables us to effectively separate and interrelate the: (1) neurocognitive processes that regulate functional awareness and overt behavioral response patterns in animals that are available from the third-person exterior viewpoint; (2) subjective conscious experiences that are only available from the first-person experiences of being; and (3) self-conscious narration in humans that is accessible via the intersubjective processes of having a shared language.

As demonstrated by the argument laid out in the first two chapters, in generating such a solution we need to not confound the ontological and epistemological considerations of these referents associated with these three domains. The solution generated by UTOK is a new map of the domains of mental processes called the Map of Mind^{1,2,3}. It identifies the first domain of mental processes as Mind^{1}, which is defined as mental behaviors exhibited by animals that are mediated by neurocognitive processes. These can be framed by a scientific exterior epistemology in a relatively straightforward manner. In contrast, the domain of Mind^{2} refers to subjective conscious experience, which is available only from the interior epistemological perspective and is much more difficult to frame via a scientific view of the world, which is behavioral and exterior in nature. Finally, there is the domain of Mind^{3}. This refers to human self-conscious narrative reflection and reason-giving. Interestingly, here the referent is intersubjective in nature. That is, it is based on a language that must be learned in a socialized context, and its content seamlessly moves across the interior subjective and exterior objective epistemological frames. Table 4.1 summarizes these points and shows how the three domains of mental processes have different ontological referents and are framed by different epistemological considerations.

The combination of the Periodic Table of Behavior and the Map of Mind^{1,2,3} gives a fundamentally new way to frame behavior and mental processes. Specifically, these tools afford us descriptive metaphysical systems that clarify the ontological referents relative to the epistemological position of the observer. The vocabulary that emerges opens a fundamentally different approach to the science of psychology, one that is based on a clear ontology of the mental that is grounded in a naturalistic scientific

Table 4.1 The three domains of mind and their epistemological vantage points

Definition	*Ontological referent*	*Interior epistemological view*	*Exterior epistemological view*
Mind^{1}	Neurocognitive functional analysis of mental behaviors		X
Mind^{2}	Subjective conscious experience	X	
Mind^{3}	Self-conscious justification	X	X

ontology. Such an approach can be labeled mental behaviorism and contrasted with the standard methodological behaviorism that is taken by mainstream academic psychology.

After detailing the Map of Mind^{1,2,3} in Chap. 11, we then move to link this metaphysical and ontological frame with metatheory. Chapter 12 explains how Behavioral Investment Theory affords us a coherent metatheory of Mind^{1}. It tracks the evolution of the animal-mental dimension of behavioral complexity across four stages of: (1) reacting; (2) learning; (3) thinking; and, finally, in humans, (4) talking. It also sets the stage for the evolution of Mind^{2}, the subjective conscious experience of being. Chapter 13 gives a model of Mind^{2} in the animal world that evolved via two primary steps. The initial, early step is characterized by the emergence of "valence qualia" (i.e., feelings like pleasure and pain) that can be thought of as residing at the base of sentience. There is good reason to believe that such core feelings may have emerged during the Cambrian explosion approximately half a billion years ago. As brains grew and behavioral flexibility associated with deliberation and thinking became more advanced, a second step emerges. Specifically, Mind^{2} grows into an "experiential self" with an inner mind's eye that operates on a global workspace that can pull from memories and anticipate future outcomes. We will review the evidence for such processes being present in birds and mammals, and perhaps other intelligent animals like octopuses.

Social processes push the evolution of the subjective conscious experience and the complexity of Mind^{2} even further in the primate line into our hominid ancestors. Chapter 14 tracks the evolution of the relational world and the human primate capacity for shared attention and intention. Here, the Influence Matrix affords a metatheoretical framework that maps the human relationship system by combining attachment theory and the Interpersonal Circumplex and shows that our capacity for mind reading and intersubjective coordination created a preverbal self–other relational field that is more advanced than that of our ape cousins. The final chapter in the section transitions into Mind^{3} and the emergence of propositional language and the Culture-Person plane of existence. It shows how JUST is a metatheory that can frame the metaphysics and ontology of human persons and the Culture-Person plane of existence.

The cumulative result is a new way to frame the metaphysics, ontology, and metatheory of the science of both mental behavior and human persons. Instead of lumping many different fields of cause and effect together and simply applying the methods of behavioral science based on the operational definitions of the researcher, UTOK gives rise to a mental behaviorism that affords metaphysical clarity regarding the ontology of the mental tied together by metatheoretical ideas that assimilate and integrate key insights and empirical findings into a coherent, consilient whole. The concluding chapter delineates the broader philosophical and societal implications of solving psychology's BM^3 problem and resolving the Enlightenment Gap.

Conclusion

The UTOK metapsychology gives rise to a new, coherent naturalistic worldview that can resolve the Enlightenment Gap and solve the problem of psychology. The next two chapters turn to the development of the Unified Theory, specifically focusing on the first two key ideas that led to its overall structure. Justification Systems Theory and the Tree of Knowledge System transform the landscape of understanding and set the stage for clarifying both the relationship between matter and mind and scientific and social-subjective systems of justification. With these key puzzle pieces in place, we will be able to better understand how UTOK is positioned to resolve the Enlightenment Gap and generate a clear ontology of the mental, defined as mental behaviors that exhibit the property of mindedness. The ultimate result is a consilient scientific humanistic worldview that can be oriented toward the cultivation of wisdom in the twenty-first century.

References

American Psychiatric Association. (2013). *Diagnostic and statistical manual of mental disorders* (5th ed.). https://doi.org/10.1176/appi.books.9780890425596

Henriques, G. R., & Sternberg, R. J. (2004). Unified professional psychology: Implications for the combined-integrated model of doctoral training. *Journal of Clinical Psychology, 60*(10), 1051–1063.

Henriques, G. R. (2011). *A new unified theory of psychology*. Springer.

Henriques, G. R., Kleinman, K., & Asselin, C. (2014). The nested model of well-being: A unified approach. *Review of General Psychology, 18*, 7–18. https://doi.org/10.1037/a0036288

Henriques, G. R. (2017). Character adaptation systems theory: A new big five for personality and psychotherapy. *Review of General Psychology, 21*(1), 9–22. https://doi.org/10.1037/gpr0000097

Henriques, G. R. (2018, May/June). The root of suffering: How to fight the neurotic loop of negative reactions to negative feelings. *Psychology Today,* 48–50.

Henriques, G. (2019). Toward a metaphysical empirical psychology. In T. Teo (Ed.), *Re-envisioning theoretical psychology: Diverging ideas and practices* (pp. 209–237). Palgrave Macmillan.

Henriques, G. (2021, October). *The "singular energy-information superforce field" that started it all.* [Blog] Medium, unified theory of knowledge. Accessed August 1, 2022, from https://medium.com/unified-theory-of-knowledge/the-singular-energy-information-superforce-field-that-started-it-all-2330bfc331d7

Henriques, G. (2022, March). President's column. *The Integrative Therapist, 8*(1), 1–5.

Koons, R. C., & Pickavance, T. (2014). *Metaphysics: The fundamentals.* Wiley.

Maxwell, N. (2019, May). Natural philosophy redux. *Aeon.*

Nirenberg, R. (1996). *The birth of modern science: Galileo and Descartes, a lecture by Ricardo Nirenberg.* Fall 1996, the University at Albany, project renaissance. https://www.albany.edu/~rn774/fall96/science2.html

Perl, E. (2014). *Thinking being: Introduction to metaphysics in the classical tradition.* Brill.

Strawson, P. (1959). *Individuals: An essay on descriptive metaphysics.* Routledge.

Vervaeke, J., Lillicrap, T., & Richards, B. (2009). Relevance realization and the emerging framework in cognitive science. *Journal of Logic and Computation, 22*(1), 79–99. https://doi.org/10.1093/logcom/exp067

Wilson, E. O. (1999). *Consilience: The unity of knowledge.* Vintage Books.

Witherington, D. C., Overton, W. F., Lickliter, R., Marshall, P. J., & Narvaez, D. (2018). Metatheory and the primacy of conceptual analysis in developmental science. *Human Development, 61*(3), 1–18. https://doi.org/10.1159/000490160

CHAPTER 5

Justification Systems Theory

Insight refers to getting a fresh perspective on the world such that one can see new solutions to old problems. Between the end of 1996 and the middle of 1997, I was fortunate to stumble upon two key insights that would plant the seeds for what would become UTOK. This chapter and the next tell the story of how they emerged. The first insight took hold over a period of about six months and became what I would ultimately refer to as Justification Systems Theory (JUST). Although the idea has matured over the last 25 years, the basic structure was uncovered during those first six months and has remained. In prior work I have referred to the idea as the Justification Hypothesis. However, as I have explained in a previous essay (Henriques & Michalski, 2018), I have come to see that this is something of a misnomer and a better label is Justification Systems Theory.

Part of the reason for the name change is that JUST consists of three interlocking claims, and only the first is rightly considered a hypothesis. Those three claims are: (1) the evolution of propositional language generated question–answer dynamics framed as the adaptive problem of justification, and this in turn shaped both the evolution of human self-consciousness and human culture such that; (2) the nature of human consciousness can be framed by the "Updated Tripartite Model" that differentiates it into three different domains of (a) the experiential self, (b) the private ego, and (c) the public persona; and (3) humans are

G. Henriques, *A New Synthesis for Solving the Problem of Psychology*, Palgrave Studies in the Theory and History of Psychology,
https://doi.org/10.1007/978-3-031-18493-2_5

transformed into persons as a function of being socialized into systems of justification, such that, relative to other animals, human persons operate in a new complex adaptive plane that can effectively be framed as Culture with a capital "C."

As suggested by this summary, central to JUST is a set of insights about what happened to the human mind and the social world as the capacity to form propositional language and "question answer dialogue" emerged. Specifically, via adopting an evolutionary lens, it asks the question: *What kinds of adaptive problems did the emergence of propositional speech result in and what are its downstream structural and functional consequences?* It then looks at modern-day humans and knits together a coherent picture of how all that came to be. JUST posits that there was a tipping point framed as the problem of justification that started a complexity building feedback loop that resulted in what some scholars refer to as "the human mind's big bang" (e.g., Rolston, 2010). More specifically, it posits the hypothesis that the problem of justification gave rise to the structural and functional organization of the human ego (i.e., Mind[3]) and the development of large-scale systems of justification that coordinate groups, specified as the emergence of the Culture-Person plane of existence. It then demonstrates how our understanding of human consciousness and the Culture-Person plane of existence can be deeply enhanced by this perspective.

Although it is anchored to an evolutionary analysis, JUST starts with our everyday lives. It highlights that the problem of justification remains very much with us and functions to shape and constrain much of what we do. It refers to the dynamics and difficulties that come with determining what is justified, both analytically regarding what is true or accurate and pragmatically in terms of our own desires and our various relationships with others placed in the sociocultural contexts in which we reside. That is, it refers to how we explain and legitimize our claims of what is and what ought to be, both to ourselves and others, as we attempt to take into consideration the current and future states of affairs. It also includes the flip side of this process, which is that we must also work to determine the justifiability of other people's actions, claims, and accounts.

Take a moment and think about how often you are describing and explaining yourself to others or listening to the accounts of others and pondering their justifiability. When you look for such processes, you will see that they are omnipresent in human activities. As I put it in my prior writings (Henriques, 2004, p. 1216):

> Unlike all other animals, humans everywhere ask for and give explanations for their actions. Arguments, debates, moral dictates, rationalizations, and

> excuses all involve the process of explaining why one's claims, thoughts or actions are warranted. These phenomena are both uniquely human and ubiquitous in human affairs. In virtually every form of social exchange, from warfare to politics to family struggles to science, humans are constantly justifying their behavioral investments to themselves and others.

We can step back and look at this book from the frame of reference provided by JUST. When we do, we can see that the process of justification is happening right now. Via this book, I am attempting to justify why you should consider my account of UTOK. Consistent with JUST, it is laid out as an interlocking set of ideas that constitute a justification system, and you are reflecting on it accordingly. Via highlighting this book as an example, it becomes clear that the problem of justification is much more than simply rationalizing one's actions or defending one's biased claims, the way a lawyer might do for a client. To be sure, this is an aspect highlighted by JUST; however, it is only one angle on it. There are also crucial aspects that include authentic social persuasion, rhetoric, and argumentation. Justification can be considered to include processes of logical and analytic reasoning that include induction, abduction, and deduction about factual states of affairs. As this description implies, the problem of justification and the process of justifying and determining what is justifiable can be placed on a broad continuum that stretches from defensive rationalizations on the one hand to honest, pragmatic social reason-giving and ultimately to logically derived conclusions by analytic processes on the other.

JUST posits that the problem of justification has several elements that need to be disentangled. First, aligning with some of the key insights of the philosopher David Hume, we can make a division based on issues of fact or accuracy, as opposed to values or pragmatic interests. In the real world, the elements of facts and values are closely knitted together. However, as Hume famously argued, questions of fact can be conceptually separated out from questions of value. The accuracy problem of justification refers to determining whether propositions carry truth value or can be said to accurately correspond to the current situation. For example, consider the time when I told my daughter Sydney not to eat the cookie before dinner. She was about four, and I went upstairs to do a few things and came down to find the cookie gone. I asked her if she ate the cookie. She said "no." I then asked her to explain the cookie's absence and account for why there was chocolate smeared on her lips. In that moment she experienced how propositional claims such as "I did not eat the cookie" can be aligned with states of affairs to raise issues of honesty. I then proceeded to mildly scold her for both disobeying and lying. However, knowing that lying is an important phase of cognitive development, I was hardly dismayed. Rather, it simply reflected an important moment in teaching her how to be a person.

The value aspect of the problem of justification pertains to employing the justification at that moment and what it implies about what ought to be, both for the person and the relational or social levels of analysis. At the most basic pragmatic level, we can ask: *Why is attention being brought to this propositional claim at this moment and what are the implications of it for those involved with it, believing it, or acting on it?* We can break this up further and say there are issues of value pertaining to the individual who is generating the proposition and the social implications for others who are considering its implications. If we continue with the example pertaining to Sydney, questions might be raised, such as: *Why don't we eat cookies first? Why do parents get to set the rules? Or why is negative attention being directed here rather than elsewhere?* As Hume's analysis makes clear, the questions about the existence of cookies and whether Sydney ate one are different than whether we ought to eat cookies after dinner rather than before.

The overall point is that the problem of justification at a minimum involves a tangled set of issues pertaining both to factual aspects of what is and value-based assertions of what ought to be. In addition, how and why propositions are novel forms of meaning-making units that carry significant implications for the world of humans. Indeed, it is the primary frame that sets the constraints of what we recognize as "question and answer dialogue." In this way of thinking, "answers" refer to propositional statements that carry "positive" meaning about what is the case or values about what ought to be. Questions, in contrast, can be thought of as pointing to a kind of "negative" space that raises issues of fact or value, and the social process of justification refers to the dynamic back and forth that takes place on the Culture-Person plane of existence. In addition, networks of propositions get tied together to frame Q & A processes, such that dialogue between people is framed by the large-scale context of justification.

As a network of interlocking ideas, JUST ties together three different lines of consideration that are rarely woven together in a coherent way. First, there is the evolutionary line of thinking. JUST starts with the evolution of propositional language and points out that as language evolved to include both propositions and the ability to question them, humans became the first animals that had to *justify* their thoughts, feelings, and actions to others initially and then to themselves as they developed a stable inner self-concept. The "Justification Hypothesis" is that this was a novel adaptive problem that shaped the human mind, specifically giving rise to the domain of mental processes UTOK labels as Mind^3. As we will see, structurally and functionally, Mind^3 is organized as propositional networks of justification systems. The second line of thinking maps the Justification Hypothesis to the major domains of human consciousness. Specifically, JUST clearly delineates the relationship between subjective conscious experience (i.e., Mind^2) and self-conscious justification (Mind^3), such that it affords an "Updated Tripartite Model" that is consistent with both modern human psychological science and Freud's central insights regarding the tension between our animalistic impulses and the judgments of our social world. The third line of thinking provides a way to frame human Culture as large-scale systems of justification. This bridges JUST to the Tree of Knowledge System, and together they clearly delineate the Culture-Person plane of existence as a separable dimension of complex adaptive behavior. Prior to explicating these three aspects of JUST in greater detail, we can enhance the justification for JUST by sharing the narrative that gave rise to it.

The Development of JUST: A Missing Piece in Understanding Human Psychology

By the mid-1990s, my training in psychotherapy had demonstrated to me that although the various schools of thought had value, they were incoherently organized in relationship to each other. This created a central dilemma, as I wanted to know how to draw on human psychological theory grounded in science in such a way to guide my practice in a more coherent and consilient way. By 1995 I had found much to like in the emerging field of evolutionary psychology. This perspective had caught fire in the early 1990s, and it offered an exciting new view of the human mind. Specifically, these scholars advanced the claim that human nature could be delineated by combining modern evolutionary theory with cognitive science. For a few years I saw in evolutionary psychology the outlines of a perspective that seemed to have the potential to organize the science of psychology at the metatheoretical level. However, by the end of 1996, I was seeing its limitations and realized it was going to be just another mid-level paradigm in psychology, as opposed to a truly effective metatheory that could ground the field in its entirety.

There were several reasons for this conclusion, including its overly strong emphasis on what its founders (i.e., Barkow et al., 1992; Pinker, 1994) called "domain modularity" (i.e., specific neurocognitive algorithms and heuristics that had developed to solve specific evolutionary adaptive problems). Their commitment to these essentialist computational forms meant that evolutionary psychology would struggle to effectively incorporate the key insights from behavioral science that suggested a more general, contextual, and environmentalist view of learning. In addition, evolutionary psychology did little to clear up the distinction between the behaviors of animals relative to socialized persons, and it did not mesh well with social science viewpoints pertaining to culture and the social construction of reality as large-scale systems of justification.

In the language of the UTOK metapsychology, we can say that evolutionary psychology led to important insights, but it was not up to the task of addressing the Enlightenment Gap or to solving key aspects of the BM^3 problem. My move away from evolutionary psychology accelerated as I developed the ideas that would lead to JUST. This is ironic because JUST is a deeply evolutionary idea. It involves an evolutionary "reverse engineering" analysis of human self-consciousness, such that it provides a clear account of both Freud's central observations about the nature of the human ego and subconscious socioemotional forces, as well as the emergence of human Culture as delineated by social constructionist theorists.

A Case Example Where the Exception Proves the Rule

In 1996, I had embarked on my doctoral training at the University of Vermont and was being supervised in my clinical work via a modern psychodynamic perspective. I also had started doing my dissertation on Aaron T. Beck's concept of cognitive distortions and the role they played in how people reacted to events. As such, I was immersed in both thinking about how people make interpretations, explanations, and attributions, and how people avoid, defend, and repress unjustifiable thoughts and feelings. In retrospect, my quest for a big picture view to frame psychotherapy and the intersection of evolutionary, cognitive, and psychodynamic perspectives combined in such a way that I was primed to be able to see JUST.

By late fall of that year, I had landed on the idea that propositional language led to a novel adaptive problem for our hominid ancestors. Specifically, propositions and the inevitable question and answer dialogue in a social environment that they would spark meant that people had to be able to justify themselves on the social stage. This insight allowed me to quickly organize many findings in social and cognitive psychology. For example, self-serving biases, cognitive dissonance, so-called myside biases, and other kinds of interpretive and attributional processes showed a strong connection to the dynamics of social influence, in that people generated attributions and explanations to afford themselves being placed in more justifiable space. Indeed, I recall reviewing Taylor and Brown's classic 1988 paper on positive illusions and mental functioning that concluded that people *tended to interpret and narrate events with a self-serving bias that was positively shifted in accordance with what the data could justify.* Viewed via the lens of JUST, we can say that there was an accuracy dynamic and a personal–social value dynamic, and the justified conclusions about the self in the social world that people arrived at could be thought of as emerging in the space between the joint pressure of accuracy and values.

The culminating insight that served as a catalyst for JUST was a case where an exception to the general tendency of self-serving biases would prove the rule that the justification systems people develop are tied to social influence and context. I was doing a comprehensive psychological evaluation with a patient who was a Caucasian woman in her late 30s, who was very depressed and extremely shy and socially anxious. Although she had an above average intellect, she had barely graduated from high school, worked as a teacher's aide, and lived in almost complete isolation on the brink of poverty. Perhaps the most striking feature about her character was her remarkably low self-esteem, and how she had inwardly turned against herself in a brutal and degrading way. She conceived of herself as worthless, ineffective, stupid, and incapable of doing anything.

On the surface, her self-deprecating tendencies seem remarkably inconsistent with the idea that people will tend to justify themselves in a way that affords them social influence. However, her developmental history clarified the situation so that what appeared at first as an exception ended up demonstrating the principle. As I did the evaluation, I learned of her difficult childhood. Her father was a raging alcoholic who dominated her timid, submissive mother, and would verbally abuse the patient, calling her weak and stupid. Crucially, he never hit the patient. However, he would be physically abusive to her older brother, who was far more defiant of his power. She distinctly remembered several episodes of her father beating her brother, while yelling that her brother needed to be more like his obedient sister. Given this developmental background, a new frame of understanding emerges regarding the way she thought of herself.

To see why, we can ask: *What would have happened to the patient if she had tried to assert her competence, her rights, and her power?* Her father likely would have turned on her, and she may well have been physically assaulted. After all, this is what happened to her older and physically stronger brother. Framed this way, her self-talk and self-concept becomes much more understandable. Placed in this social and relational context, we can see that her self-justifications were tied to how she navigated her father's actions. That is, her self-abnegation almost certainly functioned in part to avoid being abused by her father. Via aligning with her abuser and shaming herself, her critical introject (i.e., the inner voice in her private narrator that judged and critiqued herself) functioned to justify submission and deference in a context where any form of defiance was severely punished. This formulation is analogous with what is known as Stockholm Syndrome (Kuleshnyk, 1984). This is where individuals who are threatened or held hostage over long periods of time (days to years) will sometimes adopt the justification systems of their captors. Seen through the lens of JUST, this occurs because it orients such individuals to cooperate and submit, which in turn would be potentially protective from a survival or social influence perspective. With this narrative backdrop in place, we can now turn to the three lines of thought that weave together to make up JUST, with the first being how the evolution of propositions and questions gives rise to the problem of justification, which in turn becomes a major adaptive force in human evolution.

Propositional Statements and the Evolution of the Problem of Justification

The exact nature of how language evolved and the extent to which there is an innate and universal "language acquisition device" versus a more flexible structure that enables language to be learned during development continues to be debated by scholars (e.g., Greenspan & Shanker, 2004). Although these debates are important, JUST largely sidesteps the specific areas of contention. It adopts the idea that there is the basic architecture that allows for language acquisition and at least minimally structures the way language is learned, but it does not differentiate between the various accounts of exactly how this unfolds. JUST also presumes that such a symbolic-syntactical communication system would have conferred many adaptive advantages, both in sharing information cheaply and effectively and in fostering greater capacities for abstract thinking. For example, justification capacities shift the landscape of cognition in humans because humans can rapidly develop what Cecilia Heyes (2018) calls "cognitive gadgets," which are models of the world developed via the web of interlocution that allows for the sharing of novel ideas.

Although many other animals have highly sophisticated communication systems, it is nevertheless the case that human language is unique in the animal kingdom. At its core, human language involves the capacity to symbolically tag objects and events held in working memory and place them in a semantic structure. Like other cognitively sophisticated animals, humans can mentally manipulate scenes in their minds. However, humans can learn to quickly associate sounds and symbols with aspects of the mental imagery and place them in a grammar. Human language involves symbolically tagging at least three different elements that go into the represented scene. These are central entities in representational thought: (1) objects; (2) changes in the scene over time, often caused by functional actions; and (3) differences between objects and kinds of functional change patterns.

It is no accident that these three elements are the fundamental units of language. Nouns are the "things" in the scene, verbs are the patterns of activity or change, and adjectives are the differences found among entities or events. Human language also includes the capacity to place these different kinds of words into a grammatical sequence to convey propositional meanings, usually in the form of subjects, objects, and verbs. We can fill in the picture with other key aspects of grammar. For example, there are pronouns that place the observer in relation to the events in position, and there are adverbs that qualify the nature of change. There are also prepositions that place things in relation to each other on the dimensions of space and time (e.g., "over" or "across"). Thus, if I say, "The blue crab ran across the sand," you are able to conjure an image of this scene, one that is different than if I said, "The sand ran across the blue crab."

JUST takes this basic evolutionary and structural view of human language as a given. It picks up the story as language moves from a partial symbolic system into a full, open system of symbolic-syntactical communication. A partial system would be one where one could reference specific objects, events, or difference, with single words such as "antelope" or "there." However, such disconnected individual word-signs are not organized into meaningful propositions. JUST highlights the idea that propositions have truth claims and these truth claims can be questioned, which creates a crucial feedback loop. More specifically, the transition into a full semantic syntactical system creates a tipping point because such statements convey meaning that can be analyzed and argued about. Unlike "antelope" or "there," the full sentence, "The antelope are over there," carries a specifiable meaning that can then be determined to be accurate or not. According to JUST, the emergence of propositional speech then gave rise to the capacity to question the validity of such propositions. *JUST posits that it was the capacity to ask questions about propositional statements that fully set the stage for the emergence of the problem of justification, as well as the development of shared systems of justification.*

JUST frames the structure and function of linguistic thought as arising from a dialectic between questions and answers. By "answers" I mean affirmative claims or propositions that confer meaning. By "questions" I mean queries that function to do things such as elicit propositional knowledge (e.g., What is going on here?) or challenge existing claims (e.g., Why do you believe that?) or highlight the lack of knowledge (i.e., How did we come to be here on Earth?). By attending to the dynamic relationship between questions and answers, JUST recasts the basic unit of linguistic meaning—usually characterized as a proposition—as a justification. The justification for doing so is that, functionally, propositions operate in the world as justifications that serve legitimizing functions and to make knowledge claims. To help make this transition from abstract propositions to justifications in a social context, it is central to keep in mind that any proposition given in the social context can be questioned by self or others regarding, among other things, its: (a) logical or empirical content; (b) purpose in being emphasized at that moment; and (c) implications for what ought to happen, given the claim being made.

With this basic structure outlined, we can ask why the emergence of propositional speech and questioning results in an evolutionary tipping point. One reason is that asking questions is, cognitively speaking, relatively easy. To see why this is the case, spend some time with a curious four-year-old. They are likely to pepper you with questions, such as "Why don't we eat cookies before we eat dinner?" "Why are you bald?" "Why is the sky blue?" As children readily demonstrate, asking questions is much easier than answering them. That is why exasperated parents eventually say, "That is just the way it is!" This observation suggests that the capacity to ask questions necessarily motivated the search for answers. In the language of UTOK, Culture with a capital "C" is the network of claims, narratives, ideologies, and laws that have developed in response to such processes of justification. This argument becomes clearer when we dissect the three key problems that drive justification processes and note how these problems show up in the structural functional organization of both human consciousness and the Culture-Person plane of existence.

The Analytic, the Social, and the Personal Aspects of the Problem of Justification

As previously mentioned, there are three distinct elements associated with the problem of justification, which can be labeled as: (1) the analytic; (2) the social; and (3) the personal. The analytic problem can be thought of as stretching from basic accuracy (i.e., Does the statement accurately correspond to the situation?) to deeper truth claims about the nature of existence. This is a key and pervasive problem, which you should be able to see all around you. Consider, for example, whenever an individual is making a claim and you find yourself doubting its veracity, you are encountering the analytic-accuracy problem. In addition to having everyday relevance, the problem of valid knowledge ultimately becomes central for philosophy and science as justification systems evolve and become more refined in the modern era. Trying to accurately explain why the sky is blue is an analytic problem of justification. This kind of problem closely corresponds to how we think about scientific thinking. However, it is first and foremost crucial to realize that the problem of justification emerges in a social influence context of everyday living and negotiating the world in tightknit relational groups.

The emergence of the personal and social dynamics associated with the problem justification becomes clearer when we consider what happens as the capacity to justify our actions evolves. First, *human language can be thought of as a newly emergent explicit intersubjective highway of information that moves through the skin and into the world and back again without losing its informational form.* As such, linguistic justification sets the stage for much more direct interface between human subjective domains than is the case for other animals. In the language of UTOK, this is highlighting both the connection to and the jump from Mind^2 to Mind^3. In this regard, we can think of language as a kind of mental telepathy. That is, it allows for many mental processes that were formerly fully contained within the animal (i.e., Mind^2 subjective conscious experiences) to be translated and then transferred between minds in a much more direct way via the information highway that is symbolic speech (i.e., Mind^3 and the Culture-Person plane of existence).

By emphasizing the way language opens a more direct mental connection and gives rise to an *explicit intersubjectivity*, we can see both the affordances and some of the problems that also emerge as this system evolves. The affordances are clear when the group shares interests, values, and goals. Consider, for example, a tribe getting together, sharing its information about the hunt and current situation, and wondering about what the group should do the following day. The collective intelligence that is afforded by language to share information and coordinate group behavior is limited to nonexistent in the animal kingdom and is the central reason why human evolution takes a radical turn, such that we come to dominate the planet.

Conversely, problems emerge with explicit intersubjectivity when interests, goals, and investments between individuals diverge. To see why this is the case, imagine that you have a private room and that you stored valuable things in it, including a diary. Now imagine a communication device placed in that room that can be turned on and provides a direct connection between your private room and the private rooms of others. This line of connection can obviously be helpful in that now you can learn about other people's private rooms, and they can share in yours. Nevertheless, there is a problem. Or, rather, at least two broad problems. We can put these in the form of questions: (1) What information do you want to share with others and when and how do you want to share it? and (2) How do you know that the information that is coming to you from others is valid and useful?

We can imagine that there might be a fair amount in the private room that you might not want to share in an "unfiltered" way. To bring this point home, consider that you likely would not want the whole world to have unfettered access to your diary. In addition, you can wonder about the validity and veracity of information that others deliberately share with you. These problems of evaluating the validity of incoming information and filtering what is shared inevitably emerge as human language develops. The analogy is not perfect in that the private room metaphor suggests that the room was already there fully formed, and the information medium was just added to it. As the philosopher Ludwig Wittgenstein made clear, this is not exactly right because the emergence of language plays a crucial role in how the private room in your mind develops. For now, however, the point is that although it can be very helpful to construct an information highway between private rooms, it also means that one is confronted with problems of analyzing and regulating both outgoing and incoming information.

This analysis allows us to start to differentiate between the personal and social aspects of the problem of justification. The relationship between the two pertains to interests and how propositional descriptions and explanations impact social relations. Consider the following extreme example to highlight this point: Next time you are at a dinner party, and you spill a glass of red wine all over the hostess' expensive carpet, try justifying that occurrence by giving reason (a) "Oh my, I am so sorry. That was a total accident; let me help clean it up!" or (b) "I have resented you for a long time, and my aggression built up to the point where I wanted to show you the hatred in my heart."

We can make the radical prediction that the responses of the social group to Reason B would be radically different from those to Reason A. This highlights how important the reasons we give for our behavior are in terms of social influence. With this example, we can start to see how JUST's formulation of the evolutionary problem of justification aligns closely with Freud's key insights pertaining to the structure of the human mind. Specifically, it helps clarify why the ego must rationalize the selfish animalistic impulses of the id to fit into polite society. The reason is that there are personal interests that need to be protected from the judgmental eyes of society and the social interests it advances.

We can also see that this aligns with Erving Goffman's (1959) key insights regarding how we work to publicly save face. Specifically, he argued that we must wear masks or develop personas with different audiences to play particular social roles and manage the impressions we make on the social stage. What this means is that the Justification Hypothesis and the manner in which it highlights the personal and social aspects of the problem of justification sets us up to understand human consciousness in a new light. Specifically, it enables us to place, in clear relationship, the experiential portions of human consciousness that emerge as a function of us being primates, the private narrating egoic portion that generates stories of why we do what we do for ourselves, and it highlights why we have a public mask or persona, with the different domains emerging as a tension between personal and social interests. With these insights in hand, we can now shift gears from explaining how JUST gives us an evolutionary hypothesis for how and why the problem of justification created a new adaptive problem to show how it gives rise to a model of human consciousness that updates Freud and places his key observations in the context of modern empirical findings and frames in human psychology.

The Updated Tripartite Model of Human Consciousness

In our taxonomy of the domains of mental processes, we identified two distinct domains of human consciousness, namely subjective conscious experience (i.e., Mind^2) and self-conscious reflection mediated by language (i.e., Mind^3). According to UTOK, JUST gives us a metatheoretical frame for delineating the structure and function of Mind^3. This, in turn, enables us to develop a more comprehensive map of human consciousness. Specifically, it gives rise to an "Updated Tripartite Model" (UTM; Henriques, 2011) of human consciousness (see Fig. 5.1), which consists of: (1) an experiential self, which is the felt embodied sense of being in the world coupled to a relational model of the self that sits at our emotional core; (2) a private narrator or ego, which is a self-reflective justifying

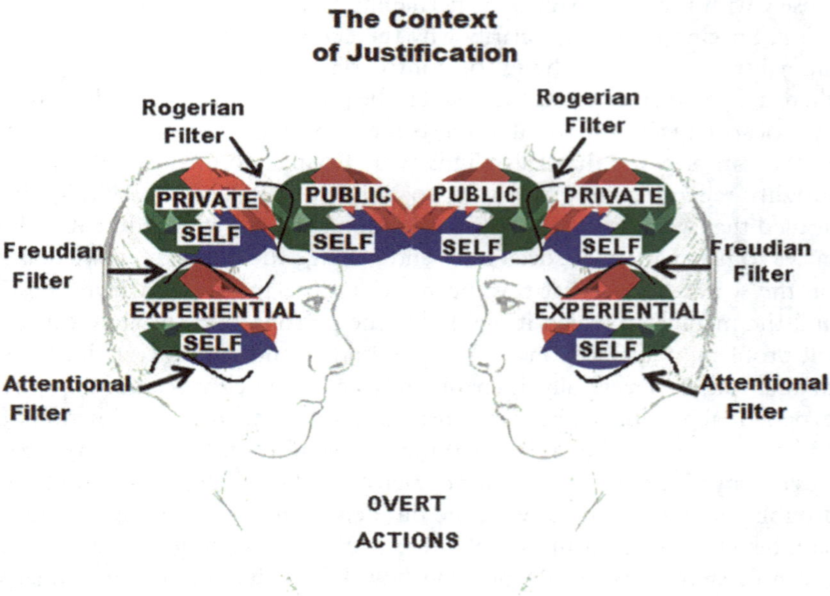

Fig. 5.1 The Updated Tripartite Model of human consciousness

position that internally narrates what is happening and why; and (3) a public self or persona, which works to manage social roles and impressions. The experiential domain corresponds to Mind^2 and the Animal-Minded plane, whereas the public and private domains correspond to Mind^3 and the Culture-Person plane of existence.

A core feature of JUST is that it provides a metatheory of the structure and function of Mind^3. The Updated Tripartite Model makes this explicit with the two domains of self-consciousness, the private ego and the public persona, which are shaped by the adaptive problem of justification, both phylogenetically and ontogenetically. The private self-consciousness system, listed as the Private Self, is the center of reflective awareness in adults and can also be thought of as the "I" or the ego. *Grounded in JUST, we can consider the ego to be the mental organ of justification* (Henriques, 2003). It is made up most immediately of the internal dialogue that weaves a narrative of what is happening and why. It is a second-order awareness system, one that translates events and feelings into language and feeds those thoughts back into the experiential system. Although other animals have the rudiments of self-awareness, this domain of explicit self-reflective awareness mediated by language is qualitatively different in humans, both as a function of our capacity to engage in self-conscious justification and as a function of our being socialized into cultures that have cumulatively evolved over thousands of years.

The public self consists of the explicit articulation to others of what one thinks and feels, along with the image one tries to project. Consistent with Carl Jung's work, we can call this the "persona," which is the "mask" that humans put on in their social interactions with others. In addition, this orients us to the fact that learning about the dynamics of giving and receiving social accounts that involve reasons and the allocation of responsibility is part and parcel of how humans become persons. In *The Presentation of Self in Everyday Life,* Goffman (1959) examined face-to-face interactions through the lens of stage acting. He articulated how interpersonal interactions could be considered performances as actors learned to manage their impressions on others in both the structured and improvised roles of everyday life. Specifically, Goffman suggested that actors work to convey a positive, predictable impression to be perceived as justifiable in the eyes of the audience.

Similarly, the psychologist Dan McAdams (2013) has described the first phase in the development of self-consciousness as the "social actor" phase. It runs from approximately age 2 to 10, and it consists of how children are thinking about the rules and roles that they are performing and whether they will be applauded or disciplined for their actions. As the example I gave with my daughter Sydney suggests, early in the phase, young children learn the rules and scripts in specific contexts (i.e., we do not eat cookies before dinner). Then, as children mature into ages 6 and 7, they see those scripts functioning across many situations and can much more readily give accounts and take responsibility. This eventually grows into what McAdams calls the agent, whereby the ego becomes a more stable force, and the child carries an identifiable self-concept across situations and recognizes their capacity to reflect and make choices about what is and ought to be. Finally, McAdams argues that in full adulthood, it becomes a narrator that weaves threads together to generate a story of the self, filled with themes, turning points, and an arc of development.

As will be laid out in later chapters, the experiential self (Mind^2) emerges out of Mind^1. Behavioral Investment Theory and the Influence Matrix provide the metatheoretical frames for understanding the structure, function, and evolution of the experiential self. These chapters will make clear why it is named the experiential self and will lay out the model of how it evolves and why it functions the way it does. As the name suggests, we can divide the experiential self into two separable domains. One is the domain of sensory and perceptual awareness. This is the seat of the witness function, which gives rise to our experience of being aware of being in the world (i.e., consider that you can simply open your eyes and the world is presented to you). The second domain pertains to the way the primate self reacts to those perceptions. These are the motivations and emotions that drive the individual to attend to what perceptions are relevant and react to them with a valence and desire to move toward or away from certain outcomes. With this division, we can see why sometimes it is helpful to refer to the UTM as giving us the "ESP-A" model of human consciousness, which refers to the Ego, the (primate) Self, the Person, and (pure) Awareness (Henriques, 2022, February).

In addition to dividing the experiential self into the domains of perceptual awareness and motivational and emotional reactions, it is also useful to divide it into two layers that correspond to our evolutionary history. The first layer corresponds to our history as animals in general and can be called the "animal body layer" of the mind. The UTOK frames this in terms of what is called the P – M => E formulation, which stands for perceptions, motivations, and emotions. We will explore this formulation in greater detail later. For now, we can note that it states that perceptions model and predict where the animal is in the world, motivational states orient the animal to approach and avoid certain animal–environment relations, and emotions energize efforts into perceptual-response sets to obtain valued outcomes and move away from threats. We can also connect the P to the witness/awareness aspect, and the motivation and emotion to the self aspect of the experiential self.

We can further frame the animal-body layer as corresponding to the first two levels on Maslow's hierarchy of needs, which include physiological and safety needs. Thus, the animal-body layer relates to things such as pleasure and pain, hunger, and thirst, one's basic place in the world, and the felt sense of either safety and trust or danger and threat. While we can frame this as the animal-body or animal-organism layer, we can then move up in our evolutionary history to our "primate-heart layer." Evolutionarily, this corresponds to our being both mammals and social primates. In Maslow's hierarchy, this layer is framed in terms of belonging and esteem. As we will see in later chapters, the Influence Matrix, the fourth key idea in the UTOK system, gives a rich map of the architecture of the social motivational processes and emotional response sets that organize the human relationship system. In short, the experiential self can be framed by two divisions. First, there is the division between the sensory-perceptual awareness of the exterior world relative to the feeling, motives, and emotions about those perceptions. Second, there is a division in terms of the animal-body layer associated with basic needs for safety, drives, and felt experiences of pleasure and pain, and the primate-heart layer associated with needs for love, power or social influence, and being seen, known, and valued by important others.

The UTM is named as a specific reference to Freud's work, as the domains of the experiential/primate self, ego, and persona carry clear parallels to the tripartite model of id, ego, and superego. In Freud's model, the id (i.e., the "It") represents the seat of animalistic energies and experiences. It is impulsive and demands gratification and pleasure. According to Freud, the primary energies that drive the id are sex and aggression. Bridging this to UTOK, we can say that the id can be located in the domains of both Mind^1 and Mind^2. In noting this, it is important to be clear that the word "conscious" for Freud is mostly associated with Mind^3 processes, meaning self-referential awareness. Raw sensations and drives for sex that are not put into language are generally considered unconscious in the psychodynamic language system. However, the language game is slightly different in UTOK. That is, sensations, perceptions, and drives are considered subjective conscious experiences and placed in the domain of Mind^2.

Consistent with Michael Gazzaniga's (2008) work on the interpreter function of the left hemisphere, JUST frames the human ego as the mental organ of justification. That is, the structural and functional aspects of propositional networks operating on the social stage (aka the Culture-Person plane of existence) require the capacity to engage in explaining events and giving accounts of one's actions on the social stage. Consistent with Baumeister's (2005) frame that humans are "cultural animals," the human ego can also be considered the mental organ of Culture, in that it is the portion of the human mind that allows humans to learn the language games of the cultures they are born into so that they can be socialized into persons. This personal–social linkage allows us to bridge into Freud's concept of the superego. This translates into "Above I," and reflects the standards of society that get internalized and idealized. It overlaps significantly with the public self, which can be thought of as the image the individual attempts to project and regulate to manage how they are seen by others. The overlap with Freud continues when we reflect on the dynamic tensions that exist between the experiential self, the ego, and the persona.

The Filters Between the Domains

Along with the experiential, private, and public domains of the human self, the UTM also depicts three filters that are operating between the domains to regulate information interface and attempt to maintain psychic equilibrium. First, there is the *Attentional Filter.* This refers to the neurocognitive filtering process regarding what gets onto the subjective screen of awareness. Placed in the domains of mental process language, the attentional filter refers in part to how Mind^1 processes are translated into Mind^2. The attentional filter involves processes of directed attention and perception, which consist of an orienting and matching process between bottom-up sensory information and top-down sense-making schema that attempt to identify the entity or event and predict subsequent changes.

To get a flavor for how we can direct the attentional filter, take a moment and shift your focus to any noises that might be in the background. This prompt likely resulted in an attentional search, and those background sounds likely became salient in your experience in a different way than before I mentioned it. The point highlights how the attentional filter can be directed so that it shines a light on specific aspects of your experience. Much more will be said about the relations between subconscious neurocognitive processes and conscious experience in subsequent chapters. For now, we can simply state that the attentional filter is the term for the interface between Mind^1 and Mind^2, and how information moves onto the stage of conscious experience.

The second filter is between ego and the experiential self, and is called the *Freudian Filter*. It is so named because it pertains to Freud's foundational insight that there is a dynamic relation and filtering process between the more basic, animalistic perceptions and urges and the self-conscious narrative process that navigates the explicitly conscious world with other human persons. The Freudian Filter works via the process of inhibiting, avoiding, or repressing disruptive, disturbing, or problematic feelings, images, and impulses and engaging in rationalizing and cognitive dissonance reducing strategies to maintain a sense of psychic equilibrium.

The model of the filter aligns with what David Malan identified as the central process of psychodynamic defense in his Triangle of Conflict (Malan & Osimo, 2014). The triangle is depicted upside down. On the bottom of the triangle there are impulses, images, or feelings, on the right side there is anxiety, and on the left side defenses. The idea is that certain impulses, images, or feelings begin to emerge on the stage of awareness, but trigger signal anxiety, which in turn activates a defense to manage the disequilibrium. To help my students see this process, I like to make the analogy with the way the old polaroid cameras worked. When you take a picture with a polaroid, it initially shoots out a gray image and then, after about 30 seconds or so, the film transforms into the picture.

The analogy is that as the image, impulse, or feeling emerges on the stage of the experiential self, the filtration system detects its implications and then signals whether the image should be brought forth. If the implication of the image or impulse is identified to be troublesome, there is a "signal anxiety" that is released, which in turn activates an inhibitory attentional shift away from the emerging image. This unconscious attentional shift away from threatening material is the classic psychodynamic defense mechanism of repression. It is an example of how the Freudian and attentional filters function to keep problematic ideas out of Mind^3. Suppression and denial are additional kinds of inhibitory defense mechanisms. There are other defense mechanisms like rationalization, moralization, or sublimation that transform the impulse or image or one's beliefs about it to make it less painful or more acceptable or congruent with one's ego and persona.

The UTM also includes a private-to-public filter, which is a filter between the private and public domains of Mind[3]. It is called the Rogerian filter because of the profound insights Carl Rogers made regarding our self and the social world. He realized that much of human consciousness and misery relates to how the judgment of others shapes our actions and the sense we have of ourselves. When you look for it, you realize that we are constantly navigating the private-public space and filtering information accordingly. Any time that you are thinking about whether to share a piece of information with another person or group of people, the private-to-public filter at work. The filters become very real whenever they are breached. Think of how vulnerable you might become if someone read your diary without permission, or, even worse, shared it publicly. The lock on the diary is a physical manifestation of the Rogerian filter. The nature of justification and influence dynamics are such that many of our thoughts are designed to be shared only with specific audiences.

It should be noted that there is a more Machiavellian side to the private-to-public filter. This refers to the fact that individuals often consciously manipulate, lie, and deceive each other through the filter. Indeed, when my daughter Sydney said "no" to the question about eating the cookie, the private-to-public filter was operative. Of course, many people lie far less innocently than my daughter. The Rogerian reference is to give a more humanistic and positive slant on the human condition, but the more rivalrous, competitive, and deceptive aspects of the private-to-public filter should not be ignored. Research demonstrates that people regularly engage in lying. Most children learn to lie by the age of three or four, and doing so earlier is generally associated with greater cognitive development. And although most are socialized into the value of honesty, estimates suggest that approximately .5 to 1% of the population are psychopathic, meaning that they have little or no concern about the feelings of others and will willfully manipulate, lie, and deceive to get what they want.

A Real-Life Example Applying the UTM to Patterns of Human Mental Behavior

A concrete example can help see how to apply the UTM and the various domains and filters to understand the dynamics of human mental behavior. When I was a teenager, there was a group of us who, like many heterosexual boys, started talking about girls and sex when we hit puberty. One friend, however, was only lukewarm when he engaged in these conversations. Later, in his early 20s, he "came out of the closet" and shared that he was gay. His narrative is instructive in that we can use it to highlight the elements of the UTM. My friend's father was critical of homosexual behaviors, and thus my friend felt he would lose social influence in our teenage group if he voiced his private desires. As such, when he started to have homosexual urges, they would quickly be seen by the private narrator as unjustifiable in the public sphere of influence.

Applying the UTM, we can say that when my friend's attentional filter started to turn toward homoerotic images, his Freudian Filter would signal anxiety and would try to block those thoughts and generate a defensive shift onto other thoughts. In addition, he did not have "being gay" as part of his self-concept. That is, his ego tried to convince himself he was not gay because the public consequences would be so severe, and so he repressed his homosexual tendencies, and they sat in his subconscious for many years. Eventually, his attitude changed, his narrator opened to the feelings, and he realized, in the explicit self-conscious sense of the word, that he was gay. That is, his experiential feelings broke through his Freudian Filter and he "came out" to himself. However, via the private-to-public Rogerian filter, he kept these feelings from others for another several years to avoid negative judgments and the loss of social influence. Finally, he let that filter down and came out of the closet to his friends and family. That convention uses "coming out of the closet" to describe this process points clearly to the filtering aspects and the dynamic relationships between the subconscious, experiential, egoic, and public aspects of human consciousness.

JUST Weaves Together Many Paradigms in Human Psychology

Thinking about verbal cognition in terms of justification systems allowed me to make connections between the major psychotherapy paradigms in a new way. Consider, for example, that the key focus of Beckian theory and therapy is the beliefs characterized by forms of self-talk. This is the portion of the mind that uses propositional language to make sense of one's self, the world, and the future. This is the domain of Mind[3] in UTOK, and with JUST, I could frame cognitive psychotherapy as being centrally concerned with self-conscious justifications and the functional role they play in navigating one's real or imagined place in the social arena.

We can apply this to the example of the self-critical woman who led to the development of JUST. Her story suggests that there is a process by which justification narratives emerge, become functional, and then "stick" in how they operate to filter information and potentially constrain options and create self-fulfilling prophecies. Indeed, with this example, we can understand the logic of cognitive therapy, because it shows how justification systems can become maladaptive. Cognitive therapy suggests that to help this individual, we would help her see that she no longer lived under the oppressive thumb of her father, and that many of her beliefs about herself were inaccurate and unhelpful, even though they were understandable given the context of her development.

JUST helped me understand the functional organization of propositional thought. Justifications were not, as several cognitive theories presumed, simply errors of logic or inappropriate inferences shifted in the negative direction. Rather than focusing on the deductive logic and inferences, the justification angle focuses on what the thought legitimizes in the social context, and what it implies about things like power, love, and social influence. This insight allowed me to share with clients where their thoughts might come from, which often is key in fostering understanding and motivating them to work to learn new ways to engage in self-talk. For example, in the avoidant client, it would likely have been much more useful in therapy to frame her negative self-talk as emerging because it justified avoidance of abuse than by arguing that she engaged in cognitive errors of inference.

As suggested by the structure of the Freudian Filter, the idea of justification systems also directly plugged me into modern psychodynamic theory. A central focus of modern psychodynamic theory is on the complex relation between self-conscious reasoning and the underlying or subconscious images, motivations, and emotions that guide those reasons. For example, individuals will frequently make excuses or rationalize their behavior to save face or feel better about themselves. Psychodynamic theorists are concerned with these kinds of rationalizing, moralizing, or intellectualizing defense mechanisms. In addition, as is made clear by the story of my friend and his homoerotic images, unacceptable thoughts (i.e., aversive feelings or desires that cannot be justified personally or socially) are not infrequently denied or repressed.

Via the JUST formulation, I saw the twin processes of repression and rationalization in a new light. I realized that the modern psychodynamic formulation of human consciousness pointed to a filtering process that centrally related to the problem of social justification. Others had made similar observations. For example, in his book *Ego Defenses and the Legitimization of Behavior*, Swanson (1988) explicitly argued that we can think of all the psychodynamic defenses as functioning to help individuals maintain a consistent, relatively stable justification narrative via blocking the unacceptable and legitimizing that which is allowed through the filtering process. In the UTM, this filtering process gets formally framed as the Freudian Filter that sits between the experiential self and the narrating ego.

Central to JUST is the idea that the explanations we develop of ourselves, others, and the world are intimately tied to our sense of being valued, our social influence, and our place in the relational matrix. This connection oriented me to pay special attention to the kinds of relational needs and social motives that moved people and influenced their reason-giving tendencies. The ultimate result was another major element in the Unified Theory, called the Influence Matrix, which provides a map of the intrapsychic and interpersonal self–other processes that guide people in the relational world. The Matrix maps the various ways people track their experience of relational value and degree of social influence, both intuitively and explicitly. This understanding is directly consistent with the relational turn in modern psychodynamic thinking, as well as the work of neo-Freudian luminaries, like Alfred Adler, Karen Horney, and Eric Erikson.

The psychodynamics associated with JUST became especially salient to me because of a sequence of events that happened directly following my insight regarding the self-deprecating patient. In developing the patient's conceptualization, I had made the connection between her prior abuse and her self-concept, and I was explaining to my fellow classmates and supervisor why I was excited about this insight. The conversation continued such that I ended up leaving work about 20 minutes later than I had told my wife I would. On the ride home (which was a short trip), there was traffic stemming from construction, which increased my delay by about 10 minutes. This was prior to the days of cell phones, and so I entered the house half an hour late.

Not surprisingly, my wife asked me to give an account of where I had been. Because I had been late earlier that week, she was primed to be annoyed with me. As soon as she asked about my tardiness, without hesitation or conscious self-reflection, I immediately blamed the traffic. That is, it effortlessly and automatically rolled off my tongue that the primary reason I was late happened to be the elements that were outside of my control and thus outside of what I should be blamed for. Of course, this account was basically a lie, or at the very least a misrepresentation. For over 20 minutes, I had eagerly discussed the idea that the patient's system of self-understanding functioned historically to justify submission and deference. And here I was, engaged in the same basic process. That exchange seared into my mind just how powerfully the human ego is shaped by such processes and functions as the mental organ of justification.

JUST not only connects with theories at the individual level of analysis, but it also aligns well with social perspectives and cultural theory. Prior to developing JUST, I was well versed in feminist theory and saw clear linkages with its central message and the implications that stem from JUST. In the late 1980s, I had taken a social psychology course on gender and women's studies. There were ten students in the class, me and nine women. It was a fascinating and eye-opening experience. I learned the central insight associated with feminism (and other critical theory, social justice, and social constructionist approaches more generally), which can be summarized via the lens of JUST as being the idea that societal institutions are constructed and legitimized by individuals and groups in power. If those in power all belong to a particular social category (e.g., White, Christian, male), then the narratives, norms, and roles that define key aspects of society will be skewed in accordance with the interests, tendencies, and attitudes held by that group. Those in the outgroup are then either explicitly or implicitly held in lower esteem (i.e., social influence and attention) and are marginalized.

Framed this way, feminism can be essentially grounded on the insight that, since the dawn of civilization, men have generally dominated power roles and institutional structures and consequently built justification systems in accordance with their interests. Feminism thus becomes the assertion that society needs to wake up to the fact that there are underlying masculine forces that are structuring the grammar of our thought, and to recognize that there are other ways to understand the world and create different social relations that are more just and equal. My point here is that my familiarity with feminism allowed me to connect JUST to how large-scale knowledge systems function to provide a context of justification in which human persons navigate and negotiate their place.

As this summary of various perspectives suggests, I came to see that JUST could function as a "hub" of an idea that connects many different aspects of the human condition and many different domains in human

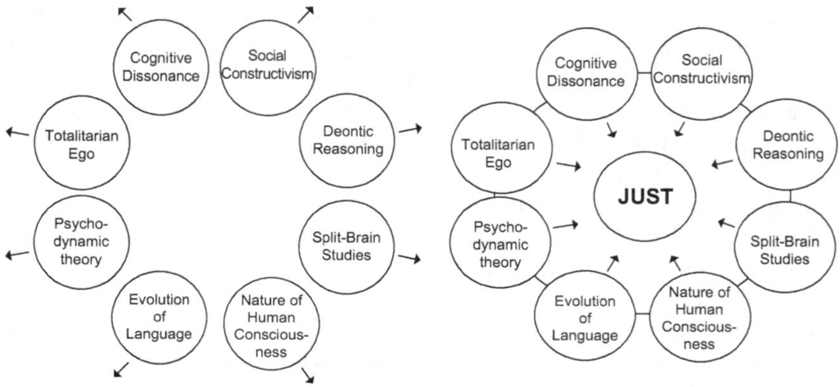

Fig. 5.2 Justification Systems Theory coherently connects many domains in the literature

psychology, social science, and philosophy together into a coherent account. It aligns evolutionary psychology with developmental, cognitive, personality, and social psychology on the evolution of language, hemispheric specialization, reason-giving, and the structure of human consciousness, and it frames the cumulative evolution of human Culture in a way that is consistent with modern sociology, feminism, and critical theory. This is visually depicted in Fig. 5.2.

The power of JUST as an account of human mental behavior becomes even clearer when we shift from framing it as a theory that explains how we justify to a foundational, metaphysical description of the structure and function of propositional thought.

From Theory and Explanation to "Back" to Description: A New Way to Describe the Structure and Function of Propositional Thought

As JUST took shape in my mind, I started to envision plausible experimental designs that could test the many predictions that resulted from the model. Consider, for example, that JUST posits that humans should seek a mental state that is justified, and that such a state would be, in part, a function of self and other evaluations. This meant that humans would be what psychologists call "motivated reasoners." In combination with the

Influence Matrix, JUST suggests that those motivations should be related to social goals like one's status or self-concept and thus the prediction that people will often reason in a biased manner, framed by the desired outcome in mind. Motivated reasoning is exemplified by lawyers and politicians who want to influence people toward specific ends. According to JUST and the Influence Matrix, motivated reasoning should be tied in a significant way to the individual's interests and place in the social field and their perceived levels of relational value and social influence. As such, it followed from JUST that if we were to manipulate social influence variables, then we should see changes in how people justify what is happening and what they are doing and why. With such a formulation, we could develop predictions about how situational variables, shifting motivational variables, and variables associated with the person's justification system all might influence reasoning processes.

JUST could also be used to make specific predictions about the kinds of biases we should see in how people tend to give accounts for their actions. For example, it predicts that people should narrate stories differently depending on the eventual outcomes associated with the events. That is, when people's actions are followed by good outcomes, then they should be more inclined to explain what happened as a function of their character, efforts, and intentions. In contrast, when the outcomes are negative, people should be more likely to seek justifications that would mitigate loss of social influence and thus tend to blame external factors. For example, when getting an "A" on a test, the attribution is more likely that the individual studied hard or is smart, whereas getting an "F" would result in more claims that the teacher or test was not fair or that the individual did not care. In addition, the same basic pattern should follow if the person affiliated or shared interests with other individuals or groups whose narratives and related outcomes were being considered. Much empirical research has found that this is exactly what happens. Self-serving or "myside" biases represent one of the most robust findings in social and cognitive psychology.

JUST also predicts that people should attempt to be somewhat consistent in their explanations because if they are inconsistent then others can accuse them of being unpredictable, unreliable, or illogical. This consistency motive should be especially strong when it pertains to their perceived social status or sense of themselves as being a justifiably good person. Consider, for example, the classic finding in cognitive dissonance where subjects in a staged experience were asked to convince other people that the task was fun and interesting, when it was in fact boring and annoying. Festinger and Carlsmith (1959) found that participants who were paid $20 (a large sum of money at the time) to misrepresent the task developed different beliefs about the task than those participants who were only paid a dollar. Those who were paid only a dollar later reported that they really believed the task was not so boring, whereas those who had been paid $20 had significantly lower opinions about the task. This was remarkable because standard learning theory would say that a larger reward should have been associated with more enjoyment. Festinger and Carlsmith argued that the key to the finding was related to the dynamics of justification. The participants who lied for only a dollar felt more uncomfortable with what they did than those who were given $20. The reason was that they lacked sufficient justification for their misrepresentation of the task. And so they altered their beliefs about the task, justifying to themselves and others that it was not so bad after all, thus maintaining a justified state of being.

JUST also predicts that people should alter their explanations for their actions depending on the audience. That is, people should explain events and their actions differently depending on: (a) the nature of the message being communicated and its implications for social influence; (b) the power relations between parties; and (c) the kind of alliance or social distance between the parties. For example, if someone had a harsh judgment about their boss, it was very likely that they would talk quite differently about that judgment to their boss directly, in contrast to how they talked to a friend who shared their dislike of that boss. One could easily envision manipulating audiences as independent variables and then exploring the nature of the justifications as a dependent variable. This highlights how much reasoning is really motivated by or at least shaped by social influence goals. Indeed, since the 1990s, the fact that human reasoning is powerfully driven by motives that are often not conscious has been one of the field's major discoveries.

I had done my master's thesis using an experimental design involving social influence and feedback, and thus I could have readily transitioned it into testing predictions. Indeed, a recent book by Hugo Mercier and Dan Sperber (2017) called *The Enigma of Reason: A New Theory of Human Understanding* directly supports the validity of this claim. In it the authors argue for an evolutionary and social model of human reasoning that has remarkable parallels with JUST. In their words from the book summary, "what reason does is to help us justify our beliefs and actions to others, convince them through argumentation, and evaluate the justifications and arguments that others address to us." The authors make the case in part by reviewing the empirical work that has been done in the last two decades, and the general patterns of findings strongly support the notion that human reasoning functions to justify and persuade others in the social matrix and that reason-giving is very much about argument and persuasion.

There are several reasons that I did not go the experimental route. *The most basic reason has to do with the fact that a transformation took place in how I was thinking about the kind of idea JUST was.* I initially thought of JUST as an explanatory hypothesis that made predictions about how explanations and attributions were offered in social settings and connected to social influence. Because I was thinking like an empirical research scientist, I initially hypothesized about ways this prediction could be explored via experimental designs. Although bringing the predictions into the research lab might have been a useful line of inquiry, three "problems" emerged that fundamentally altered the way I was thinking about the kind of idea JUST was.

First, as suggested by my summary of the Festinger and Carlsmith study, there was already an enormous literature on cognitive dissonance that had empirically demonstrated that humans generally sought a justified state of being. There was an even larger literature on the broader claim that humans were motivated reasoners that engaged in an enormous number of self-serving biases and other heuristic attributional processes that were sensitive to the social context. Consider, for example, that some of the founders of evolutionary psychology had shown that human reasoning was organized in part to detect social cheating (see, e.g., Cosmides & Tooby, 1992). That is, there was already good evidence that humans were adept in thinking about the dynamics of unjustifiable behavior that involved broken social contracts. There was even literature demonstrating that there was a recently evolved portion of the human brain in the left hemisphere that functioned as an interpreter that developed socially acceptable rationales for one's behavior. As I organized these insights, I started to see that JUST was more than a hypothesis that offered novel predictions. Rather, it was a metatheoretical idea that allowed me to zoom out across the landscape of human reasoning, self-consciousness, and culture to assimilate and integrate large bodies of existing research across several major schools of thought.

A second problem involved a change in my thinking regarding the emphasis on what I meant by justification. Much of my early focus was on the psychodynamic and cognitive dissonance angles that emphasized the ego defensive and rationalizing side of justification processes. That is, reason-giving served a primary purpose connected to underlying social motives and one's place in the social environment. However, as the concept deepened and broadened, I began to connect processes of justification to philosophy. As alluded to earlier, the concept of justification plays a central role in analytic knowledge and epistemology. That is, justification is deeply involved in what we determine to be true in the objective or analytic sense of the word and how we validate knowledge claims. In terms of development of the idea, I had started with the evolutionary problem of justification, which evolved into the problem of social justification, which shaped the evolution of the ego. I then was looking at all the psychodynamic, cognitive, and social psychological aspects of the concept in everyday life.

I then realized it moved into the refined aspects of philosophy and theories of knowledge itself. Consider, for example, the concept of epistemology and the theory of what constitutes authentic knowledge. Historically, the primary formulation for epistemology dates to Plato. It is called the *justified true belief* formulation. It is the idea that knowledge can be defined by three components: (1) an external reality that exists as a specifiable state of affairs; (2) representations or beliefs about that reality; and (3) justification, which characterizes the relation between the two and the grounds for the belief. That is, true knowledge can be defined by beliefs that accurately correspond to reality and are justified by sound logic and good evidence. What is particularly relevant to note here is that the meaning of justification in philosophy (i.e., that which is analytically true or most aptly defensible against skepticism) is essentially the opposite of what it means in the psychodynamic or social conflict contexts (i.e., defending one's self, often by generating rationalizations that hide important truths). This showed me that the concept of justification extended across a huge dimension of propositional beliefs as they functioned in the world. It was a concept that ranged from those that are the most analytically profound to those that are the most associated with self-deception, lying, and rationalizations.

A third problem that emerged was that I realized my actions that were focused on legitimizing JUST were evidence of the idea itself. That is, as I started to think of the ways I could justify the model, I realized that I was enacting the very processes that the idea contained. Seeing myself attempting to justify the idea of JUST and seeing justifications everywhere ultimately resulted in an even deeper shift. I began to experience the idea less as a potential explanation of some hidden mental process and more as a *description* of propositional thought and human verbal mental behavior that could be observed. This resulted in a profound shift in my perceptual field. In looking back, I can say that the first problem involved me shifting in my thinking from framing JUST as a theory into seeing it as a metatheory for existing paradigms, whereas the second problem shifted me into a descriptive metaphysical system for characterizing the ontological structure and function of propositional thought.

This resulted in a shift in my embodiment of being. *I could now simply observe that people were justifiers who lived in a sociocultural field of justification systems.* It simply was the case that people deliberated on their actions and experiences and developed explanations to themselves and others that justified such things. It simply was the case that people lived in social systems that functioned to both explicitly and implicitly determine what actions were justifiable. And it simply was the case that these socially constructed facts functioned to frame human interaction. Furthermore, it simply was the case that if people behave in unjustifiable ways and others had the power, influence, and inclination to do so, they were held to be accountable for their actions. Of course, these observations seem obvious. But what is new is that these obvious observations are now able to be placed in a naturalistic, causal explanatory framework in a way that had never been done before.

Although I did not realize it at the time, this shift in perspective from a mediational and explanatory view to a more foundational descriptive view would ultimately result in my discovering some profound connections with Skinner's radical behavioral perspective. To provide just one example, I could now simply observe how verbal behaviors operated in and on the social environment as a function of justification dynamics. Indeed, this rather mechanistic view was at times so strong that I started to call people "verbals," which is a direct reference to gerbils and reflects the view that we are just talking apes. And, given that justification processes operate as strange recursive loops, we can then wonder if this characterization of humans as verbals is justifiable.

Conclusion

JUST is the first key insight that started an unfolding that would lead to UTOK. It unlocks the mystery of the emergent evolutionary forces that gave rise to the Culture dimension of complexification and our transition from primates to persons. I first framed it as an empirical idea that was useful for making novel predictions about how people would engage in reason-giving on the social stage. The validity of this assertion is found in the approach taken by Mercier and Sperber as they proposed the idea that human reasoning evolved as a function of justification. But, as I had seen 20 years prior to the publication of their work, JUST was pointing to something bigger than an empirical hypothesis about the origins of human reasoning.

In making the name change from the Justification Hypothesis to Justification Systems Theory, we can now say that the former refers to the more narrow and specific idea that the evolution of propositional language resulted in the adaptive problem of justification. Specifically, statements with propositional meaning could be challenged in question–answer dialogue, and this dynamic gives rise to the problem of justification, with its analytical, social, and personal dimensions. This analysis in turn affords a clear way to frame the domains of human self-consciousness. The Updated Tripartite Model divides human consciousness into the three domains of the experiential self, the private egoic narrator, and the persona, which interface via three filters. We saw how this formulation is consistent with many lines of thought and how it is directly congruent with the dynamics that emerge with the problem of justification.

The final section of the chapter highlighted an even deeper and broader shift that emerges with JUST. This is the move from metatheory into a descriptive metaphysics and ontology for propositional networks. JUST allows for the shift from explaining human persons as working to justify their actions to simply describing them as doing so. The more common move in science is to go from description to the deeper processes of explanation. However, there is a time that you get to the ontological "rock bottom" of one's system of knowing. This is the layer of descriptive metaphysics, the layer of foundational concepts and categories that are used to understand the world. Via JUST I came to see humans as operating on a different plane of complex adaptive behavior, such that unlike other animals, we are persons navigating the sociocultural space of justification systems. It was this shift in mindset that would set the stage for the second key insight that grounds UTOK, the Tree of Knowledge System. Building from JUST, it provides a descriptive metaphysical system for both science and reality that sets the stage for a new vision of knowledge and wisdom in the twenty-first century.

References

Barkow, J., Cosmides, L., & Tooby, J. (1992). *The adapted mind: Evolutionary psychology and the generation of culture*. Oxford University Press.

Baumeister, R. F. (2005). *The cultural animal: Human nature, meaning, and social life*. Oxford University Press.

Cosmides, L., & Tooby, J. (1992). Cognitive adaptations for social change. In J. Barkow, L. Cosmides, & J. Tooby (Eds.), *The adapted mind: Evolutionary psychology and the generation of culture* (pp. 163–228). Oxford University Press.

Festinger, L., & Carlsmith, J. M. (1959). Cognitive consequences of forced compliance. *The Journal of Abnormal and Social Psychology, 58*(2), 203.

Gazzaniga, M. (2008). *Human: The science behind what makes us unique*. Ecco.

Goffman, E. (1959). *Presentation of self in everyday life*. Double Day Books.

Greenspan, S., & Shanker, S. (2004). *The first idea: How symbols, language, and intelligence evolved from our primate ancestors to modern humans*. Da Capo Press.

Henriques, G. R. (2003). The tree of knowledge system and the theoretical unification of psychology. *Review of General Psychology, 7*(2), 150–182.

Henriques, G. R. (2004). Psychology defined. *Journal of Clinical Psychology, 60*(12), 1207–1221.

Henriques, G. R. (2011). *A new unified theory of psychology*. Springer.

Henriques, G. R. (2022, February). The ESP-A map of human consciousness. *Psychology Today*. https://www.psychologytoday.com/us/blog/theory-knowledge/202202/the-esp-map-human-consciousness#:~:text=The%20ESP%2DA%20Model%20is,of%20awareness%20and%20the%20self

Henriques, G., & Michalski, J. (2018, December 1). From JH to JUST: Why a name change is justified. *Psychology Today*. https://www.psychologytoday.com/us/blog/theory-knowledge/201812/the-jh-just-why-name-change-is-justified

Heyes, C. (2018). *Cognitive gadgets: The cultural evolution of thinking*. Harvard University Press.

Kuleshnyk, I. (1984). The Stockholm syndrome: Toward an understanding. *Social Action & the Law, 10*(2), 37–42.

Malan, D., & Osimo, F. (2014). *Psychodynamics, training, and outcome in brief psychotherapy*. Butterworth-Heinemann.

McAdams, D. P. (2013). The psychological self as actor, agent, and author. *Perspectives on Psychological Science, 8*(3), 272–295.

Mercier, H., & Sperber, D. (2017). *The enigma of reason*. Harvard University Press.

Pinker, S. (1994). *The language instinct: How the mind creates language*. Harper.

Rolston III, H. (2010). *Three big bangs: Matter-energy, life, mind*. Columbia University Press.

Swanson, G. E. (1988). *Ego defenses and the legitimation of behavior*. Cambridge University Press.

Taylor, S.E., & Brown, J. (1988). Illusion and well-being: A social psychological perspective on mental health. *Psychological Bulletin, 103*(2), 193–210. https://doi.org/10.1037/0033-2909.103.2.193.

CHAPTER 6

The Tree of Knowledge System

The Tree of Knowledge (ToK) System is the most important idea in UTOK. It grounds the philosophy in a new descriptive metaphysical system that allows us to transcend the Enlightenment Gap. This is because it affords a much clearer picture of natural science ontology and does so in a way that allows us to solve the problem of psychology. Specifically, the logic by which it frames the emergence of the complex adaptive planes of existence, along with how it places scientific knowledge in relationship to social and subjective knowledge, gives rise to a missing but necessary map of the terrain. This chapter begins by sharing the original diagram that popped out of my consciousness on the heels of the insights that JUST afforded me. It then explicates in greater detail how the ToK System maps the evolution of behavioral complexification in nature, specifies the relationship between the natural and social sciences, elucidates the place of psychology's subject matter in the stratified layers in nature, and bridges our past understanding of the Great Chain of Being with modern empirical natural science.

G. Henriques, *A New Synthesis for Solving the Problem of Psychology*, Palgrave Studies in the Theory and History of Psychology,
https://doi.org/10.1007/978-3-031-18493-2_6

The Original ToK Diagram: Four Dimensions of Existence Linked by Four Joint Points

Figure 6.1 is the original ToK diagram that appeared in that quick sketch one evening in the middle of 1997. Sharing the original allows me to highlight several key aspects of the ToK System, as it contains much "vision logic" that is useful to unpack in its original form. Ken Wilber coined the term vision logic to refer to a kind of "post formal" logical thinking, and I am borrowing it here to make the connection in two senses. First, I think vision logic captures the way the ToK System works to synthesize geometric and propositional representations. Second, the ToK represents a different kind of sensibility that aligns with how Wilber characterizes post formal thought.

Over the years I have come to appreciate just how much intuitive vision logic is present in the original diagram. The most salient feature of the diagram is that rather than there being a single cone representing a continuous rise in complexification, there are four separable cones that represent different dimensions of complexity in nature. This four-dimensional view of existence is crucial to understand how the Unified Theory generates a new consilient, naturalistic worldview. Indeed, it is this structure that enables it to provide a clear descriptive metaphysical system and ontological frame for both the domain of the mental and of human persons that ultimately solves psychology's BM^3 problem. This is because the ToK System enables us to divide Mind from Life from below and Culture from Mind from above, and separate the general, nomothetic scientific worldview situated in a third-person exterior epistemological vantage point from a first-person, idiographic, qualitative experience of being in the world. These three distinctions (i.e., Mind from Life, Culture from Mind, and scientific behavioral from subjective qualitative) are necessary if one is to achieve a proper metaphysical, ontological, and epistemological solution to the problem of psychology.

The cone at the bottom emerges out of a circle labeled Physics. Inside the circle are the terms matter, energy, and gravity. This maps onto the notion that energy and matter and gravity are some of the most fundamental concepts in physics, and they emerge at the very beginning of the observable universe, which is known generally as the Big Bang. At its broadest outline, the Big Bang is the idea that the universe begins as a kind of "primordial atom," which is a super-heated, hyper-condensed state that goes through a rapid inflationary phase shift approximately 14 billion years ago,

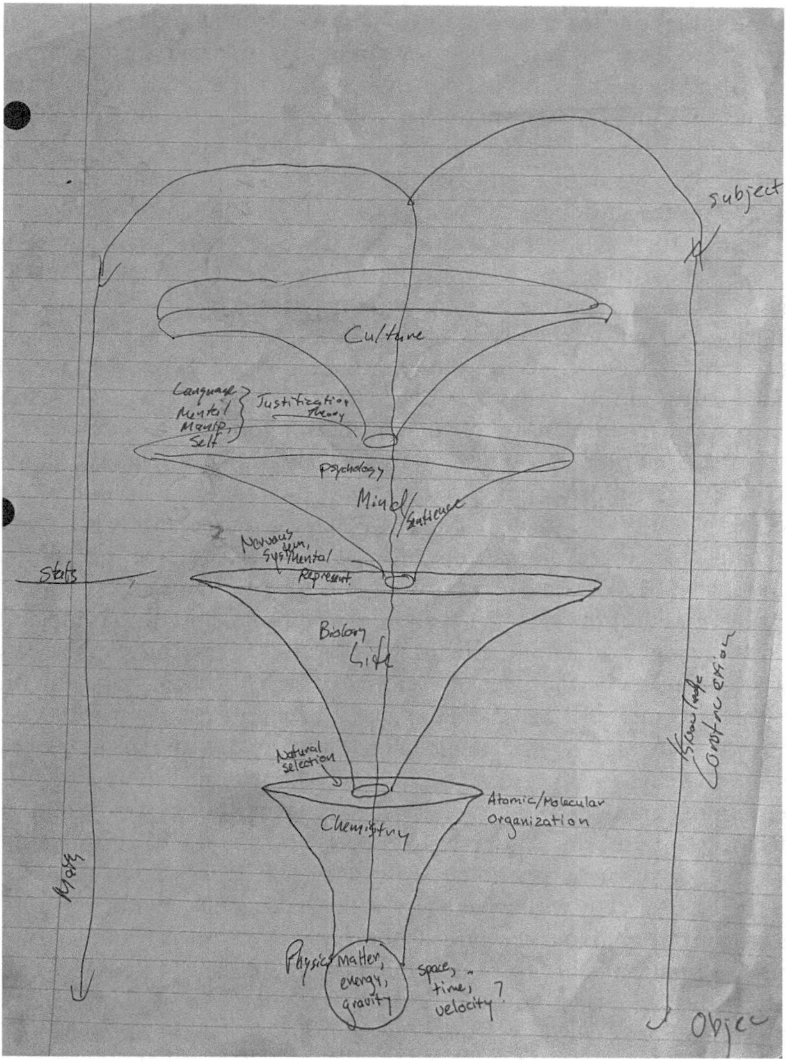

Fig. 6.1 The original Tree of Knowledge diagram

such that the four fundamental forces (i.e., electromagnetic, gravitational, strong, and weak nuclear), the various kinds of subatomic particles (i.e., bosons and fermions), and the dimensions of space and time differentiate, emerge, and unfold into what we now experience as the observable universe, and its classical matter in motion behaviors. The circle can also be thought of as symbolizing the odd world described by quantum field theory, and resides at or near the base of the UTOK naturalistic ontology.

Next to the circle at the bottom are the concepts of space, time, and velocity. These concepts connect to the foundational grammar of modern science as framing the universe in terms of matter in motion. Although not explicitly labeled on the diagram, the first dimension of behavioral complexification that emerges from the Big Bang is called "Matter." Like Life, Mind, and Culture, when capitalized in the UTOK language system, Matter refers to a dimension of complexification or a plane of existence. In the parlance of modern physics, Matter can be thought of as the domain where Energy-Information quantum field fluctuations "decohere" into measurable or observable entities that behave in accordance with classical mechanics on three dimensions of space and one dimension of time. Cahoone (2013) frames this as the difference between the material (Matter) and the physical (energy and the quantum world) orders in nature.

In the original diagram, chemistry is located at the top of the Matter cone. Chemistry is the science concerned with the behavior of matter, especially atoms, molecules, and the chemical energy transformations associated with them. The placement of chemistry makes an important point about what the graphic represents. The vertical dimension represents a continuum of structural and functional complexification. Complexification refers to the number of differentiated parts and the way their integration coheres to generate a functional form that has reliably specifiable properties. It aligns directly with what Tyler Volk (2017) calls "combogenesis" in his map from quarks to culture. Chemistry is thus placed higher in the stack because the behavior of molecules is a higher level of combogenesis than the behavior of atoms, which in turn is a higher level of organization than subatomic particles. *It is thus important to note that scale across aggregates is not being represented in the graphic.* That is, the behaviors of stars and galaxies, for example, are not explicitly mapped by the vision logic of the ToK. In terms of the natural sciences, we can trail the mapping of complexification or combogenesis from inorganic chemistry into organic chemistry and then into molecular biology.

With Life, we see the first jump from one plane of behavioral complexification into another. (Note that the Energy-Information field beneath

Matter is not considered a dimension of complexification, but is simply framed as the ultimate substance and fundamental common denominator.) Life is the dimension of existence that emerges out of mechanical and chemical energy transformations over time. However, this emergence occurs in a way that gives rise to fundamentally different behavioral patterns. Biology is, of course, the corresponding science. Natural selection is placed at the "joint point" between the Life and Matter dimensions. Although great strides have been taken in our understanding, it remains the case that the specifics of exactly how Life emerges from a "pre-biotic soup" remain a great, open scientific question (e.g., some theorists have put forth the idea that life on earth was seeded by an asteroid).

Nevertheless, despite this gap in our knowledge of exactly how the inanimate material world gave rise to the Life plane of complexification, the biological sciences successfully achieved a broad, meta-paradigmatic integration in the twentieth century. In terms of its ontology, biology is clearly defined as the science of living organisms that exhibit complex adaptive patterns of activity that are fundamentally different from inanimate objects. Moreover, there are big picture frames that organize our biological knowledge. The intersection of cell theory, natural selection, and genetics framed by molecular biology gives biology its metatheoretical organization that makes its knowledge "hard," meaning that it is grounded in a coherent ontology that is consensually shared by biological scientists. Cell theory is an odd-sounding way of saying that cells are the fundamental units of life. However, it only sounds odd because it has been so broadly confirmed that it now stands as essentially a fact. Of course, the idea began very much as a hypothesis. In addition, the modern evolutionary synthesis informs us that natural selection operating on genetic combinations of organisms, groups, and ecologies (i.e., multi-level selection) across the generations is a complexity building feedback loop that gave rise to the multitude of varied living processes we see today.

The vision logic of the diagram gives rise to the question: *Why is Life a different cone, and what, exactly, does this represent?* The answer to this question is a key aspect of UTOK's metaphysics and ontology. The explicit logic underlying the graphic was not immediately apparent to me, and the answer to this question emerged over the course of the next several years. As I noted in the original diagram, the vertical dimension clearly pertained to complexity (or what I would now call processes of nested complexification, spelled out by Volk as combogenesis). But for a time, I found myself struggling to specify exactly what the different cones captured.

We can start by noting that molecules exhibit properties and behave in ways that are quite different than single atoms or particles. In other words, patterns of emergence obviously happen *within* the material plane. Indeed, the growing size and width of the cone is meant to capture the growing complexity, variation, and new properties associated with novel arrangements of matter and energy as we move from particles to atoms to molecules across various scales of space and time. *The separate cones on the ToK graphic highlight that the difference between living processes and chemical processes is a different kind of difference than the difference between chemical processes and more basic physical processes.* This turns out to be a central point with many ramifications. As we will see in a later chapter, mapping emergence as both a function of leveling processes taking place within a dimension (e.g., atoms to molecules) and a function of novel information processing and communication networks that give rise to wholly new dimensions of complexification (i.e., from Matter to Life) is a central ontological insight that relates deeply to our understanding of both emergence and reality more generally.

For now, we can give the short answer that gets at the essence of what the higher-order cones represent. As noted, the key ingredients are information processing and communication networks that give rise to novel complex adaptive planes. Put simply, cells translate, input, store, compute, and communicate information with other cells in ways that inanimate material objects like water molecules do not. And it is the cause-effect sequences of information processing and communication networks that result in the new complex adaptive plane of existence. These networks of information processing and communication systems give rise to novel complex adaptive landscapes that have novel causal consequences that cannot be reduced to the dimension beneath them. This is an argument that will be elaborated upon throughout this work.

The third cone is "Mind/Sentience," and "Psychology" is the corresponding science. How did this dimension of complexification emerge? The base of the Mind cone connects to the Life cone via the presence of the nervous system and the capacity for "mental representation" via neuro-information processing. That I wrote Mind/Sentience is telling here. As we have seen, these terms relate to two of the three key referents for mental processes. The most general conception of the mind is the neurocognitive functionalist conception, which is the idea that the nervous system is an information processing system that coordinates the overt behavioral investments of animals. In our taxonomy of mental processes, this is the domain of Mind[1]. Sentience corresponds to the domain of Mind[2] and refers to core feelings and the subjective conscious experience of being that emerges at some point during the evolution of the Animal-Mental plane of existence.

The fundamental cause-effect processes that differentiate the Life plane from the Matter plane apply to the processes that differentiate the dimension of Mind from Life. In this case, what emerges are animals with brains and complex active bodies that engage in neuro-information processing to regulate the animal–arena relationship. Of course, just as we saw with cells and cell groups, there are novel patterns of information processing within animals and communication between them. Thus, in terms of the logic of the cones, we can see that once again it is information processing and communication networks that drive the emergence of a novel complex adaptive landscape. Behavioral Investment Theory (BIT) is the third key idea in UTOK. It is a metatheoretical formulation for this process and functions to assimilate and integrate major perspectives in the mind, brain, and animal behavioral sciences. BIT is not listed in the original ToK diagram because the specific formulation was developed several years later.

The identification of Mind with a capital "M" as the third dimension of complexification constituted by the set of mental behaviors is one of the central insights of UTOK. It is an insight that grounds a key aspect of its solution to the problem of psychology by bridging the mentalists with the behaviorists. It also clarifies why the basic science of psychology should be aligned with the Animal-Mental plane. This is a key point of confusion because many scholars, especially in Europe and in other parts of the world, identify psychology as being only concerned with human behavior. To give just one example, the Yokohama Manifesto explicitly defines psychology as the science of human being (Valsiner et al., 2016). In UTOK, the basic science of psychology corresponds first to animal-mental behavior, rather than to humans. Even more specifically, it corresponds to the sensory-motor loop that includes animals with brains and complex active bodies and can be described as the property of mindedness. These are minded creatures and basic psychology can be framed as the science of minded behaviors. As such, human psychology is not redundant, but rather is properly framed as a specific and unique branch of the larger discipline.

Last, there is the Culture cone. It connects to the Mind cone via the "Justification Theory" joint point, which as we have seen, is now called JUST. The rationale for why Culture is a different plane of complex adaptive behavior follows the same logic. Just as cells are coordinated by genetic/epigenetic information processing and communicate via chemical signals, and animals process neuronal information and engage in sensory-motor activities and communication, human symbolic language affords a novel kind of information processing and communication system that resulted in yet another emergent complex adaptive plane of existence. The essence of the Culture cone is the large-scale systems of justification that coordinate the behavior of persons. In terms of the unique mental behavioral processes that we see in humans, this is framed by the domain of Mind[3], the structure and function of which is framed by JUST. Finally, I must point out the lines emerging out of Culture on the right and left sides of the diagram. As described below, these lines represent what will become natural scientific knowledge evolving out of Culture to map the behavioral patterns in reality using objective analytics and quantification procedures (i.e., the data collection and experimental methods and logical argumentation of scientific empiricism).

The Four Joint Points

The separable dimensions of existence enable us to orient to another crucial aspect of the ToK System, which is the metatheoretical *joint points* that link them. In UTOK, a joint point refers to a complexity building feedback loop that is associated with the emergence of a new plane of existence. Quantum gravity is the first joint point. It is best considered as a placeholder rather than any kind of complete theory. It refers to the twin pillars in modern physics, general relativity and quantum field theory. As was mentioned, a truly effective synthesis between them remains elusive. For now, we can simply say that the first joint point can be framed by quantum field theory and general relativity (captured by "quantum gravity"), as well as the phase transition from the Energy Information base into the dimension of Matter at the hot inflationary Big Bang.

The modern evolutionary synthesis serves as the outline that frames the joint point between Matter and Life. According to the modern evolutionary synthesis, Life evolves as a function of natural selection operating on organisms that carry different genetic combinations across the generations. This complexity building feedback loop resulted in the metaphorical Tree of Life, and the functional adaptations that we see in cells and organisms. It is referred to as an "outline" here because many questions remain about how life began and how epigenetic and cellular physiology emerges and evolves. Indeed, several scholars are calling for significant revisions to this framework (see, e.g., Huneman & Walsh, 2017). Nevertheless, the intersection of genetics, cell theory, and natural selection placed in evolutionary ecological perspective represents the key ingredient that affords biologists a working metatheoretical paradigm to understand the Life plane of existence.

As this summary suggests, both quantum field theory coupled to general relativity and the modern evolutionary synthesis are well known and deeply ingrained in the established scientific lexicon for framing the behavior of material objects and living entities, respectively. However, as this summary notes, it is also the case that neither is fully complete. This allows us to see that the ToK System functions as a sense-making tool. It frames the outline of what we know in the physical and biological sciences and generates a clear vision logic of the relationship between these domains. That is, it places the insights of Newton's matter-in-motion paradigm (i.e., classical mechanics), chemistry, Einstein's general relativity, and quantum mechanics in geometric relation to the Matter dimension and the evolution of the complexification of material behavior starting with the beginning with the Big Bang. It then adds the dimension of Life as being characterized by Darwin's theory of natural selection, genetics grounded in biochemistry, and cell theory. It does this while providing a vision logic for how living processes exist at a higher dimension of complexification.

This overview allows us to see how the ToK can be thought of as a descriptive metaphysical system that frames the relationship between our major scientific theories and the domains of reality they map. It then builds on this general understanding to make novel assertions about the two additional joint points. It is useful to pause here and absorb the implications of adding two additional joint points to our scientific understanding. The magnitude of the first two joint points for our knowledge of the world is hard to overstate. It is readily arguable that general relativity, quantum mechanics, and evolution by natural selection represent the broadest, greatest, and most impactful scientific (meta)theories we have. The ToK System places those theories in relation and then proposes that there are two equivalently large metatheories that can be added to the mix to generate a naturalistic scientific ontological picture of the whole.

Returning to the diagram, the absence of BIT on the original is a point worth highlighting. *The vision logic clearly depicts that there should be a metatheoretical joint point between Life and Mind, just as there is a joint point between Energy and Matter and between the Matter and Life planes.* This is not an obvious claim. Indeed, many have offered big picture views of cosmic evolution that include the idea of phase transitions, but only a few include the equivalent of Life to Mind transition. To give just one example, in *Three Big Bangs: Matter–Energy, Life, Mind*, the philosopher Holmes Rolston III (2010) argued that we can obtain a grand narrative of our place in nature by framing it in terms of three "explosions" that generated: (1) matter-energy; (2) life; and (3) the human mind. The first two big bangs line up with Matter and Life, and the third corresponds to the Culture-Person plane of existence on the ToK. As he explains in his Preface (p. ix):

> There have been three big bangs: generating matter-energy, generating life, generating the human mind. These explosions form no simple continuum but a complicated, diffracted, exponential story. 'Big bang' is here a metaphor for critical, exponential, nonlinear bursts with radical consequences for exploring new state spaces with novel combinatorial possibilities. Using another term, there have been three 'big singularities.'

From the vantage point of the Unified Theory, Rolston is correct in that we need a new, coherent naturalistic picture of cosmic evolution. He is also correct in proclaiming that, following Plato, we can "carve nature at its joints" by using the metaphor of big bang singularities. Despite these positives, there is nevertheless a major difference between Rolston's view and the ToK System. Most obviously, his formulation misses the big bang of the Mind-Animal dimension that corresponds with the Cambrian explosion. By contrasting the 1997 ToK depiction of the Life-to-Mind joint point with its absence in other frameworks, we can consider this to be a kind of novel prediction that arises out of the vision logic of the ToK. Specifically, it predicts that there should be: (a) unique patterns of behavioral complexification and properties in animal-mental evolution that follow the same basic logic as the other great differentiations such that (b) it consists of a complexity building feedback loop that includes variation, selection, and retention that is (c) associated with a new information processing and communication system and that (d) the new view can generate a metatheoretical framework that can assimilate and integrate many scientific theories that attempt to map this terrain (i.e., the cognitive, behavioral, neuro and ethological sciences).

Much of my work on the Unified Theory over the past two decades strongly validates these predictions. We can see that some aspects of what this theory should look like are already captured in the original ToK diagram. It lists "nervous system" and "mental representation" as being some of the key ingredients that go into this idea. This can be translated into referencing the neurocognitive aspects of the mind, namely, the idea that, broadly construed, the nervous system functions as a computational control system that guides the behavioral investments of animals. At the time of the diagram, I still held a standard neurocognitive view and had not evolved into what might be called the "integrated 4E cognitive behavioral neuroscience" conception of mental behavior as mindedness that is given by the ToK and synthesized by Behavioral Investment Theory and recent developments in 4E cognitive science.

BIT would develop over the next three to four years. As a metatheory, BIT builds explicit bridges between the following three domains: (1) evolutionary economic approaches to animal behavior, as advanced by behavioral ecologists and sociobiologists like E. O. Wilson; (2) broad information processing computational conceptions of the nervous system (i.e., a neurocognitive functionalist view, similar to Donald Hebb's view of the conceptual nervous system and modern versions such as Karl Friston's active inference; and in recent years, BIT has been explicitly joined with John Vervaeke's metatheory of cognition as recursive relevance realization); and (3) B. F. Skinner's operant approach to behavioral selection. Directly paralleling the logic of the modern evolutionary synthesis, BIT is also structured as a complexity building feedback loop that links an information processing system with a selection process that builds a new complex adaptive landscape. Specifically, it links the idea that the nervous system functions as a recursive relevance realizing information processing system that interacts with the environment, such that the transactional consequences of an animal's action investments reinforce or extinguish subsequent paths and patterns.

In addition to offering a metatheoretical formulation for animal behaviors, when BIT is placed in the context of the ToK System, it also functions as an idea that allows us to resolve crucial metaphysical problems associated with the notion of "behavior and mental processes." Specifically, rather than separating behavior and mental processes based on the epistemological vantage point of the scientists, when viewed through the lens of the ToK System, one sees a complex active plane that can be framed as "mental behaviors." Mental here is an adjective that describes the kind of behavior patterns animals exhibit that make them so different than other living creatures. Framing the behavior of the animal-as-a-whole that produces a functional effect on the animal–environment relationship as mental behaviors is a key aspect of the new vision of scientific psychology being afforded here. It allows for the necessary shift from a methodological behaviorism based on scientific epistemology to a mental behaviorism based on a natural scientific worldview that can frame the ontology of mental behaviors that exhibit the property of mindedness.

Finally, as reviewed in the previous chapter, JUST is the outline for the theoretical joint point between Mind and Culture. Through "justification theory," I could see the human mind's big bang as emerging through the singularity that was formed via the complexity building feedback loop that arose with propositional language and question and answer dialogue, and could trace its evolution to networks of justification processes and systems that could be framed as the Culture-Person dimension of existence. This, in turn, allowed me to frame human language games and then proceed to factor them out of the equation to yield the big picture vision that is the ToK System. It would also allow me to see why human psychology needed to be sharply differentiated from basic psychology, such that the former is a social science, and the latter is a natural science.

A New Theory of Reality and Our Scientific Knowledge of It

In 2003 in the first paper that described this vision, I argued that the ToK System is a "proposal for a universally agreed upon representation of scientific knowledge" (Henriques, 2003). Although this is accurate, I lacked the proper understanding of the philosophy of science to effectively explain exactly what the ToK System achieved and how. This lack of understanding is apparent in how I began the paper with the argument that psychology's struggles were due to a "problem of epistemology." Here is how I framed the issue (Henriques, 2003, p. 151–152):

> I believe that the current approaches are not sufficient because they fail to provide a broad, clear ***epistemological framework*** that sets the stage for defining the discipline and coherently unifying the major paradigms in the field. When one asks basic questions of these proposals such as 'How are life, mind, culture, and behavior defined?' or 'How is psychology specifically differentiated from biology from below and the social sciences from above?' or 'How are the key insights from neuroscience, psychodynamic theory, evolutionary theory and genetics, behavioral science, cognitive science, systems theory, and social constructivist perspectives retained and integrated into a coherent whole?' answers are not readily forthcoming.

Characterizing the ToK System as an "epistemological framework" is reflective of the fact that my understanding of philosophy was heavily skewed toward epistemology as opposed to ontology or metaphysics. As an empirically oriented psychological scientist, I simply was not well versed in the concepts of ontology and descriptive metaphysical systems that should have been used to frame the ToK System when it was first presented. With the benefit of hindsight, I can now see that the key insight afforded by the ToK System is located in the way it functions as a descriptive metaphysical system that can organize and align our scientific theories about reality with reality, rather than what it says about epistemology per se. Indeed, the central thesis of this book is that the key frame for psychological science needs to shift from a focus on scientific epistemology to developing a clear metaphysical depiction of the ontology of the mental realm of existence.

To see this more clearly, we can divide up the ToK System's claims regarding what is currently the standard scientific view of the relationship between reality and our knowledge of it, and the claims it makes regarding how our knowledge ought to be organized after using the ToK to solve for the problem of psychology. Figure 6.2 offers a depiction grounded in the logic of the ToK System regarding the current state of natural science. It depicts the idea that there is a broad consensus that the material and living universe has evolved over time. As we described in the previous section, the physical sciences map the material dimension, and quantum field theory and general relativity form the twin pillars that afford the fundamental explanations for how the material dimension behaves. The modern evolutionary synthesis provides the outline of how the "tree of life" evolves

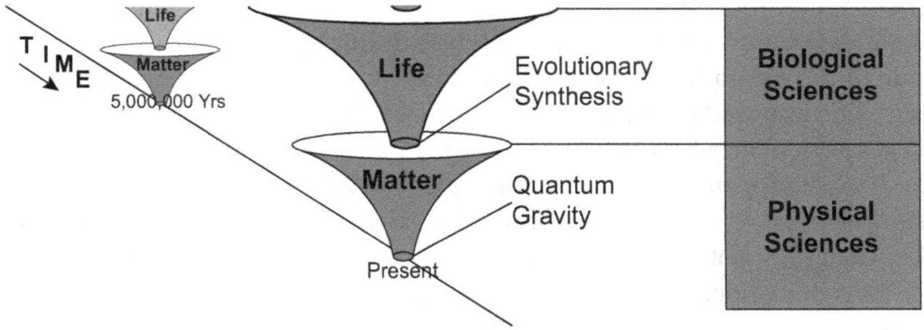

Fig. 6.2 The alignment between the physical and biological sciences with matter and life

out of the material dimension, and the biological sciences map this aspect of reality.

The UTOK argues that this arrangement is a valid and useful representation of our modern natural scientific ontology, such that most naturalistic viewpoints could be aligned with this basic depiction. To be sure, this depiction does carry some novelty, in that it is representing the material and living worlds as planes of existence that emerge out of an Energy Information implicate order. This is, admittedly, not the way many scientists would describe the current ontological picture. Despite this caveat, it is also the case that every scientist I have shown this to immediately acknowledges the logic and can follow it clearly, and agrees that it affords a reasonable representation.

The key point I want to emphasize here is what happens to the clarity of our scientific understanding as we progress up the stack of complexification. As framed by the Enlightenment Gap, UTOK asserts that the conceptual coherence of our scientific understanding of the world dramatically breaks down when we move from the dimension of Life and the biological sciences into the dimension of Mind and the domain of the psychological sciences. The problem of psychology and the missing "Life-to-Mind" joint point in many big picture schemes makes this clear. In so doing, it highlights one of the key reasons why UTOK is uniquely structured to provide a coherent natural philosophy that allows us to see the whole shebang, including the domains of the mental and of human persons. The reason is that it crisply defines them as planes of existence.

The standard ToK System representation is offered in Fig. 6.3. It is structured in a way that more clearly frames the relationship between scientific knowledge and reality than the original diagram. The left side of the diagram represents what we can call the "ontic reality," which we can define as that which exists independently of our beliefs about it. The right side of the diagram represents our scientific knowledge of that reality. Our scientific knowledge of the ontic is our scientific ontology. *The ToK System depicts this clearly because it places the position of science as being simultaneously embedded in the Culture-Person plane, and it can also be thought of as emerging out of the Culture plane.* This structure captures how scientific knowledge is both built by humans and consists of new epistemological systems of justification that give rise to a new ontological picture of reality. It is a picture of reality that arguably transcends prior systems of justification and can be defended as being more objective, accurate, and true than previous conceptions of reality. The completeness of the gestalt and the

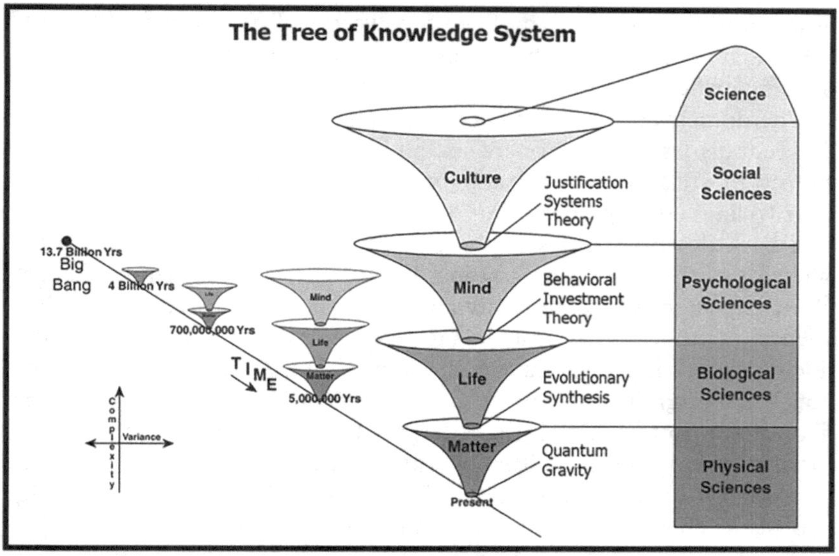

Fig. 6.3 The standard Tree of Knowledge System depiction

correspondence between the ontic dimensions of Matter, Life, Mind, and Culture and the onto-epistemological knowledge structures in the physical, biological, psychological, and social sciences is what grounds the claim that the ToK System affords a new, holistic, naturalistic scientific worldview.

Important Additions to the Original Diagram

There are three major additions that characterize the difference between the original and the current, standard ToK System depiction. We have already mentioned one major addition, which is filling in the third joint point via BIT and clarifying the concept of the Mind dimension as the set of mental behaviors. The other two major developments are the addition of the time dimension on the horizontal, and the depiction of science as a system of justification that emerges out of Culture and corresponds the four domains of science to the four dimensions. Although the time dimension was implicit in the first diagram, its explicit inclusion helps clarify some key metaphysical and ontological points. For example, the new diagram affords a way to clearly separate the "vector of general relativity" and how it leads to the Big Bang from the "vector of quantum field theory," which is at the bottom of the diagram. It also better captures the evolution of complexification across a natural logarithmic time scale.

The domains of scientific knowledge and how they correspond to the different planes of existence are also an important addition. As noted previously, this relationship is clearly hinted at in the original ToK diagram in the two lines coming out of the Culture plane that circle around and traveled back down the sides of the page. On the left side, there is a line labeled "math" and an orthogonal line labeled "stats." These concepts ultimately align with complexity and variance, and they connect to how science quantifies behavioral change across the dimensions. On the right side of the original diagram, there is a line labeled "knowledge construction." Although it is cut off, it says "subjective" on the top and "objective" on the bottom. This is the difference between subjective or pragmatic social knowledge in the cultural world and the more objective knowledge that we have acquired in science, which is exemplified by our knowledge in physics. This distinction is important in the UTOK metapsychology.

As a function of these additions, the explicit logical separation of the ontic reality and our scientific knowledge of it becomes solidified in the standard ToK System. It clearly depicts scientific knowledge emerging out of the Culture plane to develop ontological maps of the ontic territory. It took me several years to realize the importance of a descriptive metaphysics that explicitly differentiated scientific ontology from the ontic reality. That is, early in my thinking, I primarily used just epistemology to stand for methods of knowing, and ontology would be used to refer to what is real. However, the standard ToK depiction shows that when considering scientific knowledge, we need to yoke epistemology and ontology together as the scientific systems of justification that function to map the ontic reality. The next section clarifies why this is the case.

Differentiating Scientific Epistemology from Scientific Ontology and the Ontic Reality

Central to UTOK's vision for solving the problem of psychology is addressing the confusion and conflation between metaphysics, ontology, and epistemology that arise from the Enlightenment Gap. As we have shown, the ToK System identifies and differentiates (a) the ontic reality in the form of Energy, Matter, Life, Mind, and Culture from (b) scientific ontology and epistemology. Consistent with JUST, scientific knowledge of the ontic reality is framed as being a kind of system of justification. Specifically, the theories of how things in the world behave represent our scientific ontology, whereas the process by which those theories are justified to be valid represents our scientific epistemology.

For clarity on this point, we can turn to Plato's original formulation of a theory of knowledge. He argued that knowledge was best characterized as justified true belief. In this formulation, belief refers to the representation or model of the state of affairs held by the knower about the known (i.e., the knowable portion of the "external" reality). Beliefs are instances of true knowledge when they both corresponded to the actual state of affairs and are also justified by good reasons. To understand the justification qualifier, consider that if a four-year-old had heard his mother, a chemist, say that water is "H_2O" and then parroted this back when asked what water is, we would not say that the four-year-old had true knowledge about water. His mother, on the other hand, would be fully justified in making that claim because of the network of reasons that she has access to that can legitimize this claim. Although work by Geitter and others (e.g., Gettier, 1963) has demonstrated that there are exceptions to the claim that authentic knowledge can be aptly framed in terms of justified true belief, it nevertheless remains a powerful formulation.

The UTOK metapsychology uses aspects of Plato's formulation as a general framework for a theory of scientific knowledge. It does so by aligning scientific epistemology, ontology, and the ontic reality with the formulation of justified true belief. In this alignment, "justified" corresponds to epistemology, whereas "true belief" corresponds to one's scientific ontology as it represents the ontic reality. Consistent with Roy Bhaskar's critical realist view of science, which we will return to in more depth later, UTOK posits that modern empirical natural science was framed by a new kind of epistemological process that generates what Bhaskar called "transcendental realist" claims. These are truth claims that rise above subjective bias and social convention to yield an "intersubjectively constructed objective" understanding of the nature of reality. We can build from our analyses of epistemology in the previous chapter to see why this is the case.

As reviewed in Chap. 3, Gauch offered a standard view of scientific epistemology. Not surprisingly, he largely focused on the methods of science and the way science operates as a kind of epistemological system that enables greater confidence in the correspondence between one's models and maps and the territory that is reality. He paid less attention to general questions about ontology and the idea that natural science can give a worldview, noting that there was no consensual scientific worldview that scientists operated from. Gauch is not alone, and most people think about science more in terms of epistemology than ontology. For example, the physicist Sean Carroll (2017, pp. 133–134) describes science as a kind of

> methodological empiricism—the idea that knowledge is derived from our experience of the world, rather than by thought alone. Science is a technique, not a set of conclusions. The technique consists of imagining as different ways the world could be (theories, models, ways of talking) as we possibly can, and then observing the world as carefully as possible.

Although a methodological or systematic empirical epistemology is a crucial aspect of science, it is misleading to consider it all of science. Indeed, the core message of Carroll's book demonstrates why this is so. Its central argument is that physics has generated a "core theory" that provides a foundational ontological view of the ground of being, from which additional levels of nature emerge. His work demonstrates that although the epistemology of science is crucial, it is only one aspect of science. Indeed, it seems most scientists would be forced to agree that it is the knowledge that science produces about the world that constitutes its core value. However, as Gauch makes clear, when discussing the generalizable essence of science, a broad view of scientific ontology is not emphasized, for a simple reason: *There is no scientific ontology that affords a clear, coherent picture of the natural-into-human world that is consensually agreed upon by most scientists.*

We have already discussed how epistemology refers to the process by which one gathers data and justifies claims, whereas ontology refers to claims about what is real. However, we need to elaborate on the concept of ontology to afford some additional nuance. Similar to the way we divided metaphysics into the domains of descriptive metaphysics, metaphysical systems, and pure metaphysical claims, UTOK differentiates between three different meanings of the word ontology, in addition to separating ontology from the ontic reality. In particular, we can divide ontology into folk ontology, depth ontology, and scientific ontology.

Folk ontology refers to how truth claims are made or debated in everyday life. Thus, here we are talking about what people in everyday contexts use to determine the validity of a proposition. For example, if Joe is wondering if his wife Sarah is having an affair, a commonsense ontology frames this in true or false terms; simply put, she either is cheating on him or she is not. Folk ontology is crucial to getting along in the world. However, it is intuitive and not grounded in refined analyses that strive to uncover, spell out, or question the assumptions and presuppositions that undergird the claims. For example, folk psychology does not dive deeply into the relationship between the social construct of cheating and a physicalist description of material behaviors in the world.

The other two meanings of ontology are more refined and require study and analysis to ground them. In this context, depth ontology refers to the philosophical exploration that includes the complicated intersection of ontology with the ontic reality and epistemology. Many philosophers have engaged explorations of depth ontology. Perhaps most notably, Martin Heidegger plumbed the depths of ontological analyses and explored the ultimate nature of being or what he called "being-itself" (for more on recent developments in ontology, see Harman, 2018). We can see that complicated questions emerge as we collapse the experience of reality with our beliefs about reality with the realness of the moment itself as experienced by a particular knower. Consider the ToK System. On the one hand, it "maps" Matter, Life, Mind, and Culture, and on the other it calls them the ontic reality. A more proper way to frame what the ToK System offers is that it gives a new descriptive metaphysics that provides an ontological mapping of the relationship between scientific ontology and the ontic reality. The reason is that as soon as one starts making knowledge claims about the ontic reality, one has shifted into ontology.

The standard scientific epistemology is grounded in a correspondence theory of truth, and it posits a clear distinction between the ontic reality and one's scientific representations about it. Indeed, science is fundamentally about generating the best maps, taxonomies, and models to describe and explain reality; however, these concepts, categories, and theories are generally quite separate from reality itself. There are some exceptions to the claim that scientific ontology and epistemology remain clearly separable from the ontic reality. For example, both quantum entanglement and related measurement problems, as well as the problem of the double hermeneutic in the human sciences, are cases where the divisions between knower and known become blurred and require significant philosophical analyses to bring clarity. Nevertheless, the general goal of science is to develop a map that accurately corresponds to the territory, and for the most part these domains are readily separable in science.

We can apply the way UTOK frames scientific epistemology, ontology, and the ontic reality to the shape of the Earth and our knowledge about it to show how these concepts operate in a straightforward example. From a standard scientific critical realist perspective, one can say that the Earth is an ontic entity that exists independently of human beliefs about it. We can also say that for many thousands of years humans have wondered about its shape. That is, they have debated which geometric concept—spherical roundness or a disc-like flatness—accurately applies to the actual shape of the Earth. This is a basic ontological question. In the current language system, believing that the Earth is spherical is making an ontological claim about the ontic reality of the Earth. The process by which propositional claims are generated and the extent to which they are justified are questions of epistemology. Over the course of much exploration, gathering evidence and systematic processes of justification, the shared consensus is that the Earth is not a flat disc, but is essentially spherical, although it is not a perfect sphere because it bulges in the middle as a function of gravity and its rotation.

Of course, there is a small group of individuals who, to this day, believe the Earth is a flat disc. This is a different ontological picture, but one that is not justified by the epistemology or institution of science. There will always be more than one possible interpretation and, as such, the process of justification must involve comparing different justifications and developing the best explanation. This is called abduction; it refers to the kind of reasoning that generates the best explanation for the accumulated knowledge. We can thus say that the conclusion that the Earth is spherical is based on abductive reasoning. That is, it is an ontological claim that is justified by many different lines of evidence and effectively rules out other alternative explanations. This fact also allows us to recognize science as a public endeavor and that much of the fund of scientific knowledge is framed in terms of expert consensus.

The UTOK is about generating a descriptive metaphysical system that affords scientists a naturalistic scientific ontology writ large. There have been several efforts in this regard. For example, Sean Carroll's (2017) big picture view is one example. So too are E.O. Wilson's (1999) vision of consilience, Eric Chaisson's (2001) cosmic evolution, and David Christian's (2018) Big History formulation. Indeed, an argument can readily be made that the general outline of a naturalistic ontology is now available. There is a virtual consensus among naturalists that cosmic evolution can be effectively mapped on the dimensions of time and complexification. However, despite this outline, it is still the case that there is no shared, clear consilient naturalistic ontology that is generally agreed upon that includes a clear picture of matter and mind and scientific and subjective knowing. The presence of the Enlightenment Gap and its most obvious consequence, the problem of psychology, means that the view of consilience breaks down as one moves from the physical and biological sciences into the psychological and social sciences.

The vision logic of the ToK System provides us a new way to frame both the psychological and social sciences, in proper relation to the natural sciences. With its joint points between Life and Mind and between Mind and Culture and the process by which it contextualizes science from the Culture-Person plane of existence, it affords us a new descriptive metaphysics that can define the institution of Psychology and its subject matter (i.e., mental behavior), clarify the relationship between the natural and social sciences, and give us a commonsense description of the different planes of existence that update perennial philosophical positions and place them in coherent relation to findings from modern empirical natural science. These three tasks are explored in the remainder of this chapter.

The Tree of Knowledge System as a Descriptive Metaphysics and Coherent Naturalistic Ontology for Psychological Science

To effectively understand the message it is communicating, it is useful to keep in mind the origin of the ToK System. It is a map of reality and science that emerged in direct response to how American psychology was organized at the end of the twentieth century. Put in the negative, it is a naturalistic scientific ontology that did not come out of philosophy, physics, or sociology. Rather, it emerged in the conceptual space between natural science, scientific psychology, social science, and integrative psychotherapy. This means it is a map structured with specific emphases. In addition, it makes prescriptive claims about how the institution of Psychology should be defined and organized going forward in the twenty-first century. It is important to be clear about both aspects when interpreting what is being offered with the ToK System.

The original 2003 paper showed how the ToK System could align the central insights of Skinner and Freud with an emergentist physical-bio-psycho-social view of cosmic evolution writ large. It detailed how

Psychology and its subject matter could be contextualized via the big picture provided by the ToK. It further argued that BIT provided a metatheoretical frame for bridging the joint point between Life and Mind and the scientific understanding of animal-mental behavior. When viewed via the descriptive metaphysical system afforded by the ToK, it becomes clear that Skinner's key ideas were, somewhat paradoxically, about Mind. Hunting, mating, and defending a territory are mental behaviors, the set of which is represented by Mind on the ToK. BIT showed how mental behaviors could be theoretically connected to evolutionary theory, cognitive neuroscience, and Skinner's radical behaviorist formulation (but not his flawed radical behavioral philosophy).

In a parallel manner, the paper argued how JUST (at the time called the "Justification Hypothesis") filled in the joint point between Mind and Culture. It assimilated and integrated Freud's key observational insight that the human ego evolved as the mental organ of justification. It also aligned with much work in social constructionism and built a clear bridge from human psychology into cultural anthropology and the social sciences. The vision logic of the argument of the original paper is depicted in Fig. 6.4. The image on the left depicts the current organization of psychological theory. The image suggests that psychology could be thought of as

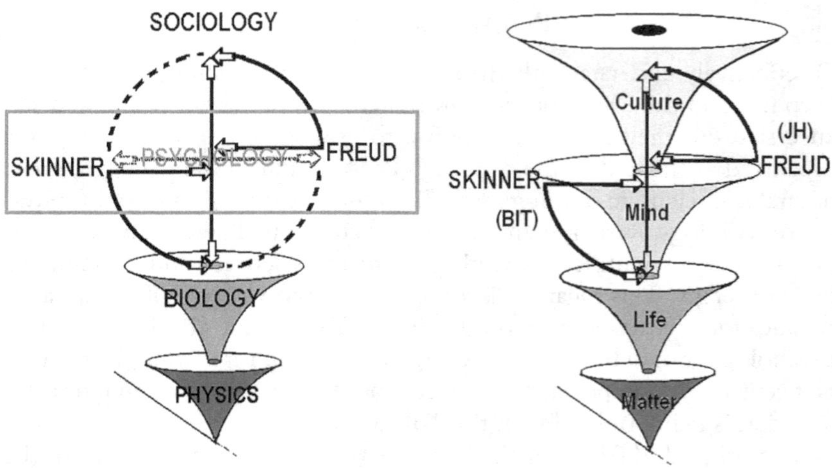

Fig. 6.4 The Tree of Knowledge aligns the key insights of Skinner and Freud on a physical-bio-psycho-social axis

existing in the poorly defined conceptual space between the ideas of Skinner and Freud on the horizontal axis, and between biology and sociology on the vertical axis. It depicts how the physical sciences map the dimension of matter and the biological sciences map the dimension of life, but consilience breaks down after that. However, consilience can be restored via the vision logic of the ToK System, by mapping two missing joint points, between Life and Mind and between Mind and Culture.

The image on the right depicts the vision logic and "conceptual rotation" I was advocating for via the ToK System, and the linkages of Skinner with Behavioral Investment Theory and Freud with JH/JUST. I chose these two icons because radical behaviorism and psychoanalysis simultaneously represented two of the historically most influential but also most diametrically opposed viewpoints in American psychology. My claim was that if these two systems of thought could be tied together in a consilient way, the rest of the field could also be assimilated and integrated in the metatheoretical architecture provided. Here was how I justified my argument (Henriques, 2003; p. 152):

> According to this analysis and in direct contrast to those who argue that unification is impossible, a unified approach can coherently unite the ideas of Skinner and Freud using the same overarching system, one that clearly spells out the errors and inconsistencies in each paradigm while retaining the key theoretical insights from both perspectives. Of course, students of psychology are not offered such a system. Instead…students are simply taught about the diversity of ideas and left to their own devices to sort out the issues. The current proposal seeks to change this status quo. The outline of a system is offered that I propose aligns the central insights of Skinner and Freud both with one another and with science at large. More specifically, I show how the science of psychology can be thought of as existing between the central insights of Skinner and Freud. In putting these pieces of the puzzle together, I offer a way to clearly define the field and provide a metatheoretical framework that can incorporate the major theoretical perspectives into a coherent whole.

A Proposed Definition for the Institution of American Psychology

In the years that followed this initial publication, my attention shifted to include claims about the institution of Psychology. Via its joint points and metatheoretical formulation, the Unified Theory very clearly demonstrated that if the science of psychology was to be coherent and retain its central emphasis on "behavior and mental processes," it had to be divided into two broad domains of basic and human psychology corresponding to animal mental and human mental behavior, respectively. I made this argument in the next major publication on the Unified Theory in a target article called "Psychology Defined" (Henriques, 2004) that served as the primary referent for two special issues of the *Journal of Clinical Psychology*.

"Psychology Defined" laid out the argument that, given the ToK System's differentiation between Mind and Culture, a major differentiation was needed in the institution of Psychology. The first branch is basic psychology, which corresponds to the scientific analysis of the animal-mental dimension of behavioral complexity. The primary task of basic psychology is to scientifically study the set of "mental behaviors" that take place both inside the nervous system (i.e., covert neurocognitive processes defined by Mind^1 and subjective experiences in animals defined by Mind^2) and between the animal and environment (i.e., overt mental behaviors). The second branch is Human Psychology, which is simultaneously an extension of basic psychology and differentiated from it. This is because it deals with human persons, who are both mental primates and self-conscious beings who can justify their actions on the social stage (i.e., they are persons who exhibit Mind^3 behaviors).

Finally, the Unified Theory leads to the obvious conclusion that there should also be a clearly identified *profession* of psychology. I made the case for a "unified professional psychology" in a separate article with past American Psychological Association President Robert Sternberg (Henriques & Sternberg, 2004). Professional psychology includes the formal practice areas of clinical, counseling, and school psychology and their various offshoots, such as neuropsychology or child and family psychology. The primary identity here is of *psychological doctors* who are health service professionals who work to assess and treat mental disorders and to foster the biopsychosocial well-being in individuals, couples, families, and communities. A psychological doctor is fundamentally different than a psychological scientist because the goals of the profession are prescriptive rather than descriptive. That is, whereas psychological scientists work to describe and explain the world, psychological doctors work to effect change in desired ways. Placed together, we can say that, according to UTOK, psychology is the science of mental behavior, the human mind, and the professional application of such knowledge toward the greater good (Henriques, 2011).

Differentiating Between the Natural and Social Sciences with the ToK System

The UTOK metapsychology is about framing the proper relationship between the science of psychology and the natural sciences "from below" and the social sciences "from above." The previous section showed how UTOK identifies a central dividing line between basic and human psychology, which corresponds to this distinction. The implication is that animal psychology can be framed as being directly continuous with and now a part of the natural sciences, and we can "follow the trail" into the human sciences, but also be aware of an important discontinuity. This continuity and discontinuity between the natural and social sciences can be seen in the pictographic representation afforded by the ToK System.

We have already noted that the ToK System depicts "Science" both as residing in the Culture dimension and emerging out of it to map the rest of the ontic reality because it affords a new and more transcendent picture that is based on a more objective epistemology than prior forms of knowing and justification. It is also important to be clear that, under the heading of "Science," the referent should be understood as the onto-epistemological knowledge structures that organize the "basic natural-into-social sciences." The specifier 'basic' refers to the difference between basic and applied sciences. The goal of the former is to describe and explain complexity and change, whereas the goal of the latter is to make a difference in the world (i.e., effect change for the better). The difference is important. As discussed above, it characterizes the primary difference between the science of psychology and the profession. Psychological scientists try to describe and explain mental behavior via research, whereas psychological doctors work to improve human well-being by carrying out assessments and interventions. It is the science of psychology that is being represented.

It is also important to be clear about what is meant by the "natural-into-social" science reference. The UTOK metapsychology frames the task of the "basic/pure/nonapplied social sciences" as being structured to describe and explain human behavior at the individual and group level, as objectively as possible. It is seen as consisting of five primary disciplines: (1) human psychology; (2) anthropology; (3) sociology; (4) economics; and (5) political science. Human psychology, which, broadly construed, would include cognitive science and linguistics in this formulation, forms the "unit base" of the social sciences. That is, UTOK posits that human persons and their self-conscious justifications are the fundamental units of the social sciences from this perspective. Of course, via its clear map of human psychology, UTOK allows the effective bridging of the system "down" into basic psychology. In this regard, we can think of human psychology's relationship to the social sciences as being akin to what neuroscience is to (basic) psychology. That is, just as neuroscience is properly conceived as the hybrid between biology and (basic) psychology, human psychology is a hybrid between basic psychology and the social sciences.

Regarding the organization of the social science disciplines, the alignment is structured as follows. First, human psychology is concerned with individual into small group behavior. Next comes anthropology, which is the general science of human behavior and is the primary social science discipline. Sociology comes next, and it can be framed as the science of human group behavior in the context of technological and industrial organizations or civilizations. Although closely related to anthropology, it brings in more focus on the macro-level social organization and societal institutions. Then there is economics, which is the study of resources, market exchange, scarcity, and wealth production at the micro and macro levels. Finally, there is political science, the science of governance and political structures in civilization. Of course, there are complicated relationships with history and philosophy here, but these two areas of inquiry do not attempt to play by the rules of the language game of modern empirical science.

In offering this description, an important gap emerges in the ToK taxonomy that must be highlighted. The ToK System does not depict human technology, which is, of course, a huge aspect of the social sciences. Societies are what Delanda (2016) calls "assemblages," and a comprehensive account of social science would need to bridge UTOK's conception of the Culture-Person plane of existence with the on-the-ground dynamics associated with societal assemblages and the evolution of technologies. More details on the nature of the Culture-Person plane are provided in Chap. 15.

The central point for our purposes is that the ToK System provides a framework for the "basic natural-into-social sciences" in that it shows that there is both continuity and discontinuity as we move across the divide from sciences of Matter-Object, Life-Organism, and Mind-Animal into the sciences of the Culture-Person plane and human societies more generally. The nature of the continuity is that (a) we can indeed apply the language game of science to human behavior and (b) via the ToK we can follow the cosmic evolutionary line from hominids as primates into modern Culture-Person activity. However, the nature of the discontinuity pertains to the fact that (a) the Culture-Person plane is of a different kind (although we should note that we see similar discontinuities with Life and Mind) and (b) as framed by JUST and many other perspectives, humans have explicit self-concept recursion capacities, which results in an added layer of complexification, and related to this, (c) there is a feedback loop between scientific knowledge as a kind of justification system that describes and explains human behavior, and how that feeds back into humans and changes them and then feeds back again into scientists.

This latter point is a massive tangle and much has been written about it. It is central to the confusions regarding the relationship between social and scientific knowledge in the Enlightenment Gap. As was described in the second chapter and with the example of ADHD, UTOK aligns with the analysis of this dynamic offered by the sociologist Anthony Giddens (2003) and his analysis of the "double hermeneutic." With JUST, we can interpret hermeneutic to mean "justification system" (Henriques, 2011). The natural sciences have, as their task, the development of an interpretive system of scientific justification that describes and explains natural phenomena that range from particles to primates. Physicists, chemists, and biologists debate and challenge each other about the valid interpretations of the subject matter. This process in and of itself does not impact the subject matter. Bacteria are not insulted if we label them as diseases. Likewise, animals do not read scientific papers or reflect on what theories of basic psychology mean about their existence.

However, the situation changes radically when we move to understanding human-social behaviors. A central reason for this is that the interpretations of scientists feedback into the human world because humans use the concepts and change their behavior accordingly. To give just one example, consider all the controversy about science and race and intelligence. The way scientists talk about race and intelligence influences how people act and think about themselves. Thus, there is a direct and iterative process between scientific justification systems and Culture at the level of how people behave. These arguments are directly present in the vision logic of the ToK diagram. It shows the continuity of Culture into Science and at the same time represents how basic Science can transcend standard cultural justifications to provide a more objective depiction of reality. Moreover, as is highlighted in Fig. 6.5, the domain of Social Science shows

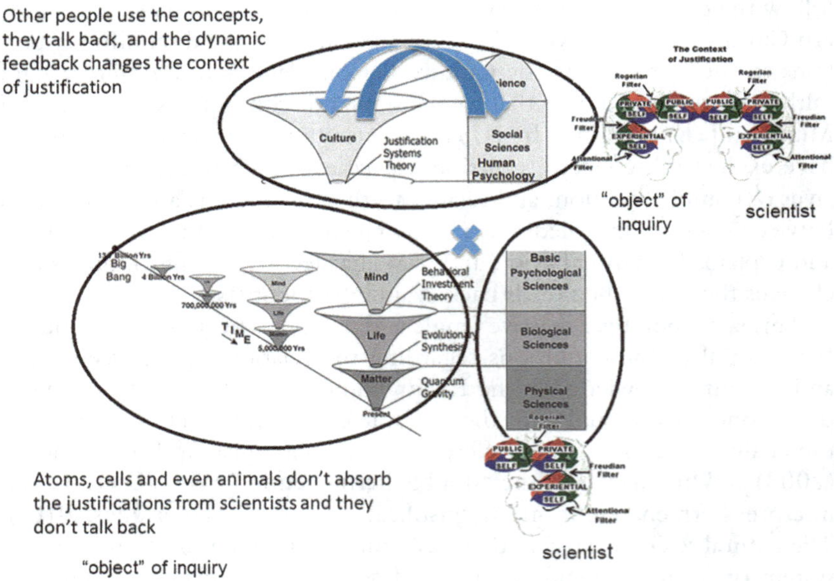

Fig. 6.5 The ToK System shows the different relation science has with the social as opposed to the natural sciences

a much more direct relationship to Culture that is present in the relationship between the natural sciences and their domains of inquiry.

The circles capture how there is an inclusive continuity of feedback between the object and scientist in the social sciences in a way that is not present between the natural scientists and their objects of inquiry. I have also framed the "object" of inquiry to make a point about language. Notice that thinking about humans as "objects" likely activates a problematic intonation, such as the justificatory retort that humans are not mere things. It is precisely the nature of this kind of iterative feedback and the conceptual tensions that arise from it that I am pointing out. This analysis enables us to see yet another angle on the problem of psychology and why the discipline of psychology sits on the fault lines of our knowledge. It shows that psychology straddles the divide between the natural and social sciences, and highlights with more clarity why basic psychology should be divided from human psychology. The summary point is to illustrate how and why the ToK System affords us a new theory of reality and science that will help frame and resolve the "science versus society" relation that is central to the Enlightenment Gap.

Commonsense Understanding and the Planes of Existence on the ToK

Although UTOK is structured to formally address the Enlightenment Gap and resolve the problem of psychology, it is also the case that the UTOK metapsychology is concerned with the bridge between science and our everyday folk understanding of the world. This is necessary to effectively address the relationship between science and society. In the final section of this chapter, we explore how the vision logic of the ToK System frames the scientific planes of Matter, Life, Mind, and Culture in a way that can be effectively bridged to how people talk about the behavior of inanimate material objects, organisms, animals, and people.

We can begin by considering a family having a conversation about what they are going to do for the weekend. Their dog is with them. We can use the ToK map to say that the family's conversation and shared systems of justification exist on the Culture-Person plane. This consists of sociolinguistic behaviors that can be legitimately described as existing "above" or "beyond" the dog in terms of complexification. This is because the dog operates on the Mind-Animal dimension. That is, as a minded entity, the dog senses the world via its brain and nervous system and moves around in it as a functional whole. As a social mammal, the dog tracks the emotional demeanor of others and engages and responds to relational cues (e.g., signals dominance or affiliation). However, the conversation about family dynamics, planning, and reflective choices involves sociolinguistic behavior patterns that operate on a different plane of existence. Even though the sounds of the words impinge upon the dog's auditory system, the meaning of the words and the nuances of the verbal exchange are not available to the dog. To say this is not being species centric. It is simply an empirical fact that dogs cannot fully participate in the Culture-Person plane of existence.

Full participation in the Culture-Person plane requires language, development, and socialization; that is, it requires growing up in a sociolinguistic environment and learning how to become a person that justifies one's actions on the social stage. If the family had a 10-year-old, a 6-year-old, a 2-year-old, and a 3-month-old infant, we would see clear differences in their stages of human development, and their varying capacities to participate on the Culture-Person plane. Indeed, much of human development and socialization is about learning how to operate in the world as a person. The 3-month-old infant does not yet operate on the Culture-Person plane, whereas the 2-year-old is starting to learn the basic elements. The 6- and 10-year-olds operate as young human persons at different stages of development and sophistication, whereas both parents operate at full adulthood.

If we put the Mind-Animal and Culture-Person planes in relationship to each other, we can thus see that there is: (1) a primate pattern of behaving in the world that corresponds to the neurocognitive and experiential mental domains (i.e., Mind^1 and Mind^2); (2) a developmental period through childhood where self-reflective language becomes operative so that the individual can start to participate in justification dynamics (i.e., a developing Mind^3); and (3) an adult stage where individuals are considered fully participating agents who both give accounts and are considered by the community to be accountable for their actions under normal circumstances. Both the 3-month-old and the dog operate solely on the Mind-Animal dimension of existence, although they do so in different ways. Each clearly demonstrates functional awareness and responsivity in a way that corresponds to Mind^1. And we can assert with confidence that both also have a subjective conscious experience of being (Mind^2). It is also the case that the dog will operate in different territories on the Mind plane than the humans. For example, the dog's remarkably superior sense of smell gives access to an olfactory sensory-motor landscape of engagement that is not available to anyone in the family. The socialized humans operate on the Mind-Animal plane in the territory of primates (i.e., hominids), and they operate on the Culture-Person plane. This is their capacity to plug into the systems of justification and engage in linguistically mediated self-reflective processes in the explicit intersubjective world of other justifiers.

For the dog and humans, as is the case for all animals, complex adaptive dynamic processes are also taking place at the Life-Organism dimension of existence. The flower on the table that the 10-year-old son brought his mother for her birthday also exists at the Life-Organism dimension. So too do all the bacteria and other single-celled life forms in the room. However, lacking a brain and nervous system and a complex active body, neither bacteria nor flowers exist at the Mind-Animal plane. Thus, in the language of the ToK System, the dog exists at a dimension of behavioral complexity "above" the flower. Once again, we must note that cells and organisms like flowers demonstrate self-organizing capacities and engage in dynamic responses to shifts in the environment. The flower might bend toward the sun, for example. We would not refer to this functional behavior as existing in the domain of the mental because flowers lack a nervous system. But cells and plants are self-organizing creatures that clearly exhibit complex adaptive behavioral patterns.

Beneath the Life-Organism plane is the Matter-Object dimension. This is the dimension of molecules, atoms, and particles at the smallest scale, and of planets, stars, and galaxies at the largest. Objects in the Matter dimension differ in their degrees of complexity compared to those of Life, Mind, and Culture. Material objects can be complicated. There can even be some complex self-organization, such as found with whirlpools, hurricanes, and tornados. However, even these physical systems do not exhibit the kinds of complex adaptive behavior patterns that organisms, animals, and people do. Consider, for example, that cells, dogs, and persons can all adjust to injury in many ways. Cells engage in repair; dogs can display submissive or affiliative cues; and humans can feel guilty and explicitly apologize. These are all examples of dynamic, autopoietic self-organizing responses operating on a complex adaptive plane. When living creatures die, their complex adaptive organization breaks down, and they drop into existing at just the material plane.

A New View on the Great Chain of Being

As a descriptive system, we can note that the ToK System affords us a basic and well-known taxonomy for dividing things up in the world. Consider, for example, that at the age of four, my daughter Sydney effectively captured the essence of this taxonomy as: "rocks, plants, animals, people." The reason this system of categories came naturally to her is that these are intuitively natural categories for all human beings. Indeed, every known culture has developed a taxonomy that corresponds to these four domains. In addition, the four domains are represented in what is arguably the most influential system of knowledge in the Western canon, namely the Great Chain of Being (Lovejoy & Stanlis, 2017). This is the idea that one can move upwards from minerals to plants to animals to humans to angels and finally to God in terms of spiritual sophistication. In the traditional Christian conception, the latter two domains reside in the supernatural plane of existence, with humans having a soul that is given by God that breathes spirit and rationality into the human being as he or she lives on Earth.

From a more naturalistic scientific vantage point, it is important to realize that the Great Chain of Being was rooted in a conception of nature that dates to Aristotle, and his *scala naturae*, or the scales of nature. Aristotle saw that living entities exhibited functional forms of behavior that inanimate objects did not. Unlike minerals, plants exhibited the functional form of metabolic growth and reproduction above and beyond the properties found in inanimate matter. In a similar fashion, Aristotle saw that animals included the physiological properties of growth and reproduction, but also demonstrated coordinated movement and sensory and perceptual experiences beyond those of plants. Finally, humans included these functional properties of both plants and animals and had exhibited symbolic language and were capable of rational thought in ways that animals were not. Table 6.1 captures this basic set of insights.

As Ken Wilber (2007) has lamented, this basic scheme fell out of favor as modern science advanced in the centuries following the Enlightenment. There are many reasons why the scheme collapsed. One reason is that the Enlightenment ended up creating a problematic and dichotomizing split between matter and mind that left out living organisms and mental animals. However, it is also the case that the original taxonomy simply could not assimilate and integrate the set of findings that science uncovered regarding how the world works. For example, there are many small-scale

Table 6.1 Aristotle's scales of nature

	ability to grow and reproduce	ability to move	ability to think rationally
Humans	X	X	X
Animals	X	X	
Plants	X		
Minerals			

and intermediate forms of being that early scholars had no way of understanding. Consider, for example, that a virus does not fit neatly into this scheme. It is neither a plant nor a mineral. However, what should have been done is that the scheme should have been reorganized and updated. The ToK System sets the stage for a return of the scales of nature as mapped by the various levels and dimensions of behavioral complexity that the modern scientific method has uncovered. The rest of this book shows how, with it, we are now in a place to develop a descriptive metaphysical system for mapping the Great Chain of Being that is consistent with modern science and solves the problem of psychology.

Conclusion

Two decades into the twenty-first century, human knowledge resides in a state of chaotic, fragmented pluralism. This is a consequence of several connected factors. Conventional conceptions of science are too often framed in terms of suggesting a physicalism that reduces everything to the material dimension of existence. In addition, too much of our thinking is siloed into separate channels of expertise. Moreover, the emergence of postmodern critiques powerfully demonstrated how much of human knowledge, even scientific knowledge, is shaped by social and contextual forces. All of these forces have combined to tear down any apparent hope for a consilient naturalistic worldview that includes the human knower.

As good clinicians know, one of the keys to adaptive behavior is insight regarding the nature of the problem. The chapters in this section have outlined how UTOK diagnoses the Enlightenment Gap and its epicenter with the problem of psychology and posits that there is a solution. JUST was the initial insight that properly framed the joint point between primates and persons. In so doing, the stage was set for the ToK System and its novel insight that Life, Mind, and Culture were all emergent complex adaptive dimensions of behavioral complexity meditated by novel information processing and communication systems, linked by complexity building feedback loops. This chapter has shown why and how the ToK System affords us a new descriptive metaphysical system that effectively maps the relationship between the four domains of ontic reality and the major domains of scientific inquiry.

The next section deepens the argument by connecting it first to natural science and then to systematic metaphysics. In so doing, it lays out the case more explicitly as to why the ToK System succeeds in providing us with a new and better map of big history. Specifically, it gives us a coherent descriptive metaphysical system that clarifies a naturalistic ontology that can solve psychology's BM^3 problem and clarify the nature of human persons.

References

Cahoone, L. (2013). *The orders of nature*. Suny Press.

Carroll, S. (2017). *The big picture: On the origins of life, meaning, and the universe itself*. Penguin.

Chaisson, E. J. (2001). *Cosmic evolution: The rise of complexity in nature*. Harvard University Press.

Christian, D. (2018). *Origin story: A big history of everything*. Little, Brown Spark.

DeLanda, M. (2016). *Assemblage theory*. Edinburgh University Press.

Gettier, E. L. (1963). Is justified true belief knowledge? *Analysis, 23*(6), 121–123.

Giddens, A. (2003). Social science as a double hermeneutic. In *Philosophies of social science. The classic and contemporary readings* (pp. 400–404).

Harman, G. (2018). *Object-oriented ontology: A new theory of everything*. Pelican.

Henriques, G. R. (2003). The tree of knowledge system and the theoretical unification of psychology. *Review of General Psychology, 7*(2), 150–182.

Henriques, G. R. (2004). Psychology defined. *Journal of Clinical Psychology, 60*(12), 1207–1221.

Henriques, G. R. (2011). *A new unified theory of psychology*. Springer.

Henriques, G. R., & Sternberg, R. J. (2004). Unified professional psychology: Implications for the combined-integrated model of doctoral training. *Journal of Clinical Psychology, 60*(10), 1051–1063.

Huneman, P., & Walsh, D. (Eds.). (2017). *Challenging the modern synthesis: Adaptation, development, and inheritance*. Oxford University Press.

Lovejoy, A. O., & Stanlis, P. J. (2017). *The great chain of being: A study of the history of an idea*. Routledge.

Rolston, H., III. (2010). *Three big bangs: Matter-energy, life, mind*. Columbia University Press.

Valsiner, J., Marsico, G., Chaudhary, N., Sato, T., & Dazzani, V. (2016). *Psychology as the science of human being. The Yokohama manifesto*. Springer.

Volk, T. (2017). *Quarks to culture: How we came to be*. Columbia University Press.

Wilber, K. (2007). *Integral spirituality: A startling new role for religion in the modern and postmodern world*. Shambhala Publications.

Wilson, E. O. (1999). *Consilience: The unity of knowledge*. Vintage Books.

PART III

A Descriptive Metaphysical System for Modern Empirical Natural Science

CHAPTER 7

A New and Better Map of Big History

Some time ago I was at a gathering where I became engaged in a conversation with a young man who had just come back from his first semester at college. I asked him what he had learned. He had a taken both a neuroscience and philosophy course and replied that through these courses he realized that he was, at bottom, "just a bunch of chemicals." Apparently, the classes had been taught by professors who had advocated for a philosophical position known as *reductive physicalism*. The college student's conclusion exemplifies what Ken Wilber (2007) aptly characterizes as the "materialist flatland" picture of reality. This view can be captured by a quote attributed to the discoverer of the atomic nucleus, Ernst Rutherford, who was reputed to have said that "all science is either physics or stamp collecting." The quip suggests that, ultimately, the only things that are ontologically valid are physical processes. Indeed, a surprising number of philosophers and scientists seem to believe that this reductive physicalist picture is the best, most justifiable, tough-minded picture of what is *really real*.

A good example of this position in psychology is called eliminative materialism, which attempts to eliminate folk psychological conceptions and reduce all psychological and phenomenological concepts to the language of neurobiology. The logic is that there is one ultimate substrate where all cause–effect relationships take place, and we can call this ultimate

G. Henriques, *A New Synthesis for Solving the Problem of Psychology*, Palgrave Studies in the Theory and History of Psychology,
https://doi.org/10.1007/978-3-031-18493-2_7

substance "physical." This worldview is a kind of substance monism, and it can be contrasted most directly with substance dualism, which is the claim that minds (or other entities, such as ideas or souls or God) are made up of fundamentally different kinds of essences than the things that make up the material world. The logic of reductive physicalism can be summarized as follows: *If the world consists of only one substance and that substance is physical, then real scientific causes are physical and mechanistic in nature.*

The philosopher Alex Rosenberg (2011) embraces this view and calls it *scientism*. His work is a good example of what a strong reductive physicalist view of the world looks like. Here is how he describes the process of psychotherapy (pp. 284–285):

> There are also scientifically serious approaches to talk therapy … and they might work. Stranger things have happened. Scientism has no problem with the improbable, so long as it is consistent with physics. However, if talk therapy does work, it will be like this:
>
> 'Your therapist talks to you. The acoustical vibrations from your therapist's mouth to your ear starts a chain of neurons firing in the brain. Together with circuits already set to fire in your brain, the result is some changes somewhere in your head. You may even come to have some new package of beliefs and desires, ones that make you happier and healthier. The only thing scientism insists on is those new beliefs and new desires aren't thoughts about yourself, about your actions, or about anything. The brain can't have thoughts about stuff …
>
> There is no reason in principle why the noises that your therapists makes, or that someone else makes (your mother, for example), shouldn't somehow change those circuits "for the better." Some of those changes may even result in conscious introspective thoughts that seem to be about the benefits of therapy. Of course, science shows it is never that simple. It also shows that when talking cures work, they almost always do so as part of a regime that includes medicine working on the neural circuitry. The meds reach the brain by moving through the digestive system first, without passing through the ears at all.'

From the vantage point of the Unified Theory, the strong version of reductive physicalism is both tragic and absurd. The material flatland view of the world is tragic in the sense that it essentially guarantees a sense of meaninglessness at the Culture-Person plane of existence. The reason has to do with the way human belief-value networks are linked together to form systems of justification. Returning to the young college student, the idea that he was, at bottom, just a bunch of chemicals could easily operate like a kind of Mind[3] virus, infecting the rest of his belief-value propositional nodes and modes of justification. Indeed, the conclusion collapses our loves and dislikes, our hopes and dreams, our desires and feelings into a mechanistic unfolding from which it would be hard to derive any sense of purpose and meaning. That is, if what is really real are chemicals (and presumably atoms and the particles and the quantum fields that make them up), then the life we experience in all its richness essentially becomes a meaningless, epiphenomenal illusion.

If it were just tragic but also true, then we would need to learn how to cope. And yet this statement about coping belies the rather obvious fact that there are ontological realities above the bottom floor of subatomic physics. Indeed, as the physicist Sean Carroll (2017) notes, cats are real entities, but they do not exist at the fundamental level of quantum fields. This brings us to why the materialist flatland view is absurd. To begin with, it carries the performative contradiction that implies that the conclusion about the world being chemicals is itself just a bunch of chemicals. *Embedded in the act of making the argument is the claim that things like concepts, ideas, and explanations are different in important ways from the subatomic floor that constitutes the physical base of our ontic reality.*

To have a knowledge system that is coherent, it must be the case that *both* subatomic particles *and* our ideas of subatomic particles are real in some ways. Thankfully, the basics of this conundrum are framed by the combination of the ToK System and JUST. Ontological claims are justifications, and via its novel depiction of evolutionary emergence, the ToK System shows how justifications and justification systems—even those that are fictional—are ontically real (i.e., they exist on the Culture-Person plane). This analysis aligns well with what the physicist David Deutsch (2011) calls "good explanations," which are the scientifically justified systems of justification that map the ontic reality.

Contra Rosenberg, there is much scientific evidence that the way psychotherapy works is not mediated by chemical mechanisms, but rather is mediated by the information processing and communication networks that operate at the dimensions of Mind and Culture. That is, two people come together, one of whom is suffering, and the other is a healer. Decades of empirical research demonstrate that much of the outcome in psychotherapy is a function of the quality of the relationship and the participatory dance between the therapist and client. Much of the dance happens in the shared, implicit intersubjective relational space that the two develop together, which is framed by the Mind dimension of existence (i.e., the nonverbal communicative and interactive patterns of primates). The shared narratives that frame and constrain the procedures, the narrative sense-making, and the actual words that form the "talking cure" take place at the Culture-Person dimension. The repeated findings in medicine more broadly that "placebos" generate major effects show without a doubt that much change happens as a function of what is believed to be the case rather than the Newtonian impact of some mechanical force.

There is a crucial difference between the claim that, in some sense, everything is energy and matter, which is consonant with UTOK, and the reductive physicalist claim that everything is *just* energy and matter. It is the "just" that is key. To understand why we need to focus on this subtle but crucial distinction and why it is so often overlooked, we need to understand the history of the large-scale systems of justification that have given rise to this modern understanding in the West. Indeed, the fact that we need to place this set of understanding in the West shows that different cultures will come at these fundamental issues of matter and mind via different ways of understanding.

The Origins of Reductive Physicalism and Its Impact on Meaning-Making

Many of the problems that gave rise to the Enlightenment Gap and its inability to generate a coherent map of matter and mind and social and scientific knowledge are found in the tensions between the major worldviews that were operative at the time when modern science became established in the seventeenth and eighteenth centuries. To map this territory of beliefs, one must have a basic understanding of three different worldviews, which are: (1) Cartesian dualism; (2) Christian dualism; and (3) Newtonian matter in motion physicalism. Cartesian dualism stems from **René** Descartes' philosophical reflections that resulted in him positing the existence of two different worlds of substance and causation, the physical and the mental. In the next chapter, we will return to Cartesian dualism and the role it played in the problems the Enlightenment had in developing a coherent philosophical understanding of the relationship between matter and mind.

Our task here is to understand the Newtonian matter in motion view of the universe, the concept of reductive physicalism, and what a modern conception of naturalistic emergent evolution looks like. To frame this history, we need to place the emergence of the Newtonian matter in motion "substance monist" view with the Christian belief in substance dualism and the separate domains of Heaven and Earth. Specifically, we need to be aware of their core justifications, and the tensions and polarities operating between them regarding the relationship between matter and mind and the fundamental nature of the world. We then need to contrast reductive physicalism with emergent naturalism.

To understand the *existential meaning* of reductive physicalism (i.e., the experienced meaning of concluding that one is "just" a bunch of chemicals), we need to place it in the Western sociohistorical context and the large-scale systems of justification that have shaped the broader cultural frames of understanding and meaning-making. To do so, we need to keep in mind that the meanings of arguments and claims are defined in part by what they are contrasted with and defined against. To see this with an everyday example, imagine Julie is a small-time actress who has recently put out a movie. She is happy with her work and hopes it gets noticed. Let us further imagine that she receives the very surprising notification that her work has been nominated for an Oscar. She tells all her friends and family and feels honored.

Now imagine that there had been a strange clerical error, such that her name was confused with the name of the woman who was actually nominated. The error is corrected, and she receives a phone call letting her know of the error. It is likely that the experience will leave Julie with a sense of disappointment and she may end up feeling worse about her acting career than if it had never happened. Why? As demonstrated by many experiments in psychology, we derive meaning from contexts and contrast effects. When Julie believed her work was elevated, she experienced the joy in the unexpected elevation. However, when this was removed, she then had to go through an emotional correction. The point of this example is to drive home that our meaning systems are defined in relationship to other meaning systems involving hopes, expectations, and past experiences. If, prior to receiving the errant message, Julie had been told she had not been nominated, she would not have been surprised, and it would not have carried much meaning. This contrast effect is important to keep in mind when we consider what a substance monist, Newtonian matter in motion worldview would have meant to a culture that was almost entirely Christian.

The Enlightenment thinkers valued the power of reason, and leading intellectuals argued that the natural world could be understood using logic, mathematics, and the empirical method. Especially relevant to the current analysis was the publication of Isaac Newton's *Principia* (*Mathematical Principles of Natural Philosophy*) in 1687. This is arguably the single most important scientific publication in history. It is one of the central works that resulted in a transition from pre-modern thinking into a modernist sensibility. What did Newton do in *Principia*? He developed a mathematical framework that described matter in motion (sometimes called "classical mechanics") that connected processes on Earth with those in the heavens under one set of laws. He did this so well and so completely that the Newtonian paradigm was the foundation of physics and astronomy (and thus all of science) for almost 225 years, up until the development of modern physics in the beginning of the twentieth century.

Although Newton himself was deeply Christian and held a dualistic worldview, by the time the Enlightenment was in full swing, several increasingly skeptical and influential scholars had emerged who were using the new science to challenge the Christian worldview. Most were deists like Voltaire (1694–1778), but some were full-fledged atheists like David Hume (1711–1776). The Enlightenment intellectual Pierre-Simon Laplace (1749–1827) is an exemplar of someone who advanced the physicalist worldview. He believed everything was completely determined by the laws of matter in motion. There is a famous (but likely apocryphal) incident in which Napoleon, upon hearing of Laplace's strict determinism about how the world works, asked about the place and power of the Creator in determining events. Laplace boldly answered that he "had no need for that hypothesis."

In contrast to the Christian worldview, the Newtonian physicalist posits that the world only consists of matter in motion, that matter obeys strict laws, and everything is determined by these laws. It also posits that matter has always existed and can never be created or destroyed; only its form can change. Finally, there is no higher reason for the way matter is the way it is. Human beings are just complex arrangements of matter, and they exist because they just happen to be how matter is organized right now. Of course, this worldview is in deep conflict with the fundamental message of Christianity.

To this day, the competition between the Christian theistic and impersonal Newtonian worldviews reverberates throughout the West. In a paper titled "Measuring the Horizon: Objectivity, Subjectivity, and the Dignity of Human Personal Identity," Frank Ambrosio and Elisabetta Lanzilao (2013) argue that the "impersonalist" vision of modern empirical natural science and its matter in motion frame of reference has never been effectively squared with the theistic and "personalist" views of the pre-modern period. They argue that this conflict has generated an ongoing state of cultural warfare and lay out how a secular spiritual sensibility might engender a productive way to resolve the tensions between the theistic and impersonal matter in motion worldviews. The UTOK metapsychology very much aligns with their call and envisions a new bridge between science and spirituality in the twenty-first century.

One key element of this bridge pertains to a necessary shift away from the matter in motion worldview. Specifically, a shift needs to be made toward an emergent naturalism that affords a clear, accessible stratified ontology at the appropriate levels and dimensions of analysis and needs to take root and grow like a tree in the minds of the populace writ large. Thankfully, this shift toward a more stratified ontology has been under way for more than a century.

From a Physical Reductionism to an Emergent Naturalistic View of Scientific Knowledge

Over the past several decades there have been an increasing number of calls for a consilient naturalistic worldview grounded in emergence. The zeitgeist is now such that most scholars, philosophers, and scientists would concur that a mechanistic flatland view of the world is inadequate. Indeed, the current worldview adopted by most scientists is probably best characterized in terms of a nonreductive physicalism or emergent naturalism. There have been several developments in both science and the philosophy of science over the past century that have resulted in this shift.

One major shift in the twentieth century was a change from thinking about the universe in terms of matter to thinking about it in terms of energy. That is, energy now shares with matter what we might call foundational ontological status, that is, both energy and matter are fundamental concepts in physics. Some of this change began with the physicist James Clerk Maxwell, and his remarkable work in unifying the electrical and magnetic forces into the modern understanding of the electromagnetic force. An even more foundational development in our understanding of the relationship between energy and matter was made by Albert Einstein.

His famous $E = mc^2$ equation depicts how energy (E) is equivalent to mass (m) multiplied by the speed of light (c) squared. Many physicists now consider energy as the more fundamental concept and conceive of the universe beginning in a state of pure energy.

There are many implications that follow from this shift. One is that the focus moves from primarily an "object" view to a "process" view. This process view of reality can be seen in the philosophy offered by Alfred North Whitehead (1929). It shifts one's conception from thinking of the universe as a set of mechanical things bumping into each other like billiard balls into seeing the universe more as a series of unfolding and emergent events mediated by fields and flow. A way to appreciate the difference is to go from thinking of the universe in terms of nouns into thinking of it in terms of verbs. In addition, the energy view has different implications for the ground of being. Consider, for example, how the conclusion "I am a bunch of chemicals" has a different feel to it than "I am an unfolding wave of energy."

A second major development was that the realization that the universe evolves, and that complexity in the cosmos has increased over time via natural processes. Of course, the most significant contribution to this vision was made by Charles Darwin. His theory of natural selection allowed scientists to understand the basic ingredients regarding how life could evolve over time and be shaped by its relationship to the environment across the generations. The evolutionary view was extended into matter with the emergence of the Big Bang, by which the universe has a specifiable beginning and thus a documentable history. This resulted in a view of cosmic evolution (see, e.g., Chaisson, 2001).

There have also been important changes in our understanding of the foundations of physics. In the first few decades of the twentieth century, developments in quantum mechanics blew up the strictly deterministic picture that people like Laplace had of how matter behaves. It is now largely understood that the most basic elements of the universe (i.e., particles behaving at the smallest units of scale) have a fundamentally indeterministic and objectively random character. *That is, there are theoretically unknowable variations that make the future unpredictable, even in theory.* Most physicists believe that this indeterminacy is such that, even if the universe began with the exact same initial conditions, it would unfold in radically different ways. This, in turn, means that the Laplacian idea of it unfolding as a pre-determined complicated clock-like set of mechanisms is wrong.

Another major change in science and philosophy involved the emergence concept of information. There were at least two big developments here, one in information theory and the other in information science. Working in Bell labs in the 1920s and 1930s, Claude Shannon developed information theory, which would revolutionize our understanding of how information can be transmitted, and how messages can be stored and encoded. In parallel, the science of information processing grew into a full-fledged field that would give rise to massive developments in cybernetics, systems science, artificial intelligence and computer science, the cognitive revolution in psychology, and in the last 30 years the emergence of complexity science, with its analysis of complex adaptive systems, attractor states, and chaos theory. These developments provided scientists with a new perspective on causation. Rather than causation being purely mechanistic in terms of exchange of forces, there are many systems whose causal properties are described in informational terms of inputs, computational processes, and outputs.

A sixth big change was in the philosophy of language and the sociology of knowledge. The twentieth century saw the rise of the view that human knowledge was always contextualized in sociohistorical movements and framed by specific language games that could never be justified as revealing the transcendent truth. These developments were so significant that there was a fundamental shift in epistemology, which is represented by the move from the modernist conception of science and rationality leading to ever more progressive and accurate depictions of reality into postmodernist sensibilities that emphasize context and immanence. Exemplified by scholars like Foucault and Derrida, central to the development of a postmodern sensibility is the idea that human language and knowledge were contextually embedded in cultural and power relations such that no grand meta-narrative for the universe and our place in it was possible.

The combined point to be made here is that any worldview that reduces our conception of knowledge and reality to a reductive physicalist, mechanical, "just chemicals" material flatland is nonsensical. And, consequently, a broad evolutionary emergentist view has gained increasing traction in our understanding of the material-into-living-into-mental-into-cultural world. The difficulty has been in generating a picture that is simultaneously coherent, incorporates theories and findings from modern science, and effectively manages the enormous diversity of opinion and empirical findings about how to make sense of the picture-as-a-whole. In short, up until this point there has been no big picture scientific worldview that is consilient at both the metaphysical and metatheoretical level of analysis. However, despite the difficulties, there is a general, shared big picture view that is coming into focus, and the evidence for this is found in the popular Big History movement.

Big History: A Scientific Worldview Mapped on the Dimensions of Time and Complexity

Big History is an integrative, big picture view that has attracted significant attention across the globe. Launched by the world historian David Christian (2018), Big History (BH) is a broad, interdisciplinary movement that attempts to generate a map of reality that weaves everything we know—from the creation of the universe to the advent of agriculture and the modern technological revolution—into a coherent tapestry of understanding. It is an interdisciplinary view that maps cosmic evolution on the dimensions of time and complexity. It has a journal and a society and there is now a series of books written explicitly from a BH vantage point that have been endorsed and supported by Bill Gates.

As a formal academic discipline, history refers to the systematic study of the past, as recorded by humans in written documents. Thus, professional historians typically study human events that stretch back approximately 5000 years, corresponding to the earliest written records. Although historians start with human written records, it is of course the case that the form and function of humans 5000 years ago was the result of a long evolutionary process. When the longer arc of evolution is attended to, the formal edges of the discipline begin to fade away. The somewhat oxymoronic "prehistoric" period now becomes the context from which ancient civilizations and formal human history emerged. Congruent with UTOK's metapsychology, Big Historians argue that this context is crucial for understanding how the last 5000 years have unfolded.

Once one steps into prehistory, a long line of continuous succession appears. Following the trail back 100,000 years ago, one encounters our ancient paleolithic ancestors who spoke and lived in hunter-gatherer tribes. If we increase that period by a factor of ten, one million years ago we see our pre-human hominid ancestors who were making stone hand axes and likely did not have full linguistic capacities. Moving our gaze back by another factor of ten, ten million years ago we see a world populated by our great ape ancestors. The trail then winds into mammals that evolved on the scene about 100 million years ago, and then into "simpler" animals 500 million years ago. A billion years ago and we are at the earliest plants and other multi-celled creatures. At four billion years, we find the beginning of life and then the history of the solar system. At approximately 14 billion years ago, we find the beginning of the universe itself. The point here is that one can follow the trail of time all the way back and start from the beginning. This continuity brings us to one of the central insights of the BH tradition, which is the recognition that we can scientifically situate human history in the larger cosmic evolutionary narrative.

The humanistic impetus for embracing the larger cosmic picture afforded by Big History can be justified in part via the broad emergent naturalistic view it affords humanity. The argument is that the picture of emergent naturalism is much more conducive to meaning-making than the reductive materialistic flatland vision. Big Historians argue that, when modern reductive visions are combined with globalization and the post-modern state of fragmented knowledge, the consequence is high levels of narcissism and nihilism. In *Origin Story*, Christian (2018) characterizes the situation and justifies the need for a BH as follows:

> Globalization and the spread of new ideas [have] corroded faith in traditional knowledge ... Some people responded with aggressive or even violent, defenses of their own religions, tribal or national traditions. But many simply lost faith and conviction, and along with them, they lost their bearings, these sense of place in the universe. That loss of faith helps explain the pervasive anomie, the feeling of aimlessness, meaninglessness, and sometimes even despair that shaped so much literature, art, philosophy, and scholarship in the 20th Century. (p. ix)

Christian does not think that if we adopt a scientific worldview we are inevitably "doomed to a chronic state of fragmentation and meaninglessness." He believes that modern scholarship needs to be woven together into a new "global origin story" that gives rise to a collective project regarding how we can live sustainably and do so with both scientific accuracy and humanistic values.

Christian (2017) further spelled out his vision of the BH movement in the inaugural issue of the *Journal of Big History*. He argued that BH aspires to provide a "universal understanding of history" (p. 12) that works "to link the findings of specialist scholarship into a larger unifying vision" (p. 13). The first explicit goal is that BH would empower humanity with a greater understanding of who we are and our place in the cosmos. The second goal is to offer a vision of existence that is both universal and consilient. Third, BH was structured to advance interdisciplinary collaboration via an integrative reference point that is not just located on "the individual islands and continents of modern scholarship" (p. 14) but includes a big picture frame of reference that supports the many links between them. Finally, with an eye toward the dialectic between the sciences and humanities, Christian emphasized the point that BH can provide a meaningful narrative for the modern age, one that is grounded in science and can help humanity understand our place in the cosmos and how we might use that to chart a course toward a wiser future.

Although the ToK System and BH were developed independently, there is much overlap in their vision, both scientifically and in relation to human values. For example, in direct alignment with Christian's (2018) articulation of a need for an origin story, I characterized the ToK System as "a picture of the universe story, as presently mapped out by scientific inquiry that potentially provides us with a shared origin myth" (Henriques, 2011, p. 259). It is named in explicit reference to the Tree of Knowledge of Good and Evil in the Bible, to forge a linkage between our modern scientific "logos" and the pre-modern foundational "mythos" that has provided archetypal frames for understanding the human condition. The metaphor further embraces the idea that we need both scientific sensibility and moral clarity structured in a way that allows us to eat heartily from the Tree of Knowledge and flourish in doing so. Concordant with BH, the overall UTOK is positioned in a manner that proclaims that "scientific knowledge does have a story to tell about humanity, and it is crucial that we convey such knowledge in the context of a meaningful narrative that explicitly emphasizes a moral component" (Henriques, 2011, p. 259).

The Thresholds of Complexity in Big History

BH is an emergent naturalist vision that tracks cosmic evolution on the dimensions of time and complexity. The field identifies eight "thresholds" that represent the major shifts in emergence that led to modern human life: (1) the Big Bang and Origins of Matter; (2) Stars and Galaxies; (3) Chemistry; (4) Earth and Solar System; (5) Life; (6) Homo Sapiens; (7) Agriculture and the Rise of Human Civilizations; and (8) Modern Technological Revolution. A ninth threshold has been posited to be emerging in conjunction with the digital revolution. A summary of each threshold will help clarify the time by complexity picture of energy, matter, and information flow provided by the Big History view of the universe and our place in it.

Threshold 1: The Big Bang and Origins of Matter

BH starts at the beginning of our universe approximately 13.8 billion years ago. The BH scholar Barry Wood offers a useful analysis of this threshold. First, he argued that we should think of the Big Bang as the "Big Beginning," in that it is the point from which the universe as a whole—which includes all the matter, space, and time that exists—sprang forth and has grown in complexity since. As such, we can consider the Big Beginning the point at which space and time are set at as close as possible to zero. Then, there was a massive inflationary big bang process that resulted in a phase shift from a state of pure energy-information into matter particles. As the universe expanded and cooled, particles organized into atoms like hydrogen and helium, and the more familiar Matter plane of existence had fully come into being.

This transformation into Matter allows us to be on more familiar footing. At the atomic level, we have a basic understanding of entities (i.e., particles and waves), fields, and the force interactions that take place between them, as well as change processes that occur in space and over time. Beneath the atomic level, the Standard Model of Elementary Particle Physics provides the foundational base of the material dimension of existence. From this, we can trace the emergence of increasingly complex material objects into atoms and molecules. In *The Origin Story*, Christian (2018) recounts the process by which the atoms in the early universe began to arrange themselves into large gas clouds. These clouds then began to collapse in on themselves as a function of gravity, which in turn set the stage for the next threshold of complexity in the BH narrative.

Threshold 2: Stars and Galaxies

Gravity pulled the gaseous clouds together in denser and denser arrangements. However, as more and more material entities become concentrated in a region, more and more collisions happen, and things heat up. That creates an internal pressure, as the heat causes things to expand. However, if there is enough mass, the pull is greater than the expansion, and things will get denser and denser, which also means things get hotter and hotter. At approximately 10,000,000 degrees the heat is so extreme that violent collisions between protons can cause them to overcome their repulsive tendencies (because they are positively charged, they tend to repel each other) and fuse together because of the strong nuclear force. As proton particles become fused at high temperatures, a small portion of their mass is converted into radiation. As Einstein's famous equation shows, even a small amount of mass can produce a huge amount of energy. These dynamic forces work together to give us the basic elements for the formation of a star.

In some ways that parallel how atoms coalesced together to form stars, the gravitational pull of stars coalesced into the large-scale patterns and clusters we call galaxies. We can think of the universe being a bit like a loaf of raisin bread, such that each raisin represents a galaxy of stars. Galaxies vary in size. Some have "only" a few hundred million stars, whereas others are massive and include as many as 100 trillion stars. The number gets even more fantastical when we consider the fact that the latest estimates suggest that there are hundreds of billions and maybe even trillions of galaxies. Although the universe is perhaps not infinite in the absolute sense, to our human minds, it might as well be.

Threshold 3: Chemistry

Most of the atoms in the universe—about 75%—take the form of hydrogen, which normally consists of one proton and one electron. Helium, which typically consists of two electrons orbiting a nucleus with two protons and two neutrons, makes up just under 25% of the rest of the atoms. Given our anthropomorphic position in the world, this might seem odd. After all, the human body is made up of mostly oxygen, nitrogen, carbon, calcium, and phosphorous, along with hydrogen. If 99.99% of all the atomic matter in the universe is in the form of hydrogen and helium, where did these other chemical arrangements come from? The answer is that, except for lithium, which appeared at the Big Bang, all the naturally occurring atoms mapped by the Periodic Table of the Elements were forged in the bellies of stars. This is a good reminder of how odd Earth is relative to the rest of the universe and how odd life is in terms of its complex chemical makeup.

The basic process that gives rise to the heavier elements is that, as stars both age and grow, they eventually run out of free protons and start fusing helium nuclei together. However, the fusion of helium is not indefinitely sustainable. There is a frenetic drama that takes place between the collapsing power of gravity and the expansion of heat and the formation of heavier nuclei, such as carbon and oxygen. Eventually, the star may explode into what is called a "supernova," which is a gargantuan release of energy that might, for a second, equal the radiation energy of a whole galaxy. It is during these moments that the heavy elements that make up the naturally occurring atoms in the Periodic Table are dispersed into the universe. This is why Carl Sagan said we are made of "star dust."

Chemistry is the science of these atomic elements, and how they form into molecules and the behaviors they exhibit and the changes they undergo during a reaction with other substances. Chemistry is a central player, for it is the forces of chemistry, the forming and breaking of chemical bonds, that rest at the heart of all complex matter. Thus, whereas physics deals with the foundations of the material dimension, chemistry is where all the complicated transactions occur between atoms and molecules. It is from the bed of these complex energy exchanges that Life eventually emerges.

Threshold 4: Earth and Solar System

The next threshold in the Big History scheme is planet Earth, which is, of course, part of the solar system. The study of our sun is called "helioseismology." Geologists and other Earth scientists study the behavior of things like oceans, mountains, and other rock formations. They divide the history of the Earth into different time scales, the largest of which is the eon. The Earth was created about 4.5 billion years ago, and the Hadean Eon is the period of its first half-billion years and stretches from the beginnings of the Earth's formation to the next era, the Archean Eon. The Earth looked very different in these eons, as it had a hot molten surface early on, and then was regularly pummeled by asteroids. The current Earth-Moon arrangement is thought to have been the consequence of a spectacular collision of Earth with a hypothetical Mars-sized body called Theia over four billion years ago, prior to the earliest forms of single-celled life.

One of the most important ideas in Earth science is that of plate tectonics. This notion was first proposed by the meteorologist Alfred Wegener in 1912 under the name continental drift (Spaulding & Namowitz, 2005). Like many before him, Wegener noticed the remarkable way that various continents seemed to be shaped like pieces of a puzzle that had come apart. However, Wegener marshalled a significant amount of physical evidence that extended beyond the shapes of the shorelines and included both fossil evidence and geological evidence that supported the idea that the now very distant landmasses were at one point conjoined. The theory did not receive large-scale acceptance in Wegener's day in large part because there was no good evidence that the continents could move in the way the continental drift proposed.

Over the next 50 years, however, a wide variety of evidence, from clarifying the basic structure of the Earth's outer layers to discovery of mid-oceanic rifts to patterns of magnetic striping to careful measurements of distances and movements over time, coalesced to result in the theory, now called plate tectonics, being accepted as scientific consensus by the 1970s. Along with quantum mechanics, general relativity, and the discovery of the structure of DNA, the theory of plate tectonics represents one of the truly great scientific achievements of the twentieth century.

Threshold 5: Life

Although it is certainly possible—even probable—that life exists elsewhere in the universe, science maps our current knowledge of the empirically documented universe. The best scientific evidence suggests that living entities, in the form of simple single-celled organisms, were present on Earth by 3.7 billion years ago and may have started as early as four billion years ago. Exactly how life originated remains a bit of a mystery, although there are many clues and several plausible models. Biologists have a basic framework for understanding how life evolved after it started. As noted in the previous chapter and the discussion of theoretical joint points, biology is organized via cell theory and the modern evolutionary synthesis. It provides the overarching meta-paradigm that biologists use to frame the emergence of living entities. Although it is not the final word, it has provided biologists with a unified conceptual system from which to operate to understand organisms and living processes.

In the language game of the ToK System, the cell is to Life what the atom is to Matter in that it is the fundamental unit of organization that operates at this dimension of behavioral complexity. Cells provide the basic organizational structure that differentiates the organism from the outside world and allows the component parts to engage in information processing, along with energy intake, metabolism, growth, and reproduction. Both the ToK System and BH emphasize the important role that information processing and communication play in understanding the key features that define Life. In contrast to the inanimate world, life exists as a collection of information processing systems that have stored information across the generations and are shifting in response to ongoing experiences. In the words of Dave Christian (2018, p. 79), living organisms are "informavores … they all consume information, the mechanisms they use for reading and responding" to their environments. Whereas cells are the fundamental units of self-organization and information processing, genes are the fundamental units of information storage on the Life dimension.

Just as atoms are more complex than particles and molecules are more complex than atoms, living entities became more complex when single cells merged to become multi-celled creatures. For over a billion years, cellular life maintained a relatively basic structure (Lane, 2015). Then, at about 2.7 billion years ago, a massively important structural change happened when there was a remarkable jump in cellular complexity. That jump was the emergence of eukaryotic cells, meaning cells that had a nucleus contained in a membrane. Eukaryotic cells were a quantitative game changer in terms of behavioral complexity at the biological dimension. Such cells are much larger and far more structurally complex than simple cells like bacteria. Even more important, they set the stage for the emergence of multi-celled creatures. This is what Christian calls "Big Life," which is a "sub-threshold" in the BH system. Multi-celled creatures like plants exhibit many emergent properties that are not present in single-celled creatures. And they set the stage for the emergence of the biosphere over the next several billion years, including the animal kingdom.

Threshold 6: Homo Sapiens

Approximately 250,000 years ago, a different kind of animals emerged on the scene. According to BH, the appearance of Homo sapiens represents a new threshold separate from Big Life because not only do they have capacities to adapt to the environment, they also have remarkable capacities for language and technology that allow them to adapt the environment to suit their needs in an unprecedented way. This capacity to shape and control the environment stems from the human capacities for cooperation, coordination, and invention. It means that humans are not simply at the mercy of their ecology but can migrate into virtually every corner of the globe. It has also meant that, when they do migrate into new territories, they usually end up dominating the landscape, killing off other big game and domesticating the land around them.

Consistent with the ontology of the ToK System, Christian identifies the key difference in our species as being found in the way we process information and communicate. He writes (p. 171), "We don't just gather information, like other species. We seem to *cultivate* and *domesticate* it, as farmers cultivate crops. We generate and share more and more information and use it to tap larger and larger flows of energy and resources." The consequence of this information processing and communication is what Michael Tomasello (2016) calls "cumulative cultural evolution." This is a process unique to humans, and it is what warrants our species being denoted a novel threshold in the BH formulation.

Threshold 7: Agriculture and the Rise of Human Civilizations

Human societies consist not just of cultural systems of justification that make up the Culture-Person dimension of existence, but also the practices and skills, the accumulated technologies and industrial infrastructure, and the bio-ecological context in which the group lives. The early technological innovations that contributed to this transformation can be seen in tools like hand axes, spears, and, perhaps especially, fire. Fire changed the way we communicated, ate, and worked together and it showed how we could tame the forces of nature. Despite the important shifts that came with fire, an even greater suite of technological and ecological changes emerges with agriculture. Over 15,000 years, agriculture laid the groundwork for dramatic changes in human behavioral practices and energy transformation, such that it set the stage for the emergence of human civilizations.

The transition from nomadic foraging to horticulture to agriculture was not smooth, uniform, or universal (Graeber & Wengrow, 2021), and it is unclear exactly why agricultural practices began to emerge between 12 and 15,000 years ago, although the shifting climate and a receding Ice Age almost certainly played a role. Nevertheless, whatever the ultimate cause or causes of agriculture, it marked a major change in human history and is associated with an accelerated pace of change. In particular, it allowed population growth that resulted in the emergence of larger, denser, and eventually more complex societies and the birth of the first human civilizations about 5000 years ago.

Threshold 8: The Modern Technological Revolution

The modern era launched in the seventeenth and eighteenth centuries, with the insights from the Enlightenment and new modes of production taking hold. This period now has an official geological name, called the Anthropocene (Lewis & Maslin, 2015). The reason is that the global world system that humans have generated is changing almost everything, including how we live, communicate, and relate to and rely on the planet. These dramatic changes are echoing through the biosphere, which is why it has been labeled a new geological age. In the mid-nineteenth century, with the Industrial Revolution and the power of the steam engine to change fossil fuels into work effort in full swing, humans generated another massive transformation in energy flow. This power, coupled as it was to modern science and ideals about liberal democracy, resulted in a transformation in the world. European powers began a process of colonialization that would impact every corner of the globe.

Threshold 9: The Singularity, the Fifth Joint Point, and the Future of Humanity

Many argue that the rate of change is accelerating, and that we may be fast approaching yet another transition that warrants designation as a new BH threshold taking place in the first half of the twenty-first century. This threshold is often characterized in terms of a social or technological "singularity." Although the concept of a singularity means different things to different theorists, it is commonly interpreted as a phase shift that denotes either a transformation in artificial intelligence or a social–technological interface that would transform virtually all aspects of human behavior. Offering an in-depth review of several similar models of the accelerating evolution of complexity, Korotayev (2018) concluded that there are data "to indicate the existence of sufficiently rigorous global macroevolutionary regularities (describing the evolution of complexity on our planet for a few billion years), which can be surprisingly accurately described by extremely simple mathematical functions." In other words, there are ways to characterize the evolution of complexity to suggest an emerging threshold that would be crossed in the next decade. That threshold is one way to conceptualize the singularity.

The ToK System offers one way to frame the concept of a singularity appearing in the relatively near future. We can start by noting that the ToK includes the insight regarding the accelerating pace of complexification and change. What it further suggests is the possibility of the emergence of a new dimension of existence, perhaps occurring in the relatively near future in a way that is consistent with the timeframe specified by Korotayev (2018). The ToK System posits that novel dimensions of existence occur when new information processing or semiotic systems emerge and then become networked together and then regulated via a centralized control system. Based on this logic, we can ask: Are we seeing the emergence of novel information processing systems and have they become networked together in a centralized way? There are some obvious technologies that present themselves in a way that results in us answering in the affirmative. Computers, the internet, and the interface between human and artificial intelligence systems are suggestive of an answer that might be "yes."

Comparing and Contrasting Big History and the Tree of Knowledge System

As suggested by this final point, there is much consistency between the BH view of the cosmos and the ToK System. Both are emergent naturalistic evolutionary visions of the universe set on the dimensions of time and complexity. Both strongly reject reductive physicalism and share the perspective that the academy needs macro-level frames to effectively organize the disparate fields of study and situate disciplinary findings in a larger picture of understanding. Both agree that such frames are necessary for good science and for developing a shared understanding that grounds humanistic values and a vision for the future.

At the same time, there are important differences between the two approaches (see Henriques et al., 2019). Although it attempts to achieve a holistic vision, BH remains anchored to a standard modernist empirical epistemology, which means that it is vulnerable to being limited by the Enlightenment Gap. Indeed, from the vantage point of the Unified Theory, BH currently lacks the necessary descriptive metaphysics for a truly coherent naturalistic ontology. This becomes especially evident when we consider the problem of psychology and the ontology of the mental. As we saw with the thresholds, BH completely overlooks the Life to Mind joint point. We can contrast this with the ToK System. Because the ToK System organizes the evolution of complexification via the planes of existence and it resolves the problems of both the mental relative to the physical and scientific relative to social knowledge, we can assert that the ToK assimilates and integrates the BH vision and upgrades it with the appropriate descriptive metaphysical conception. In short, the ToK System gives a new and better map of Big History that gives rise to a clear naturalistic ontology that solves the problem of psychology and fills in the Enlightenment Gap.

The Eight Thresholds Versus the Four Dimensions of Behavioral Complexification

The most obvious difference between the frameworks is the contrast between BH's eight thresholds of complexity and the four separable dimensions of behavioral complexification depicted by the ToK System. BH essentially characterizes the evolution of complexity on a single continuum, one that goes from particles to atoms to molecules to organisms to societies. Each step is characterized as a threshold, but there are not different kinds of thresholds. The ToK System argues for a different vision, which becomes apparent when we place the BH thresholds in relation to the dimensions of behavioral complexity on the ToK (Fig. 7.1).

As reviewed, BH has a single category for jumps in emergence (i.e., thresholds). In contrast, the ToK System differentiates levels of emergence within a dimension (particles to atoms) and dimensions of behavioral complexification that appear with new complex adaptive systems that

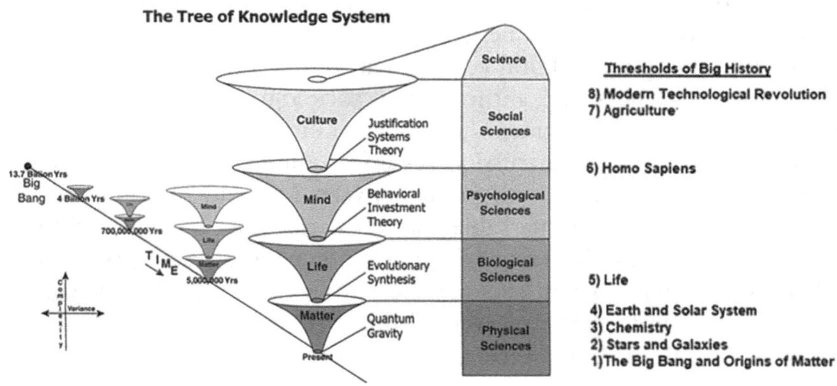

Fig. 7.1 Comparing Big History's thresholds with the Tree of Knowledge System

generate new planes of existence (e.g., the Matter to Life jump). This point highlights a key difference between the systems, one that should not be underestimated. BH has a general picture of emergence that tracks the appearance of new entities and their properties and aligns them across time and complexity and corresponds them to the various domains of science. The ToK System differentiates standard forms of emergence that occur within a dimension from emergence events that result in fundamentally new planes of existence. The ToK highlights that new information processing and communication mediums are key to the generation of these new complex planes of adaptation. Genetic/epigenetic information processing and cell–cell communication give rise to Life, neurocognitive information processing and animal communication give rise to Mind, and symbolic language and justification processes between people give rise to Culture. These kinds of emergence are different.

The Problem of Psychology and the Descriptive Metaphysics of Mind Versus Matter

The crucial difference between the thresholds of emergence and the dimensions and levels frame of the ToK System brings us to the second big difference. BH offers no identifiable place for the domain of "Mind" and its corresponding domain of science. That is, from a UTOK vantage point, both the subject matter and institution of psychology are essentially absent from the BH formulation. The thresholds in BH jump from biology—which corresponds to the emergence of life approximately 3.8 billion years ago at threshold 5—to the emergence of early modern humans studied by paleoanthropologists 250,000 years ago at threshold 6. At this point, the human social sciences (i.e., anthropology, sociology, political science, and history) become the prominent disciplines of inquiry.

As was noted in the discussion regarding the original ToK diagram, there should be a major joint point between Life and Mind. Christian (2018) does note the remarkable transformation that happened during the Cambrian Explosion approximately 550 million years ago. This is when animals with brains and complex bodies emerged and began to dominate the landscape. However, all this is placed under the BH "sub-threshold" of Big Life and receives scant attention in the formulation. In contrast, explicitly drawing a clear dividing line between the minded behaviors of animals and living behaviors of other organisms is a key feature of the ToK System. It unequivocally highlights that the complex adaptive plane of animal behavior mediated by brains and complex active bodies represents a qualitative shift in patterns of behavior in nature. Moreover, the core logic for differentiating the complex adaptive plane of Mind from Life is the same as the logic differentiating organisms from inanimate objects. The latter half of this book details the argument. The summary statement is that as we move from invertebrates like worms and butterflies into vertebrates like fish and reptiles and finally into mammals and primates, we can readily trace Mind, or mental evolution.

From the vantage point of the ToK and its central concern with the problem of psychology, the absence of mental evolution in the BH formulation is diagnostic of a more general confusion regarding how to think about mind in relationship to matter from a scientific vantage point. It shows how starting with the language game of physics and biology results in a crucial blind spot in how to think about mental behavior, subjective conscious experiences felt by animals, and self-conscious reflective capacities seen in human persons. Indeed, many, if not most, of the big picture views provided by physicists and biologists miss this dimension. Maynard Smith and Sasthmary's (1995) well-known model of major transitions in evolution also completely overlooks the emergence of the nervous system and animal behavior as a major transition. The physicist Eric Chaisson (2007) maps the epochs of evolution; however, he also misses the Life-to-Mind jump.

Ultimately, comparing the ToK System with BH yields two key take-home points. First, there is growing momentum toward big picture systems among many natural scientists and humanists. Virtually all these frameworks map the universe on the two axes of time and complexity. BH is an exemplar of this mode of thinking, and it offers a powerful interdisciplinary way to frame our scientific knowledge. It should be considered the standard reference point for big picture views, and it demonstrates their utility and capacity to assimilate and integrate many lines of thought. The ToK is highly consistent with the basic conceptual structure provided by BH, and thus there is much potential for productive synergy.

Despite this positive alignment, there are also important differences between BH and the naturalistic scientific ontology afforded by the UTOK. When we compare BH to the ToK System, we see that it completely misses Mind and thus affords no way to frame the various domains of mental processes and their relationship with the more general concept of behavior. When this fact is combined with the reality of the Enlightenment Gap and the problem of psychology, it becomes obvious that BH cannot achieve a fully consilient naturalistic scientific worldview. In contrast, with its joint points and novel conception of the dimensions of complex adaptive landscapes as qualitatively different planes of existence, the ToK System represents a new map of BH that allows us to advance past the Enlightenment Gap and toward a coherent origin story and scientific ontology for the twenty-first century.

Classifying Four Scientific Worldviews on a "Materialist" Continuum

The contrast between reductive physicalism, Big History and related naturalistic approaches like E. O Wilson's (1999) consilience, and the ToK System affords us the opportunity to reflect on the concept of scientific worldviews. Indeed, there are many debates about whether science results in a worldview or not. Many have argued that scientific materialism has been adopted without proper justification. One such critique of scientific materialism is found in the Galileo Commission Report. Entitled "Beyond a Materialist Worldview: Towards an Expanded Science," it opens with a quotation from the philosopher Mary Midgley as follows:

> This whole reductive programme—this mindless materialism, this belief in something called 'matter' as the answer to all questions—is not really science at all. It is, and always has been, just an image, a myth, a vision, an enormous act of faith. As Karl Popper said, it is 'promissory materialism', an offer of future explanations based on boundless confidence in physical methods of enquiry. It is a quite general belief in 'matter', which is conceived in a new way as able to answer all possible questions. And that belief has flowed much more from the past glories of science than from any suitability for the job in hand. In reality, not all questions are physical questions or can be usefully fitted to physical answers.

The report, authored by Dr Harald Walach (2019) on behalf of the Scientific and Medical Network, claimed that the dominant paradigm in science was a "reductive physicalist ontology." The report focused on consciousness and posited that the mainstream scientific view was that consciousness "is nothing but a consequence of complex arrangement of matter, or an emergent phenomenon of brain activity." This critique raises central questions about the nature of scientific worldviews.

It is useful to begin by noting that science *per se* does not necessarily commit one to a worldview. The most generalizable aspect of modern science is that it functions as an epistemological frame and method for obtaining knowledge. The methods of science can be applied to a specific area of inquiry or many different areas, all without rendering a worldview or committing someone to a worldview. Consistent with this analysis, the Galileo Report divides science into practices and methods and contrasts these with what they see as a dominant scientific worldview. The report frames the practices and methods of scientific inquiry as "Science 1.0." In contrast, "Science 2.0" refers to a "materialist world view which comes with a narrow methodology and an overreliance on a certain type of rationality." The report advocates for a shift in "Science 3.0," which they framed as being a "post-materialist" or "spiritually informed" science.

Given that UTOK embraces a naturalistic scientific worldview, but also critiques physical reductionism, questions emerge regarding the relationship between the Galileo Report's characterization of scientific worldviews and the one being advanced here. A close read of the Galileo Report reveals a lack of clarity regarding the kind of reductionism or materialism that it proclaims is adopted by the mainstream scientific worldview. Consider, for example, the claim that consciousness is "nothing but a consequence of a complex arrangement of matter" or an "emergent brain activity." These are substantially different perspectives that must be disentangled if we are to achieve clarity about what is being proposed. In this section, I delineate a continuum of four different scientific worldviews that offer different perspectives on materialism and emergence and help frame the kind of scientific worldview being offered by the ToK System that grounds UTOK.

We can begin our discussion with the view that the universe consists of "nothing but" the fundamental base levels of physical reality. This is the reductive physicalist view that we described at the beginning of this chapter, which is labeled here as Scientific Worldview A. This worldview characterizes consciousness in the first sense as "nothing but a complex arrangement of matter." Although, as we saw at the beginning of the chapter, this has some modern defenders, I do not see it as being the dominant scientific worldview.

In contrast to the strong reductive program, most present-day scientists embrace some version of emergence. That is, they acknowledge that there are emergent properties, and it is necessary to develop vocabularies and specific methods of inquiry to describe them. Although there are many different angles on what emergence might refer to, there are two weak versions that are adopted by most scientists and philosophers. One is property emergence, which refers to the fact that aggregate groups have properties that do not appear in the individual units (e.g., fluidity emerges with lots of water molecules, but does not exist at the individual molecule level). The second pertains to human knowledge, pragmatics, and epistemology, such that our approaches to understanding require us to talk about "higher-level" phenomena. We can consider the scientific worldview that embraces general forms of emergence to be Scientific Worldview B. It is a nonreductive physicalism.

Sean Carroll's (2017) *The Big Picture: On the Origins of Life, Meaning, and the Universe Itself* is a good example of Scientific Worldview B. His "poetic naturalism" posits that the ultimate ontology is found at the quantum or foundational level of physics. However, he also argues that emergent phenomena are both real and important. For example, he notes that cats do not appear in the Standard Model, but they are real, and we need vocabularies to describe how they behave. E. O. Wilson's vision in consilience is also a good example of Worldview B. It seems that most who subscribe to Big History would fall into this camp. However, many in BH might also be more aligned with the next two worldviews, as the ontology of BH is not fully specified.

Although Scientific Worldview B is much more sophisticated than A, it is not the perspective adopted by UTOK, which we can classify as Scientific Worldview C. The reason is that Scientific Worldview B embraces only a weak form of emergence, framed in terms of aggregate properties and epistemological necessity. The UTOK argues for more than just "weak emergence." The shape and "vision logic" of the ToK System clearly differentiates between emergence that takes place within the dimensions and the kind of emergence that gives rise to new dimensions. The ToK posits that the kind of emergence that drives Life, Mind, and Culture can be framed in ontological terms. That is, there are new causal forces that drive the behavior patterns of organisms, animals, and persons. The primary feature of Life, Mind, and Culture that I emphasize is the novel kinds of information processing and communication networks that appear and function to coordinate complex adaptive behavior patterns.

One way to characterize the new emergent processes that lead to these complex adaptive planes is in terms of epistemics, which refers to a kind of knowing. That is, Life, Mind, and Culture can be framed as kinds of knowing processes that engender new kinds of causal properties that cannot be reduced to mere aggregations of parts that generate new features. Rather, they are systematic arrangements that give rise to new complex adaptive behavior patterns. Such behavior patterns do not exist in the inanimate material plane and can be lost. For example, organisms can die, animals can lose consciousness, and humans can lose their capacity for explicit self-conscious awareness and justification. These describe shifts into lower planes of existence. This means that UTOK does not equate naturalism with either physicalism or materialism. These latter terms suggest that nature is a single material cone of complexification that moves from atoms to molecules to cells to animals to societies. In sum, the defining feature of UTOK that differentiates it from Worldview B is the idea that the kind of emergence that gives rise to the dimensions of Life, Mind, and Culture is qualitatively different than the processes that increase complexification within the dimensions of matter.

Finally, there is Scientific Worldview D, which, as framed by the Galileo Report, can be described as "post-materialistic science." It should be stated that a close read of the report reveals that the document is really a combination of two different arguments that should have been separated. One argument woven throughout the report is the criticism of the reductive materialist view of science. This is a view that is strongly supported by UTOK, especially as it is applied to Worldview A and, to a lesser extent, Worldview B. The second argument in the Galileo Report was the claim that consciousness has a kind of ontological primacy that is either chronologically prior to or independent from the material world. It is an argument that is much more contentious, radical, debatable, and uncertain. For example, at times, the report seemed to embrace the argument that consciousness does not require brain activity and that phenomena like parapsychology (i.e., clairvoyance, telepathy), near death experiences that point to life after death, and spiritual or transcendent experiences give credible scientific evidence for an ontology of an infinite consciousness or a dual world metaphysics. Such claims are interesting and worth considering, and there are many powerful anecdotes and lines of investigation that suggest that this is a possible map of reality that might be realized. I believe we should be open to and take seriously this class of phenomena. I label such phenomena "exo-naturalistic events" (i.e., processes that operate fundamentally outside a coherent naturalistic ontology). Although there are many interesting anecdotes and lines of investigation, it is also the case that extraordinary claims require extraordinary evidence, and I do not think the threshold of evidence has been met for such a radical shift in the foundational metaphysical system and ontology of a scientific worldview.

The primary point here is that the UTOK gives rise to a clear philosophy of science that is grounded in a naturalistic ontology that is different from philosophies of science that are characterized either as a reductive physicalism (Worldview A), an epistemological emergentist perspective (Worldview B), or a post-materialism view that overturns naturalism or includes extra-naturalistic phenomena (worldview D). Central to the UTOK vision of a Scientific Worldview C is the claim that there are two kinds of emergent processes, and that the emergent processes that give rise to new dimensions of existence have new ontological consequences that are not causally reducible to Matter or the language of physics. As the next chapter makes clear, this argument is strengthened when we consider the works of several philosophers of science who have ontologically mapped the different orders of nature.

Conclusion

Prior to the emergence of reductive physicalism, the Great Chain of Being provided the standard metaphysical conception of the way the world works. For many centuries, Europe was almost totally wedded to a worldview that combined the understanding the Greeks had of the scales of nature with a Christian view of the heavens. However, the modern Scientific Revolution started a radical transformation in the Western worldview. Over the course of several centuries, Christian dualism was replaced with a monist physicalist view. For many, this collapsed even Aristotle's scales of nature into a material flatland. Thankfully, the pendulum has begun to shift toward an emergent naturalism that restores the utility and credibility in seeing the whole. The Big History movement is a powerful example of such thinking. As similar movements gain general traction, thinking about the whole of human knowledge on the dimensions of time and complexity will become more commonplace.

Although BH and other movements toward a consilient, scientific view of the whole are positive developments, the failure of these efforts to attend to the problem of psychology demonstrates that a general emergent evolutionary view grounded in empiricism coupled to a commonsense naturalism does not suffice. The reason is that it leaves the Enlightenment Gap unaddressed. To dissolve the Gap and achieve a genuinely coherent scientific naturalism that includes the human subject that knows about the world, we need to develop a more metaphysically and ontologically sophisticated view of evolution and the orders of nature it produced.

The ToK System gives a new and better map of Big History. This chapter shows that it assimilates and integrates the key insights of Big History and advances the formulation considerably by providing a model that differentiates emergence within dimensions from complexity building feedback loops that result in new dimensions of existence. Doing so allows for distinctions that clarify the nature of the mental in a way that is lacking in the standard Big History framework. The differences in scientific worldviews—for example, between Worldviews B and C discussed at the end of the chapter—raise important philosophical questions regarding the validity of the ToK System's ontological frame. And, as the next chapter makes clear, recent work in metaphysics and the philosophy of science strongly supports this claim.

References

Ambrosio, F. J., & Lanzilao, E. (2013). Measuring the horizon: Objectivity, subjectivity, and the dignity of human personal identity. *Open Journal of Philosophy, 3*(04), 32–40.

Carroll, S. (2017). *The big picture: On the origins of life, meaning, and the universe itself.* Penguin.

Chaisson, E. J. (2001). *Cosmic evolution: The rise of complexity in nature.* Harvard University Press.

Chaisson, E. J. (2007). *Epic of evolution: Seven ages of the cosmos.* Columbia University Press.

Christian, D. (2017). What is big history? *Journal of Big History, 1*(1), 4–19.

Christian, D. (2018). *Origin story: A big history of everything.* Little, Brown Spark.

Deutsch, D. (2011). *The beginning of infinity: Explanations that transform the world.* Penguin.

Graeber, D., & Wengrow, D. (2021). *The dawn of everything: A new history of humanity.* Farrar, Straus, and Giroux.

Henriques, G. R. (2011). *A new unified theory of psychology.* Springer.

Henriques, G., Michalski, J., Quackenbush, S., & Schmidt, W. (2019). The Tree of Knowledge System: A new map of Big History. *Journal of Big History, 3*(4), 1–18.

Korotayev, A. (2018). The 21st Century Singularity and its Big History implications: A re-analysis. *Journal of Big History, 2*(3), 73–119.

Lane, N. (2015). *The vital question: Energy, evolution, and the origins of complex life*. WW Norton & Company.

Lewis, S. L., & Maslin, M. A. (2015). Defining the anthropocene. *Nature, 519*(7542), 171–180.

Rosenberg, A. (2011). *The atheist's guide to reality: Enjoying life without illusions*. WW Norton & Company.

Maynard Smith, J., & Szathmáry, E. (1995). *The major transitions in evolution*. Oxford University Press.

Spaulding, N. E., & Namowitz, S. N. (2005). *Earth Science*. McDougal Littell. ISBN 0-618-11550-1.

Tomasello, M. (2016). The ontogeny of cultural learning. *Current Opinion in Psychology, 8*, 1–4.

Walach, H. (2019). *Beyond a materialist worldview: Towards an expanded science*. Lulu.

Whitehead, A. N. (1929). *Process and reality: An essay in cosmology*. Macmillan Company.

Wilber, K. (2007). *Integral spirituality: A startling new role for religion in the modern and postmodern world*. Shambhala Publications.

Wilson, E. O. (1999). *Consilience: The Unity of Knowledge*. Vintage Books.

CHAPTER 8

Toward a Coherent Naturalistic Scientific Ontology

This chapter explores considerations in metaphysics and naturalistic ontology that provide evidence for why the four-dimensional view of the planes of existence given by the ToK System is justified. First, we briefly review the categorical ontology developed by the philosopher Nicolai Hartmann in the twentieth century and show that it directly aligns with the ToK System. Then, we move to a more recent analysis that explains why natural science should be bridged with descriptive metaphysics and why the result is a picture of the orders in nature that is also in direct agreement with the ToK System depiction. Finally, we end with an exploration of Roy Bhaskar's critical realist view of scientific ontology and demonstrate how it is strongly aligned with UTOK's theory of reality and our scientific knowledge about it. The conclusion is that these philosophical works provide ample justification for the descriptive metaphysical system and corresponding ontology given by the ToK System.

Hartmann's Categorical Ontology

In his *A Guide to the Perplexed* (1977), E. F. Schumacher criticized modern scientific thinking for losing sight of the Great Chain of Being. Instead of a materialistic flatland, he advanced the claim that we should divide the world into the inanimate, the animate, the animal, and the human, and

G. Henriques, *A New Synthesis for Solving the Problem of Psychology*, Palgrave Studies in the Theory and History of Psychology,
https://doi.org/10.1007/978-3-031-18493-2_8

argued that there is an ontological discontinuity between them. He summarized the point as follows (pp. 20–21):

> Is there really anything beyond the world of matter, of molecules and atoms and electrons and innumerable other small particles, the ever more complex combinations of which allegedly account for simply everything, from the crudest to the most sublime? Why talk about fundamental differences, "jumps" in the Chain of Being, or "ontological discontinuities" when all we can be really sure of are *differences in degree*? It is not necessary for us to battle over the question whether the palpable and overwhelmingly obvious differences between the four great Levels of Being are better seen as differences in kind or differences in degree. What has to be fully understood is that there are differences in kind, and not simply in degree, between the powers of life, consciousness, and self-awareness. Traces of these powers may already exist at the lower levels, although not noticeable (or not yet noticed) by man. Or maybe they are infused, so to speak, on appropriate occasions from "another world." It is not essential for us to have theories about their origin, provided we recognize their quality and, in so doing, never fail to remember that they are beyond anything our own intelligence enables us to create.

The idea that the inanimate, animate, animal, and human are ontologically separable categories was the focus of the philosopher Nicolai Hartmann. Over the course of many books, he developed a "categorical ontology" that identified four "layers of reality" as follows: (1) the inorganic layer (i.e., atoms, molecules, and spatial and temporal dimensions); (2) the organic layer (i.e., cells and plants, growth and metabolism); (3) the psychical/emotional layer (i.e., animals that could perceive and feel, act and move with purpose or intention); and (4) the intellectual/cultural/spiritual layer (i.e., self-conscious humans who reasoned in a social context). These four layers align directly with Schumacher, and they closely correspond to the Matter, Life, Mind, and Culture dimensions of behavioral complexification on the ToK.

Hartmann devoted much rich philosophical analysis to explaining the interrelationships and differences between the layers (Kleineberg 2016). For our purposes, his work can be summarized via his identification of four different "laws," which are as follows:

1. The law of recurrence: This refers to the fact that properties of the lower categories can appear in the higher levels, but never vice versa.
2. The law of modification: This refers to the fact that the properties in the higher levels modify themselves (i.e., they are causally interactive).

3. The law of the novum (which means new property emergence): The higher category is composed of a diversity of lower elements, but it is a specific set of properties that is not included in the lower levels.
4. The law of distance between levels: The different levels emerge in leaps and as a function of iterative processes, which means they can be clearly distinguished.

As these laws suggest, Hartmann made a detailed philosophical case for why there were fundamental ontological differences between the higher and lower dimensions and that the properties and cause–effect relations of the higher layers could not be described and explained via the lower ones.

Whereas Hartmann's work on categorical ontology supports the core logic of the ToK System, we can switch the lens around and show how UTOK significantly advances Hartmann's analyses. Hartmann acknowledged that he did not know what exactly constituted each plane of reality, or how they emerged. Moreover, he was rather flummoxed when confronting how to define subjective conscious experience in his system. This is not surprising, given the knowledge of his day. Having died in 1950, Hartmann was not privy to the developments in the latter half of the twentieth century. Understandably, Hartmann was most confident in the difference between the inanimate and the organic ontological layers. He identified the modern evolutionary synthesis as a basic framework for understanding this transition and the emergence of the life layer of ontology. However, he did not develop a clear picture of what constituted the other two layers, nor did he have a good metatheory for what gave rise to their emergence.

Of course, these are precisely the gaps that UTOK fills in. First, the ToK System explicates each "layer" as a dimension of behavioral complexification or plane of existence. Second, it frames the Matter dimension emerging out of Energy at the Big Bang and asserts that Life, Mind, and Culture emerge from the dimensions beneath them as planes of complex adaptive dynamic systems mediated by different kinds of information processing and communication networks. Third, via its "joint points" the ToK provides a general formulation for framing the emergence of each dimension. Thus, the UTOK formulation addresses Schumacher's uncertainty about the nature of the differences between the layers and provides frames for their emergence. In addition, via Behavioral Investment Theory and Justification Systems Theory, the Unified Theory also affords much greater metatheoretical specificity regarding Hartmann's last two layers.

The general point to be made here is that there is a synergistic relationship between Hartmann's categorical ontology and UTOK. In fact, his work makes an excellent philosophical case as to why there can be ontologically separate dimensions of complex adaptive behavior, and why the higher dimensions cannot be reduced, even in theory, to the lower ones. As such, his analyses do much of the philosophical labor that supports the different ontological planes of existence mapped by the ToK. Before leaving Hartmann, it must be noted that, as suggested by the term "psychic-emotional," Hartmann's primary reference point for the third layer of reality was what we call Mind^2, or subjective conscious experience. One of UTOK's main accomplishments is its capacity to obtain the proper relationship between behavior and consciousness, a bridge that is crucial to solving the Enlightenment Gap. As we will see in the latter half of this work, with UTOK we can effectively box in Mind^2 ontologically and properly place it in relationship to science and subjective knowledge.

A more recent philosophical analysis lends even greater convergent validity to the ToK System's depiction of reality and science. In *The Orders of Nature* (2013), the philosopher Lawrence Cahoone analyzed the ideas regarding the relationship between matter and mind that were present in the Enlightenment. He found them deeply inadequate and not up to the task of effective sense-making in the twenty-first century. To remedy the problem, he called on both philosophers and big picture scientists to bridge commonsense naturalistic approaches with systematic metaphysical analyses to upgrade our naturalistic ontological picture of the world. His philosophical critique and conclusions regarding the orders of nature yield a descriptive metaphysical system that is powerfully aligned with the map of reality and science provided by the ToK System.

On Naturalism and Metaphysics

Cahoone began his work by making the case that fruitful and necessary bridges can and should be built between scientific naturalism and modern-day metaphysicians. He justified this claim by contrasting it with the fact that there are several modernist and postmodernist forces that are positioned against this bridging project, such that there are strong institutional headwinds that resist such developments. One such headwind is that many philosophers identify naturalism with physicalism, which, as discussed in the previous chapter, is the claim that everything is reducible to or determined by the physical. Cahoone skillfully argued that reductive physicalism cannot account for the most complex features of human reality, such as "mind, culture, ethics, freedom, or art," and argued that emergent naturalism was a much more viable approach.

The second problem Cahoone identified was the fact that many scientists have a strong aversion to anything that smacks of metaphysics. It is presumed by many big picture scientists who take an emergent evolutionary view that one can simply adopt a "natural ontological attitude" without getting into the philosophical baggage associated with traditional metaphysical conundrums. As Cahoone (2013) put it (pp. 1–2), for most scientific naturalists, "naturalism is not metaphysical at all, [as] nature is what we are left with when we abandon metaphysics." Such individuals approach the world via a scientific attitude grounded in empiricism. From the vantage point of the Unified Theory, it is reasonable for scientists to eschew *pure* metaphysical questions. However, as the problem of psychology makes plain, there is a clear and obvious need for a descriptive metaphysical system for clarifying the concepts and categories pertaining to matter and mind, as well as for understanding science as a kind of justification system.

A third problem identified by Cahoone is the postmodern sensibility and the strong antipathy that emerged in the twentieth century among philosophers toward big picture synthetic philosophies. Indeed, as Lyotard (1979) noted, the defining feature of the postmodern sensibility is that grand meta-narratives are not viable. From a postmodern philosophical position, even a naturalism like Big History that implies a capacity to supersede other cultural systems of understanding is problematic. As Cahoone notes, to these poststructural philosophers, any naturalism that is framed as a general metaphysics is as illegitimate as any general metaphysics.

These attitudes combine to create strong skepticism in academic circles toward efforts that blend or bridge naturalism and metaphysics. Cahoone effectively summarized the criticisms that any such endeavor faces as follows (p. 2):

> A systematic metaphysics tries to inquire into many of these things [i.e., matter, mind, ethics, cultural objects] all at once in a coordinated way. It is this kind of general inquiry that has the worst current reputation, seeming the most obvious suspect for an inquiry still seeking an anachronistic view of the Whole, which is impossible, or a view from 'nowhere,' which is inconceivable, or claiming to incorporate its own meta-language—the language in

> which the basic terms of the theory are defined—which is illogical. Thus a systematic metaphysical naturalism continues to arouse a variety of negative responses: if it's naturalism, it's not metaphysics; if it's systematic metaphysics, it's not naturalism; if it's both, it epitomizes the errors of traditional philosophy exposed by thinkers with names like Nietzsche, Dewey, and Carnap, not to mention Wittgenstein, Heidegger, Quine, Derrida and Rorty.

In short, the current institutional sensibilities of both science and philosophy are largely defined against a descriptive metaphysical systematic approach, such as found in Cahoone's orders of nature or the ToK System.

This institutional resistance is deeply misguided and should be strongly rejected. The Enlightenment Gap unequivocally means that we are lacking in our big picture understanding, and that there is much philosophical work to be done in constructing such a vision. This was obvious in the previous chapter. BH embraces a standard commonsense naturalism; however, when placed in direct contrast with the ToK System we can see that it is not structured in a way that can address the key questions regarding the nature of matter and mind and their interrelations. From the vantage point of UTOK, philosophers and big picture scientists should be working together and actively seeking a descriptive metaphysical system for developing a coherent naturalistic ontology consistent with scientific findings from across the spectrum of empirical inquiry. Thankfully, this is exactly the task that Cahoone set out for himself. His focus was on philosophy, and thus he did not directly attend to either the Big History movement or the ToK System. Nevertheless, his analysis is highly instructive, both in how he positioned the need for a systematic metaphysics and what he concluded to be a reasonable picture based on twenty-first-century knowledge.

Consistent with UTOK's emphasis on the Enlightenment Gap, Cahoone anchored his analysis to how modern philosophy has struggled enormously with developing an adequate view of the relationship between matter and mind. He laid out the position that instead of attempting to locate a foundational set of truths, modern philosophers should first look to empirical science to develop a descriptive metaphysical picture that stretches from the base of the physical reality to the cultural complexity that involves scientific methods and explanations. That is, he took what UTOK frames as a "descriptive metaphysical-empirical systems approach" to the big picture and proceeded to generate a naturalistic ontology that has different orders of nature that closely align with the ToK System.

Cartesian Dualism and the Enlightenment's Mind-Body Problem

In the previous chapter we briefly contrasted the Christian and physicalist worldviews and alluded to the need to also consider Cartesian dualism to understand the matter versus mind problem that resulted in the Enlightenment Gap. Cahoone's analysis affords us the opportunity to do so. He laid out why much confusion in philosophy has a great deal to do with the work of René Descartes. Cartesian substance dualism is the idea that matter and mind occupy two completely separable causal planes that are mediated by different substances. If we consider the deep ideological tensions between the Christian and Newtonian mechanical worldviews, we can appreciate why Cartesian dualism plays a central role as a third system of justification. One reason is that it can be considered as a kind of "psychological middle ground" position between the Christian worldview and Newtonian matter in motion determinism. This is because Descartes' understanding allowed people to simultaneously believe that the best science of the day mapped the physical dimension and that the mental dimension of existence connected humans to God in a way that was broadly consistent with Christian dualism.

It is not just because Descartes' work afforded a pragmatic psychological resolution to the conflict between Christianity and matter in motion science that it had an impact. Descartes was also a brilliant mathematician, scientist, and philosopher who made many lasting contributions to our knowledge. Nevertheless, he is perhaps best known today for his dualist arguments for matter relative to mind. His well-known and oft-repeated dictum "cogito, ergo sum" translates as "I think, therefore I am." This maxim communicates Descartes' idea that he was a thinking and self-reflective entity. Descartes' foundational insight was considered as such because he argued that a self-conscious person could be most certain of this fact above all others. Moreover, it seemed obvious to Descartes that his self-reflective thought processes were radically different in form and function than the material world, such that his thoughts could not possibly be caused by the mechanistic matter in motion processes so apparent in the material dimension. Putting these claims together, one arrives at the inevitable conclusion that the stuff of matter and the stuff of mind are fundamentally different kinds of substances.

Descartes arrived at this conclusion in part by adopting an extremely skeptical view of knowledge. He questioned everything and sought to start his system with certain knowledge and build from that foundational truth claim. Descartes saw that he could not trust what other people told him, as it was possible that they were either lying or mistaken. He even realized he could not trust that his own perceptions corresponded to truthful representations of the external world. He noted that he had dreams that seemed very real, but were, in fact, illusions. Given this, he justified that it was at least theoretically possible that some "evil demon" was casting a spell on him, such that it was possible he was in a dream-like state of consciousness and every perception he had was a kind of hallucination. This notion still tickles the imagination, and it is referred to by more recent philosophers as the brain in a vat problem, and it was famously represented in the blockbuster Hollywood movie *The Matrix*.

Descartes noted that even if the external reality was obliterated, one thing remained. It still had to be the case that he was a thinking, deliberating entity. This was true even when he tried his best to doubt it. That is, even as he tried to argue that the world was an illusion and that he was not himself, he nevertheless remained an entity that was thinking about his existence. Thus, "I think, therefore I am" became his foundational truth claim. Additionally, he argued that thoughts did not seem to exist anywhere in the dimensions of space or behave in a way that could be caused by anything material. The combination of the foundational truth of his self-conscious existence and the fact that it was inconceivable to Descartes that his mind could be explained in terms of the mechanisms of matter were the core reasons that justified his belief in substance dualism. Importantly, Descartes thought that only humans (and, by implication, God) were imbued with this separate substance, and that other animals were not conscious in this sense.

Although his formulation was powerful, there are (at least) two key problems with it. First, Descartes offered no theoretical formulation for the nature of the mental or spiritual substance that enabled his self-conscious mind to have the properties it did. Descartes simply asserted that such a substance or field had to exist. But this assertion does not solve the problem so much as it introduces a whole new set of questions about the nature of the separate mental world. This problem is related to the second key problem, which Descartes overlooked in his first rendering of his formulation. It can be stated in the form of a question: *If the two domains of mind and matter are completely separate, then how could they possibly influence each other?* So significant are these two problems that very few modern philosophers adopt a substance dualist view of the world.

Western Philosophy's Bipolar Disorder of Matter versus Mind

In his critique of the Western philosophical tradition, Cahoone (2013) argued that two key errors have misguided the field of inquiry, both of which were present in Descartes' thinking. The first error was that many philosophers have sought certain and infallible knowledge as the starting point from which to build their systems. Cahoone rightly argued that this kind of foundationalism is misguided. To show why, he pointed toward the success of modern science, where knowledge is neither certain nor infallible, yet the modern empirical natural science enterprise has nonetheless been incredibly generative. Modern scientific descriptive taxonomies and theories can be thought of as plausible inference to best explanation, supported by logic and evidence. Via science and the methods that support it, we have achieved an enormous amount of high-quality knowledge. With a high degree of confidence, we can assert that electrons and protons are ontically real, and that the atomic theory of matter has much generalizable validity in describing the Matter plane of existence.

Although this knowledge is neither infallible nor certain in the absolute sense, it is nevertheless highly plausible, has much evidence that supports it, and generally rules out other interpretations, all while leaving the door open for more evidence and analysis. Moreover, there is little justification for having absolute certitude as the foundation from which one starts. Cahoone argued that a coherent and systematic view of the world and the truths that modern science has generated is a preferred approach to building up knowledge claims about the world. This is what grounds his descriptive, systematic approach to metaphysics, and his embrace of the notion that there are different orders of nature. It is very similar to the logic and coherence-based sensibility that grounds UTOK's theory of scientific knowledge and the reality it maps.

The second major problem with the modern philosophical tradition that Cahoone identified was Descartes' fundamental conceptual starting point that the world is (a) matter or (b) mind or (c) some dualistic combination of the two. Cahoone again turned to scientific knowledge and pointed out that the world as mapped out by science looks very different than this bimodal conception of reality. Consider, for example, that sophisticated animals like whales do not fit neatly into a matter versus mind dualism. Such creatures exhibit many of the kinds of behaviors that are "in between" the static, mechanistic picture of matter and the dynamic, self-organizing, autopoietic, complex adaptive picture of the human self-conscious mind.

Cahoone (2013) proceeded to argue that the matter *versus* mind split was the dominant "bipolar disorder" of modern philosophy, and he strongly rejected it as deeply misguided. He characterized this bipolar disorder as follows (pp. 6–7):

> [It is] the belief that reality is constituted by at most two kinds of entities or properties, the physical and the mental, a disorder shared by idealism, dualism, and physicalism or materialism, reductive or nonreductive. That disorder encouraged us to think physics is the only metaphysically interesting natural science, that human mind is the only part of nature that creates problems for a (physically oriented) metaphysics, that knowledge and mind are solely human possessions, that all the other natural sciences—chemistry, the Earth sciences, biology, engineering—are metaphysically unimportant. This dualism has been repeatedly and recursively applied, multiplying subschools (for example, between 'scientific' naturalists and 'humanistic' naturalists), but always with the same tendencies. It arguably has something to do with the congealing of twentieth-century Western philosophy into two opposed hermetic traditions, analytic and continental philosophy, one (in its metaphysics) tending to focus on highly specialized problems in the interpretation of physics and the possible reduction of mentality, the other rejecting natural science as inhospitable to whatever matters to the human prospect (there being some exceptions on both sides who, as is said, prove the rule). In a broader context, both are manifestations of the conflict of C. P. Snow's the 'two cultures.'

This passage gets to the heart of the consequences of the matter-mind aspect of the Enlightenment Gap. When one looks across the landscape of the academy, one sees many problematic dualisms that stem from it. In psychology, we see dualities such as "behavior versus mind" or "brain versus mind" or "subjective versus objective" and "mental versus physical." Cahoone's focus was on philosophy, and he showed how Western philosophy has been trapped inside a matter versus mind dichotomy for much of its history. He also traced how physics was the most successful science and tended to pull all of science in that direction. The consequence was that for many scientists and philosophers the vision of the natural world went from the scales of nature described by Aristotle down into the material flatland advocated by reductive physicalists. Cahoone, however, noted that physics is lacking in its capacity to describe and explain the self-conscious mind, and much else. He offered a forceful rejection of reductive physicalism for an emergent, naturalistic view effectively captured by a systematic metaphysics. The vision that emerged can be readily classified as falling under Scientific Worldview C as described in the previous chapter and shows striking parallels with the ToK System.

Cahoone's Orders of Nature

In contextualizing his approach and the traditions that preceded it, Cahoone (2013) placed his systematic metaphysics as an integrative approach that blended aspects of the major branches of philosophy (i.e., the analytic, continental, and pragmatic branches). He most directly identified as an emergent naturalist, and his approach connected historically with the British emergent evolutionists who were active in the 1920s. Cahoone centered his argument in the claim that, when we look broadly at nature from a scientific view, many complex entities exist between the continuum that stretches from the base of matter mapped by the standard theory of particle physics to the activities of human persons engaged in things like constructing the Universal Declaration of Human Rights. From this he argued that, in contrast to the bipolar split between matter and mind, a much more pluralistic portrait of the various domains or orders in nature becomes clear. He argued that an emergent evolutionary view was needed to capture this wide distribution of behaviors and that the proper way to ground it was in a systematic metaphysics.

Cahoone explored the difference between emergence and reduction and how they exist in dialectical tension. In a manner that strongly corresponds to the map of reality offered by the ToK, he also explored both emergence within the orders and how radically different properties emerge that give rise to entities and behavior patterns that should be identified as qualitatively different orders. He ultimately identified five separate orders of nature: (1) the physical; (2) the material; (3) the biological; (4) the mental; and (5) the cultural. As will be made clear, these orders of nature line up closely with the concepts of Energy, Matter, Life, Mind, and Culture on the ToK System. Given that they were independently conceived and derived from two disciplinary vantage points (i.e., philosophy versus psychology), the correspondence between the two visions is strong evidence for convergent validity.

The Five Orders of Nature

The result of Cahoone's analysis is visually depicted in Fig. 8.1, and there are several aspects to this diagram that need to be highlighted. We can start by noting the close alignment between it and the ToK System. The largest, dotted circle that is labeled the physical corresponds to the circle at the base of the original ToK. Then there is the material dimension that exists within the broader physical set but is demarcated by a circle which represents how it can be thought of as also being "above" the physical. The same pattern of higher dimensions being logically *both* above and within the dimensions beneath them holds for the biological, mental, and cultural orders. This is, of course, also present in the vision logic of the original ToK diagram. In addition, like the ToK System and Big History, Cahoone places these orders on the dimensions of time and complexity, as indicated on the left side of the diagram.

Cahoone's graphic traces cosmic evolution via an alphabetic notation. It starts at the physical order of nature with the quantum gravity and Planck scale beginning at the Big Bang (A) and moves into the radiation era, for which we have direct empirical evidence provided by cosmic microwave background radiation. The trail of the evolution of behavioral complexity proceeds into the quantum particle realm (B) and then atoms (C) and then the development of stars (D). This transition sets the stage for what Cahoone characterized as a new order of nature, the material, which roughly corresponds to the science of chemistry. The stars produce heavy elements (E), and the graphic follows the trail into our solar system and planet Earth, where we have immediate contact with the more familiar elements of the material world, such as minerals, water, and air (F). Then there is a jump into macromolecules (e.g., proteins and RNA) and from there we get the emergence of single cells like bacteria (G). This is the realm of the living and its ecological thickets, which Cahoone calls the biological order.

Cahoone then traces the biological order through the emergence of invertebrate animals that appeared at the Cambrian Explosion at approximately 550 million years ago when there was a dramatic "explosion" of animal life forms (H). Cahoone identified the mental order of nature with the emergence of vertebrates, such as fish and later reptiles and then into birds and mammals. He calls these "bio-psychological thickets," which is appropriate given the current institutional ambiguity as to whether we should consider animal behavior a biological or psychological discipline

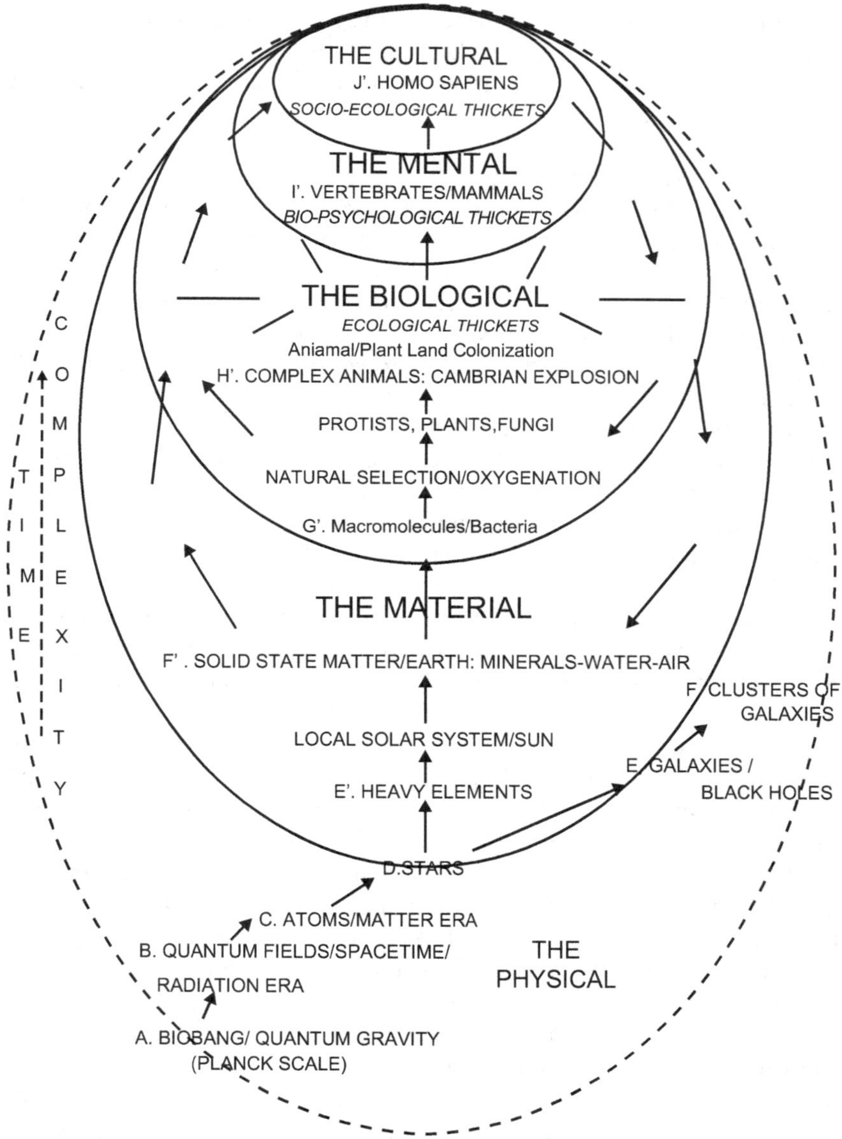

Fig. 8.1 Cahoone's (2013) depiction of the orders of nature. (Reproduced with permission)

(I). Finally, there are Homo sapiens and the cultural order of nature, characterized by socio-ecological thickets (J). With this overview of the schematic in hand, we can now move to delineate the specific orders in greater detail.

The Physical and Material Orders

Given that Cahoone was not aware of either Big History or the ToK System when he formulated his orders of nature (Cahoone, 2018, personal communication), the convergent overlap is striking. Nonetheless, there are also a few important differences with the ToK System that require some analysis. The most obvious difference resides in the fact that Cahoone differentiated the physical from the material orders of nature. In the ToK System, physical and material are considered largely synonymous, as they are in many language systems, although the physical is often framed as the broader term because it includes energy. Given that atoms are clearly matter as the word is defined in physics, it does seem odd or at least potentially confusing that, using Cahoone's frame, we should refer to atoms as being part of the physical rather than the material order. However, if one examines the diagram more carefully, entry "C" suggests that atoms do align with what Cahoone labels the "Matter Era." In addition, work philosophy has shifted the description of the ground of ontic reality from the material to the physical as a function of modern physics and findings from quantum field theory and general relativity.

Deeper analyses reveal clear parallels that suggest close correspondence in the way both UTOK and Cahoone map this aspect of the world. One obvious linkage is to connect the physical in Cahoone's diagram to how the physical was represented in the original ToK. Let us recall that it formed the basic circle at the foundation of the original ToK diagram, out of which the dimension of Matter emerged. Once this alignment is seen, then another primary effective linkage can be made by seeing that Cahoone's concept of the physical corresponds primarily to what is identified as the Energy-Information implicate order in UTOK. As we have noted, the Unified Theory posits that the universe began as an "Energy Information Singularity" and energy (as opposed to matter) is the ultimate common denominator substance of the universe. This means that, at bottom, everything is a form of energy. It also means that the Matter dimension emerges out of Energy. By aligning the physical in Cahoone's system with Energy on the ToK System, we see that the two descriptive metaphysical systems are in closer agreement in this domain than first appearances might suggest.

Cahoone's (2013) distinction between the physical and material order brings us to what is the biggest difference between Cahoone's system and the ToK System. Although Cahoone describes why the orders of complexity are separable, he does not explicitly define them as dimensions of behavioral complexity or planes of existence. In addition, he does not explicate the emergence of each new dimension via metatheoretical "joint points." In contrast, UTOK specifies that Matter, Life, Mind, and Culture are all well framed as dimensions of behavioral complexification. Moreover, the ToK System is explicit in arguing that the different dimensions that arise following the Matter dimension are different planes of complex adaptive behavior patterns that emerge as a function of complexity building feedback loops that give rise to self-organizing dynamic systems mediated by novel information processing and communication patterns. And, via Behavioral Investment Theory and Justification Systems Theory, UTOK provides two key metatheories that frame the structural and functional organization of the mental and cultural orders. As such, the UTOK can be framed as offering a very similar picture, but one that is more detailed and theoretically specified than what Cahoone lays out.

The Biological Order

The biological order in Cahoone's scheme corresponds directly with the ToK System's conception of the Life plane of existence. Consistent with the ToK ontology, he describes the remarkable, complex adaptive processes that make animate entities and behavior patterns so different from inanimate ones, and explicitly identifies life as a set of processes that cannot be reduced to the language of physics or chemistry. Cahoone (2013) tied the functional way that living entities seek certain valued outcomes (i.e., they move toward states that enable survival and reproduction) to the information processing capacities of cells. He characterized these organic behavioral processes as follows (p. 183):

> [L]iving things have form. The relation of their structure to their components is more independent than that of non-living things, for the structure remains the same through a virtually complete exchange of token molecular components. The form or characteristic set of structuring activities which accomplishes this is constituted by something utterly foreign to non-living matter: a complex macromolecule that functions as a code to dictate the

> continual reconstruction of the individual … Cells are the first things that, as far as we know, are literally constructed according to an existing detailed plan.

As suggested by this passage and its reference to "form," Cahoone made the argument that to understand the functional behavioral properties of Life, we need to return to Aristotle and invoke his notions of formal and final causation. The quotation that summarizes his analysis of the biological order makes this clear (Cahoone, 2013, p. 187):

> Biology thus gives us clear examples of final causes … Within organisms there must also be *teleomatic* material processes as described in Chapter 6. … When any encompassing biological system—society, organism, organ system, organ, tissue, cell—restricts the output of a component system, dictating any properties or behavior of the component, this too is selection. Lastly, because the organism participates in an environmental ecology and species evolution, its form has actually been selected over time for that ecology, hence its form serves a final cause, adaptive fit, given a genetic inheritance.
>
> One may say there is no reason to reintroduce final causes here. Aristotle thought a ball rolling down an inclined plane exhibited final causation, moving to rest in its 'natural place' on the surface of the Earth. We moderns say that is not telic (and we are right). So isn't the wren eating a seed doing the same mechanical thing? No. For even putting aside its greater complexity, there remains this: if the ball doesn't roll, it is still the same material object, but if the bird doesn't eat, it dies. The dead bird obeys the rules of chemistry and physics as before, and as the ball does. But it is no longer alive. And it was life, after all, that was being investigated here.

When this passage is considered in relation to Hartmann's categorical ontology and the different dimensions of complexity argument provided by the ToK, we clearly see the idea that there is a fundamentally different kind of emergence when we move from the inanimate to the animate. Many other scientists and philosophers have made similar claims. In their *Tree of Knowledge: The Biological Roots of Understanding*, the two Chilean scientists Humberto Maturana and Francisco Varela (1987) introduced the concept of "autopoiesis" to characterize a dynamic, self-organizing system to provide a frame for such processes. More recently, Stuart Kauffman (2019) has argued that the structure and information networks of cells exist in “a world beyond physics.” He characterized it thusly (pp. 5–6, italics in original):

> The evolving biosphere manifests this surging upward in complexity, from protocells 3.7 billion years ago to the millions of species now. … The biosphere literally constructs itself and does so into a biosphere of increasing diversity. Again, How and why is this? Remarkably, the answer may be ‘Because the living world *can* become more diverse and complex and in an ongoing way *creates its own potential to do so.*’

The point of all of this is that the living order is a new plane of existence with different kinds of processes and properties engendering novel cause and effect relations.

The Mental Order

After summarizing the need to consider Aristotle’s concepts of formal and final causation in explaining living behavior, Cahoone turned to analyze the mental order of nature. Rather than thinking in terms of matter versus mind, he emphasized how it is the biological order of nature—as opposed to the physical or material orders—that gives rise to the mental order. This means that when we consider the place of the mental order in nature, we should be clear that we are already dealing with complex adaptive processes and their formal and final causal processes. He made the point as follows (Cahoone, 2013, p. 191):

> If physicalist reductionism is inadequate to biology, why should we expect it to work in the investigation of mind? Our problem is the relation of the mental, not to the physical, but to the biological. Central nervous systems are living systems of cells exhibiting teleonomic behavior. The problem is not, how do [mental] properties prevail in a physical world, but how do they prevail in a living biological system in interaction with a physico-material-bio-social world?

In addition to these considerations, Cahoone clarified why the mental order of nature corresponds with the animal rather than the human. Linking his analysis to what Antonio Damasio called Descartes’ Error, Cahoone lamented the fact that most of the work in the philosophy of mind is in fact about the philosophy of the human mind. Cahoone (2013) put it as follows (pp. 190–191):

> [The] study of mental activity by rights should be an offshoot of ethology, and what is called 'philosophy of mind' should, by dint of actual practice, be called 'philosophy of human mind.' At its narrowest, the metaphysical problem of philosophy of mind lies in the relation between, say, an animal's feeling of pain and the electro-chemical behavior of its CNS. The narrowest part of the river is usually the best place to ford. Consequently, the present chapter [on the mental order of nature] is not about human mind; it is about mind in general, which is to say, animal mind.

Placed in the lexicon of mental behavioral processes provided by the Unified Theory, Cahoone is highlighting how the philosophy of mind has failed to make the crucial distinction between the domains of $Mind^1$ and $Mind^2$ in animals and $Mind^3$ in humans. Moreover, as this passage suggests, and Cahoone notes, it is here that the hard problem of consciousness resides.

In his analysis of the mental order, Cahoone follows the phylogenetic trail through the appearance of the early nervous system and into the invertebrates. He then notes the emergence of more complicated forms of mentation, which involves perception, memory, and problem solving. He argues that such higher levels of functional awareness and responsivity are strongly suggestive of subjective experience (i.e., $Mind^2$). For Cahoone, like Hartmann before him, this is the essential emergent feature of the mental order. As such, it is not surprising that he began his discussion of the mental order of nature with reflections on the hard problems of consciousness. He lamented the difficulty both in understanding how consciousness can emerge from a material brain and how mental causation can be considered consistent with the laws of the physical universe. This highlights how, for Cahoone, the mental corresponds to subjective conscious experience or $Mind^2$.

Equating Mind as an order of nature with the domain of $Mind^2$ is an error from the vantage point of UTOK for several reasons. First, with its exterior epistemological position, science is grounded in tracking behavior and $Mind^2$ is thus a tricky and ambiguous domain for science. Second, given our current knowledge regarding the neurobiological mechanistic ontology of $Mind^2$ (i.e., the hard problem), it is not clear when and where $Mind^2$ appears in the animal world. Third, UTOK's metaphysical system and metatheory makes clear that we can justifiably ground the mental order in the concept of mental behavior framed by a neurocognitive functional analysis of animals with brains as complex active bodies operating as minded animals in the world. As will be made clear in later sections of this work, UTOK's distinction between Mind^1 and Mind^2 coupled with BIT's metatheoretical formulation for understanding the Life-to-Mind joint point specifies why framing these issues in terms of mental behaviors or mindedness resolves the philosophical conundrums that trip up so many. Nevertheless, Cahoone gets several of the key pieces of the puzzle correct, including properly placing the mental in the domain of the animal.

The Cultural Order of Nature

In discussing the cultural order of nature, Cahoone starts with the important point that we did not evolve from apes, but, of course, always have been apes and still are. By reminding us that we are still primates, Cahoone rightly places the emergence of Mind^3 and the Culture-Person plane of existence in the context of the animal mind. He then proceeds to highlight that there are nevertheless remarkable differences that make our complex adaptive behavior patterns quite exceptional in the rest of the animal kingdom. He reviews the archeological and anthropological records documenting our evolutionary history and offers several candidate ideas for what makes us uniquely us. These include: (a) our tool-making ability and how this shapes us; (b) our intellectual abilities; (c) our capacity for self-conscious reflection; (d) our capacity for symbolic language; and (e) our capacity for empathy and reading the minds of others. The latter four capacities are, of course, explicitly framed and interrelated by JUST, with the latter insight supplemented by the Influence Matrix.

We find another point of correspondence when we consider how Cahoone characterizes culture. Although he was not aware of JUST, Cahoone's conception of culture has several elements that line up with the idea. For example, he argues that human culture is about constructing meaning-making systems and is closely aligned with history in the sense of the evolution of meaning-making complexes over time. Such processes involve narrative and verbal propositions, and how new members are socialized into the practices of the group. Cahoone also traced the development of oral indigenous cultures into the development of agricultural and foraging societies and, finally, modern civilization. He also differentiated culture from societal institutions and technologies. And he followed the evolution of cultural belief systems directly into science in a way that parallels the ToK System's depiction of science as a specific kind of justification system.

From Culture to Science
The parallels between Cahoone's systematic metaphysics and the ToK System continue when we consider his closing chapter on the nature and evolution of human knowledge. In it, he traces how human language, self-consciousness, and reason-giving emerged and resulted in the evolution of various kinds of epistemological practices. Situated as such, Cahoone traced the evolution of various forms and structures of justification through social and pragmatic reasoning about truth statements regarding local states of affairs into more complicated analytic proposition and finally into science and postmodern philosophical critiques regarding grand knowledge systems. Seeing science as an extension of culture and culture as an extension of nature, he sees the orders of nature coming full circle to include an epistemological, justificatory process of understanding the ontology of evolutionary emergence.

In making this last connection, Cahoone shows how an emergent evolutionary view of the orders of nature anchored to a systematic metaphysics sets the stage for framing the other aspect of the Enlightenment Gap. Specifically, we can start to generate a coherent picture of the relationship between social-pragmatic knowledge and the emergence of scientific epistemology and its ontological claims about the nature of the ontic reality. This insight connects us to the work of the philosopher Roy Bhaskar, whose critical realism provides us with a clear understanding of how the evolution of social justification systems can produce scientific methods and processes that afford us a realist view of natural science situated in a critical hermeneutic context of knowledge generation and discovery.

A Critical Realist View of Scientific Ontology

The edited volume *Metatheory for the Twenty-First Century: Critical Realism and Integral Theory in Dialogue* (Bhaskar et al., 2016) provides a powerful overview and critical analyses of the potential synergies between Ken Wilber's Integral Theory and Roy Bhaskar's Critical Realism. Both perspectives find common ground in their appraisals of the current state of the world and of human knowledge, and the need for more harmonious bridges between science and spirituality that would enable us to have wiser relationships with the planet and technology, and between different national and cultural groups. Specifically, both perspectives share an awareness of a series of profound and global problems or "meta-crises" (see Björkman, 2019) facing humanity, which include: (1) an ecological crisis, both in terms of macro-level climate change and in terms of depleting the earth's resources and changing a multitude of ecosystems; (2) a meaning and mental health crisis, such that depression, anxiety, existential angst, and confusion about what is true and good are pervasive; and (3) a deep problem with the current structure of our knowledge, such that it is fragmented, disorganized, and not amendable to orienting toward wisdom. The authors put the issues as follows (pp. 2–3):

> [M]uch of the contemporary academy remains hypnotized by either the hyper-analytic, hyper-specialized, fragmented gaze of late modernity, or the sliding scale of postmodern relativism and its antipathy to integrated knowledge and meta-level understanding. Together these two orientations offer inadequate understanding(s) of our many complex problems and their root causes, let alone the socio-ecological crisis at large. Without being able to adequately illumine such root causes, the academy remains largely impotent to address and help transform them. This point is underscored by the fact that, to date, the dominant metatheories of modernity, such as positivism, have not only failed to alter fundamental trajectories of human-induced ecological degradation (Biermann et al., 2012; IPCC, 2014) but are in fact deeply implicated as underlying causal forces contributing to such trends, as has been widely argued by philosophers and social theorists alike (Bhaskar, 2002/2012, ch. 2; Wilber, 1995).

A chapter by Paul Marshall (2016) explored the relationships between Integral Theory and Critical Realism and emphasized their complementarity and their potential synergy. He noted that the core strength of Wilber's system is found in the four quadrants, which serve to clarify epistemology and justify an integral methodological pluralism to framing and acquiring knowledge. He also identified the utility of Wilber's taxonomy of developmental lines, levels, states, stages, and types, and the fact that it is grounded in a spiritual view of the cosmos. Marshall also rightly identified that the core strength of Bhaskar's work was in ontology, which is the foundation of his philosophy of critical realism.

Bhaskar's Critical Realist Ontology

Enlightenment Common Sense: The Philosophy of Critical Realism (Bhaskar & Hartwig, 2013) was Roy Bhaskar's final book, before he died in 2014. It summarizes the key features of his work, focusing most directly on his first phase of inquiry, critical realism, which was then followed by dialectical critical realism, and then a shift into a secular spiritual vision called "metaReality" that blended science with Eastern wisdom traditions. Bhaskar's critical realist philosophy is deeply consistent with both the ToK System's theory of reality and our scientific knowledge of it, and UTOK's larger vision for a system of knowledge that is oriented toward the cultivation of wisdom.

As we have seen, central to the theory of reality and scientific knowledge advanced by UTOK are the following three metaphysical categories: (1) the ontic reality (i.e., the world referenced by the dimensions of Matter, Life, Mind, and Culture, all of which existed prior to the development of our scientific knowledge of them); (2) scientific ontology (i.e., the descriptive taxonomies and theories about that reality in the various domains of science); and (3) scientific epistemology (i.e., the frames, principles, and methods used to justify the correspondence between the ontic reality and our scientific ontology). In addition, the ToK System places science in the context of Culture and framed via the social construction of reality.

Although the terminology is slightly different, this frame directly aligns with Bhaskar's critical realist philosophy. Paralleling the UTOK's distinction between the ontic reality and scientific onto-epistemology, Bhaskar made the distinction between the *intransitive* and the *transitive* ontological domains. The intransitive domain refers to the real world that exists independently of our beliefs about it, and thus corresponds to what we have called the ontic reality. In contrast, the transitive domain refers to the theories about the world, and thus corresponds to what we have called scientific ontology, which emerges out of scientific epistemology. Bhaskar refers to the theories we have about the world as being transitive because they are capricious and in flux. In contrast, the actual history of the universe does not change with human beliefs about it, and thus is intransitive.

It must be noted that our depiction of the ontic reality is ultimately dependent on our ontology, which is one of the features that makes the relationship complicated. Consider, for example, that there is some debate regarding the actual age of the universe. When I first published the ToK System back in 2003, I placed the age of the universe at approximately 15 billion years old, which was congruent with most scientific work at the time, which generally estimated a range between 12 and 20 billion years. In the last two decades, the methods have increased in sophistication and allowed much greater specificity, such that 13.8 billion years is now an oft-cited fact. However, in 2019, reports by Adam Reiss and others (Powell, 2019, May 18) indicated that the universe might be expanding faster than previously understood. Some scholars argued that this would mean that the universe might be approximately 12.5 billion years old, over a billion years younger than what had been strongly proclaimed by many. Even more dramatically, there are a tiny minority of cosmologists who argue that the entire Big Bang model is in error (see, e.g., Lerner, 1992). The point is that, as far as human knowledge goes, there is an ongoing relationship between our transitive ontological claims and the intransitive ontic reality.

The ToK System depiction clearly frames scientific knowledge as being *both in the Culture-Person plane of existence and as emerging out of it.* This corresponds to the fact that science is a human construction, and it points to the idea that achieves a transcendent realist picture of the world and our place in it. The word transcends here refers to the fact that it extends beyond the normal subjective and socially constructed systems of justification and achieves deeper and more epistemologically valid truth claims. Transcendent realism can be framed via a thought experiment. It posits that if humans encountered another intelligent species who had detailed, logical knowledge of the universe, they too would describe the world in a way that aligned with the Big Bang, quantum mechanics and general relativity, and the atomic theory of matter. This directly aligns with the realist argument Bhaskar is making. Of course, if an alien species had a different but also valid propositional map of the foundational structures of the material universe, then transcendent realism as framed here would be false, and one would be returned to Kant's transcendent idealism and the claim that our reality is indeed framed by the categories of the human mind (i.e., phenomenology).

Another point of alignment pertains to how the ToK System and critical realism frame the arrow of time and the nature of the real. Bhaskar divides up the ontic world into three conceptual categories: (1) the real; (2) the actual; and (3) the empirical. The real refers to the set of causal generative structures and mechanisms that potentially give rise to the actual events in the world. Two examples of such generative mechanisms are gravity and the electromagnetic force. The actual refers to entities and events that emerge from the generative potential mechanisms that exist and have a history, and thus could, at least in theory, be documented to have occurred. Finally, the empirical refers to actual entities and events that have been observed by a specific observer.

Figure 8.2 visually depicts this relation, and it shows how the real is the broadest category, then the actual, then the empirical. For a concrete

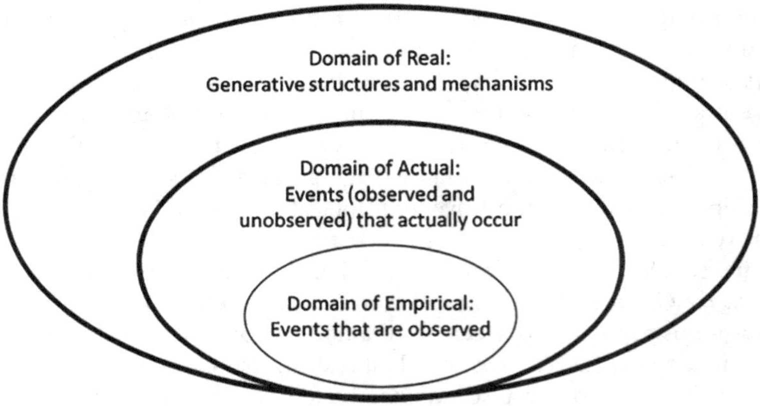

Fig. 8.2 The nested relationships between the domains of the real, actual, and empirical

example, consider the goldfish in my fish tank. They are real-actual-empirical entities, in the sense that (1) they consist of generative mechanisms (e.g., bio-chemical processes such as metabolism and energy transfer) and (2) are actual entities that exist on the "grid" of space and time and (3) have been observed by me. Stars from a distant galaxy are real and actual, but not empirical in the sense that they cannot be observed by me.

The distinction between the real, the actual, and the empirical can be aligned with the metaphysical–empirical continuum we laid out in setting the stage for developing a consilient picture of science. Specifically, the metaphysical aspect refers to the naming of the concepts and categories. For example, the differentiation between these three domains pertaining to the real is a metaphysical argument. Then there are the scientific systems of justification—the metatheories, paradigms, and theories—that provide scientific ontological frames that map the generative structures and mechanisms in the ontic world. This corresponds to Bhaskar's domain of the real. More specifically, it refers to our transitive ontological claims about the intransitive real causal mechanisms. For example, natural selection operating on genetic combinations is a theory of real generative mechanisms that play a key role in the generation of the Life plane of existence. The actual world is the unfolding of history that emerges from the potentials given by the generative structure and mechanisms. The scientific process of justification of determining the real and actual requires evidence of the correspondence between theory and reality. This is analyzed via the empirical process of data collection, experimentation, and analysis of alternative explanations. This empirical process can be done descriptively in open systems, or it can be done experimentally in closed systems.

This tripartite distinction between the real, the actual, and the empirical also aligns with the time dimension as framed by the ToK System. Specifically, the ToK depicts emergence as being generated by the unfolding wave of change, which can be framed by entropy and the arrow of time. The framing is that the real generative mechanisms create a potential that is realized into actual events. At its broadest level, this can be framed in terms of the flow of energy and information from the past into the present, such that the edge of that unfolding is the possible- into-probable future, which UTOK frames as the wave of causality. The actual refers to the trail of history. The empirical, then, refers to the domain of events that are happening in the sphere of energy and information that is available to the observer in different positions in the space–time continuum.

Bhaskar's Critique of the Epistemic Fallacy

Bhaskar also made key distinctions between philosophical and scientific ontology. His philosophical ontology is grounded in what he called "the epistemic fallacy," which encapsulates his critique of philosophers like Hume and Kant whom he argued reduced ontology to epistemology. Specifically, he argued that the implication of Hume and Kant is that what humans have access to is phenomenology (and, from more modern perspectives, socially constructed language games), such that genuine ontological claims about the ontic reality become impossible. He argued that this has been implicitly woven throughout much of modern science, which is somewhat ironic in that many argue that science is committed to realism. Bhaskar cogently argued that too much work in the philosophy of science has been concerned with how we know, as opposed to focusing on and getting clear about what is real. He argued that we can make assertive claims about ontology via what he called the "TINA" principle, which stands for there is no alternative. An example of applying the TINA principle would be the claim that the Matter dimension of complexification is made up of atoms. Placed at the proper level of specificity (i.e., acknowledging that particles make up atoms), Bhaskar argued that there is no alternative to this conclusion about the ontology of Matter.

To my knowledge Bhaskar never applied his critique of the epistemological fallacy to psychology. However, his analysis works brilliantly in this domain. In fact, the first two chapters of this work demonstrated that mainstream academic psychology essentially reduces ontology to scientific epistemology. That is, it generally defines behavior as the part of the domain of the mental that is available to scientific inquiry and methodology. This decimates Psychology's capacity to have a clear picture of the ontology of the mental that is independent of scientific methods and epistemology. The result is a chaotic, fragmented pluralistic system of knowledge such that the whole is less than the sum of its parts. The latter half of this book is devoted to untangling this knot of confusion and generating a metaphysically clear descriptive map of behavior and mental processes that appropriately and comprehensively interrelates epistemological and ontological considerations and gives rise to a coherent integrated pluralistic view of the mental and its relations to other dimensions of existence.

Bhaskar's Four Planar Model of Human Knowing and the Stratification of the Ontic Reality

Bhaskar's scientific ontology refers to his analysis of how science can make claims about the ontic reality (or, in his vocabulary, the intransitive ontology). Central to critical realism was Bhaskar's framework for understanding how humans learned about the world as they operated within it in a way that was consistent with realism, naturalism, and the social activity of human beings and the processes by which they construct knowledge systems. Toward that end, he developed the transformational model of social activity that could be represented by what he referred to as the "four planar" model of social being that posited that all human social processes occur simultaneously on the four dimensions of: (1) material transactions with nature; (2) social interactions between people; (3) social structure; and (4) the stratification of the embodied personality.

These domains are apparent when reality is viewed via the lens of the ToK System. It shows that (1) humans are embedded in the material world, (4) there are intrapsychic processes at various levels of operation playing a mediating role as people engage in exchanges in the relational world, all of which takes place in (2) a sociocultural context of justification, as well as (3) historical developments and institutional and technological mechanisms and constraints. Consider, for example, if you and I were engaged in a coversation about UTOK and its validity. We would be embedded in the material world, engaged in social processes of justification, embedded in a social institutional structure, and would each have subpersonal, nonverbal, primate mental behavioral processes that situate us as embodied participants in the exchange.

Bhaskar paired the four planar model with the idea that the ontic reality is genuinely stratified into different levels of structure, processes, and events. He argued that the stratification of reality can be seen in three different ways. First, he argued that the distinction between the real generative causes and the actual events represents a crucial ontological distinction. This feature of the ontic reality is mapped by a scientific ontology that is organized into different levels and dimensions, as well as theories of the real and empirical findings that correspond to the actual. In articulating his view of the stratification of reality, Bhaskar also noted, in direct accordance with the ToK System, the cosmic evolutionary fact of emergence, such that the universe was, at its beginning, far more simple and undifferentiated, but out of energy have come particles, atoms, molecules, cells, animals, humans, and societies. He posited this as strong evidence of the stratification of the ontically real.

The next section of this book shows how the UTOK frames this set of insights in terms of behaviors that take place on different levels and dimensions of complexity. The concept of behavior includes structures and processes that are strung together in events. Subsequent chapters will show how even greater detail can be achieved in understanding the levels and dimensions of behavioral complexity and the differentiation of the domains of science that map them. The conclusion is a vision that is consistent with and enriches Bhaskar's critical realist philosophy of science and picture of reality.

Conclusion

An attitudinal hallmark of modern natural science is that it eschews metaphysics. The early success of empirical investigations resulted in a false confidence that the scientific enterprise could simply proceed based on observation and measurement, experimentation, and a commonsense language system. Unfortunately, the vocabulary and conceptual grammars available during the Enlightenment were simply not up to the task of generating a naturalistic scientific ontology that coherently framed the domain of the mental. The failure of commonsense natural scientific approaches like Big History to even see—let alone resolve—the problem of psychology is strong evidence that such descriptive metaphysical systems are needed.

This chapter demonstrated that several philosophers have mapped nature in a way that is highly concordant with the ToK System. Given this history of the Great Chain of Being and the long-standing work on integrative levels of reality, this is not surprising. Hartmann's categorical ontology and Cahoone's systematic metaphysical description of the orders of nature allow us to grasp the metaphysical and ontological issues at play. Bhaskar's critical realism provides a vocabulary for the ontology of scientific knowledge that is directly concordant with the ToK System, and it maps the relationship between social knowledge, natural science, and the social sciences in a way that directly aligns with key insights from UTOK.

The conclusion from this section is that with its clear differentiation of the different dimensions of behavioral complexity and specification of why there are separable complex adaptive planes, the ToK System assimilates and integrates the insights from these philosophers and provides a descriptive metaphysical system that enables us to have a clear theory of scientific knowledge and its relationship to both the ontic reality and the Culture and society in which science is embedded. This affords us a new map of Big History and bridges natural science to work in systemic metaphysics and moves us closer to a coherent naturalistic ontology. The next two sections of the book spell out how UTOK enables us to advance our clarity regarding behavior and mental processes so that we can fill in the metaphysical, ontological, and metatheoretical pieces and resolve the BM^3 problem that resides at the epicenter of modern psychology's conceptual difficulties.

References

Bhaskar, R., & Hartwig, M. (2013). *Enlightened common sense: The philosophy of critical realism*. Routledge.

Bhaskar, R., Esbjörn-Hargens, S., Hedlund, N., & Hartwig, M. (Eds.). (2016). *Metatheory for the twenty-first century: Critical realism and integral theory in dialogue*. Routledge.

Björkman, T. (2019). *The world we create: From God to market*. Perspectiva Press.

Cahoone, L. (2013). *The orders of nature*. Suny Press.

Kauffman, S. A. (2019). *A world beyond physics: The emergence and evolution of life*. Oxford University Press.

Kleineberg, M. (2016). From linearity to co-evolution: On the architecture of Nicolai Hartmann's levels of reality. In K. Peterson & R. Poli (Eds.), *New research on the philosophy of Nicolai Hartmann*. De Gruyter.

Lerner, E. J. (1992). *The big bang never happened*. Vintage.

Lyotard, J. F. (1979). *The postmodern condition: A report on knowledge*. University of Minnesota Press.

Marshall, P. (2016). Toward a complex integral realism. In R. Bhaskar, S. Esbjörn-Hargens, N. Hedlund, & M. Hartwig (Eds.), *Metatheory for the twenty-first century: Critical realism and integral theory in dialogue* (pp. 140–182). Routledge.

Maturana, H. R., & Varela, F. J. (1987). *The tree of knowledge: The biological roots of human understanding*. New Science Library/Shambhala Publications.

Powell, C. (2019, May 18). The universe may be a billion years younger than we thought. Scientists are scrambling to figure out why. *Mach NBC News*. https://www.nbcnews.com/mach/science/universe-may-be-billion-years-younger-we-thought-scientists-are-ncna1005541

Schumacher, E. F. (1977). *A guide for the perplexed*. Random House.

PART IV

Defining Behavior and Its Deep Connection to Modern Science

CHAPTER 9

Behavior: The Central Concept in Natural Science

Solving the problem of psychology requires a meta-perspective that affords the capacity to coherently frame the interrelationships between three concepts: science, behavior, and mental processes. The previous chapter showed how the ToK System provides a new map of the ontic reality and our scientific knowledge of it. In so doing, the ToK System gives us a descriptive metaphysical system that potentially affords a coherent naturalistic ontology, one that can more effectively situate scientific knowledge in relationship to subjective and social knowledge. As such, it sets the stage for addressing key elements of the science versus social knowledge side of the Enlightenment Gap. However, the primary focus in the present work is on developing the proper relationship between matter and mind. Specifically, in these next two sections of the work, we turn to the BM^3 problem that lies at the heart of the problem of psychology and provide much greater clarity regarding what is meant by behavior and mental processes and their interrelations.

This chapter and the next deconstruct and then reconstruct the concept of behavior from the vantage point of the Unified Theory. What emerges is a previously overlooked relationship between the concept of behavior and science writ large. More specifically, the view afforded by UTOK is that behavior is a central concept in science, such that it can be seen to frame the metaphysics (i.e., the concepts and categories), epistemology (i.e., the ways science obtains knowledge), and the ontology (i.e.,

G. Henriques, *A New Synthesis for Solving the Problem of Psychology*, Palgrave Studies in the Theory and History of Psychology,
https://doi.org/10.1007/978-3-031-18493-2_9

how science frames reality) of modern empirical natural science. To develop this argument, we review the history of behavior in psychology and explore how and why the concept ultimately generalizes across the scientific disciplines from physics to anthropology. We then deconstruct the various aspects of behavior, focusing on how behavior frames scientific epistemology in terms of third-person observation, measurement, quantification, and change, and then proceed to show how it frames scientific ontology in terms of mapping behavioral patterns in nature at various levels and dimensions of analysis.

Although the arguments that follow are novel, it is nevertheless the case that the claim that behavior is a central concept in science is, in some ways, rather conventional. Consider, for example, that the Google Dictionary based on Oxford Languages defines science as "the intellectual and practical activity encompassing the systematic study of the structure and behavior of the physical and natural world through observation and experiment." If we agree that behavior can be defined as changes in entities and their relations across time (more on this later), then by implication the concept of behavior can be considered to include the structure of those entities. However, the reverse is not true, as structure is static and does not include the dynamic aspects of behavioral patterns over time. Following a similar logic, we can also fold physical into the broader concept of natural. Natural includes physical, but physical does not include all that is natural. This analysis gives us a definition of science such that the core of it is about the systematic observation, description, and explanation of behavioral patterns in the natural world.

While this is a fairly conventional way of framing science, it is also the case that the relationship between science and behavior is largely unexamined (although there are exceptions; e.g., see Merleau-Ponty's (1963) *The Structure of Behavior*). The need for such an analysis is obvious when we consider the fact that the concept of behavior originates not from physics or the philosophy of science, but from the science of psychology. The argument that systematic analysis is needed is greatly strengthened when we consider that, although behavior is a central concept in psychology, UTOK shows why psychologists have gotten this concept wrong. The UTOK's position is that "behaviorism" should not be associated with psychology, but rather represents a general scientific approach to nature writ large. That is, natural science operates on a kind of universal behaviorism. Psychological science, in contrast, is about a particular kind of behavior, properly described as mental or minded.

Chapter 3 detailed psychology's confusions regarding the relationship between behavior and mental processes. There we saw how behavior came to be a construct that emerged to frame the mental in a scientifically accessible way. This chapter reverses the relationship between science and behavior. Rather than thinking about behavior as being the scientifically accessible aspect of the mental, it examines how modern science can be interpreted as the systematic study of behavior patterns across all the various aggregates, levels, and dimensions of existence in nature. We can then reinterpret mainstream academic psychology's frame via this shift in perspective. A scientific approach is one that inevitably employs a third-person behavioral epistemology. Thus, it is the rules, language game, and epistemology of science writ large that results in psychological science having to approach the mental in behavioral terms.

The argument regarding the relationship between behavior and science generally can be summarized in three key claims. First, behavior is an overlooked concept in the philosophy of science. This is because the concept emerged in psychology, and it has been associated with leaders of behavioral psychology like Watson and Skinner or the logical behavioral approaches in the philosophy of mind, such as that adopted by Gilbert Ryle (1949). This focus on "mind as behavior" in behavioral traditions has blinded the academy to consider why science in general represents a behavioral approach to understanding nature writ large. The second key claim is that modern science is grounded in a behavioral epistemology, such that the methods and processes of justification in science are situated in an exterior empirical position of observation grounded in quantifiable measurements of structures and change processes. The third claim is that behavior is central to the descriptive metaphysics and ontology of modern natural science. That is, scientific statements about what is ontically real can be framed in terms of entity–field change that takes place at different levels, aggregate scales, and dimensions in nature. Embedded in this last claim is the idea that entities, fields, and change are foundational metaphysical concepts in science.

By seeing how the philosophy of modern science is framed by the conceptual grammar of behavior, several important puzzle pieces fall into place. First, greater clarity is achieved regarding what defines science as a unique kind of justification system. That is, natural science can be framed as a justification system that is based on an exterior empirical epistemology that maps the ontic reality, and that reality as framed by modern science can be described as observable behavioral patterns that are stratified across different levels and dimensions in nature. Second, when we consider the ToK System to be a map of the various behaviors of the universe, we achieve a substantially clearer relationship between: (a) the ontic reality and (b) the scientific ontological theories that function to map and model the ontic reality and (c) the epistemological justifications and methods and processes that constitute the way scientists justify their conclusions about the validity of those maps and models. In such a formulation, behavior can be conceived of as a nexus point that bridges the patterns of complexity and change that exist in the ontic world with our transient scientific theories and methods that map that world. This is represented visually in Fig. 9.1.

These insights afford us a fundamental shift in how we conceptualize the essence of a scientific worldview. The previous section demonstrated that reductive physicalist or materialist views of scientific knowledge are misguided. Instead, a much more accurate conception is an emergent naturalistic worldview that identifies the separate levels and dimensions in nature. This section adds to the UTOK scientific worldview by framing nature in behavioral rather than material or physicalist terms. The behavioral view is much more dynamic, fluid, and changing. As such, it shares much with Whitehead's process philosophy. And it is much more consistent with what we know. It aligns much better with the Life, Mind, and Culture planes of existence than either materialism or physicalism. And

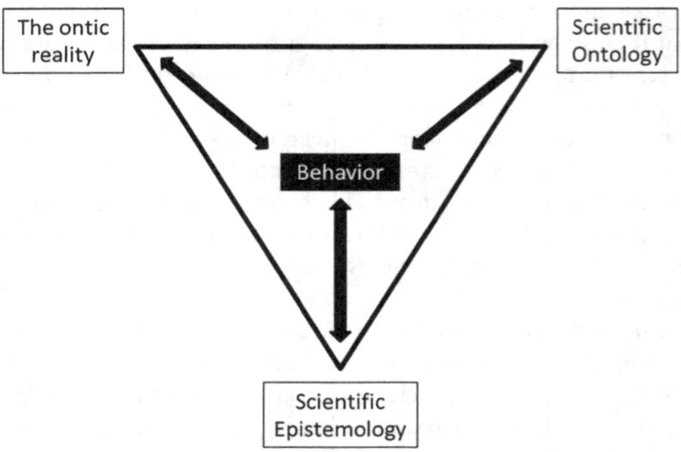

Fig. 9.1 Behavior frames scientific knowledge and epistemology in mapping reality

behavior also aligns better with the modern understanding that the ultimate "ground floor" of the natural world consists of fluctuations in quantum fields, rather than clock-like material mechanisms. With this summary of where we are headed, we can now take a step back and trace how the concept of behavior emerged in psychology, and how it spread into common usage across the scientific landscape.

The Concept of Behavior in Psychology and Its Confusions

Without a doubt, behavior is, historically, one of psychology's most central concepts. And yet, despite its centrality, there is little consensus regarding exactly what the term behavior means. Even the famed radical behaviorist Skinner (1988) said that "there is no essence of behavior." However, we can make good progress on understanding what behavior means if we apply the ToK System and use that to link behavior to the concept of science writ large. To get an effective grip on the concept of behavior, it is helpful to start by seeing how the term can be used in a general way. For example, sometimes behaviors are conceived of as movements that generate measurable effects. Other times the term is used in a more specific sense, such as the "functional responses of an animal to stimuli." This more specific meaning is present in the behavioral science and philosophical approaches to understanding mind as behavior. The fact that there is a general and specific meaning of behavior results in much equivocation and confusion.

To see why, consider the contradictory ways in which the term behavior can be used by psychologists. On some occasions, it is used *to connect what psychologists study to what other "real" natural scientists study*, as in "unlike those Freudian folks, we are a real, objective science because we study and measure observable behavior." However, sometimes it is used in precisely the opposite manner. That is, it is used to *differentiate what psychologists study from what other "real" scientists study,* as in "psychology is *the* science of behavior," which is presumably different from what physicists and biologists study. The second chapter included a quotation from Ed Wasserman justifying behavioral psychology. A careful reading of it reveals that he used the concept of behavior to simultaneously connect psychology to the more basic sciences like biology *and* to differentiate its approach and subject matter from those sciences. Yet he does not highlight the crucial shift in meaning. As any logician will tell you, if the same term is used to justify precisely opposite things without any clarification of meaning, then there is a problem with it.

The confusion between the general and specific meanings of behavior relates to the confusion in psychology between ontology and epistemology. The general meaning of observable change links psychology to the natural sciences via the epistemological meaning of the term. That is, it refers to the fact that knowledge in science is generally grounded in third-person empirical observations, and it captures the fact that scientists see the world via the epistemological lens of observable and measurable change. In contrast, the specific meaning of functional responses of animals to stimuli in the environment represents a differentiation from the more basic sciences. Here the referent is found in the ontological meaning of the term behavior. It is capturing the point that different sciences study fundamentally different kinds of behavioral patterns, and behavioral psychologists study the minded behaviors (i.e., observable, functional responses that arise via sensory motor loops) exhibited by animals, in contrast to biologists who study the behavior of cells and physicists who study the behavior of inanimate objects like atoms. The reason this important difference in meaning has been hidden is that early behavioral psychologists were confused about the epistemological and ontological frames of reference in sorting out how to scientifically study mind as a kind of behavior.

The term behavior emerged in the scientific lexicon largely through the writings of John Watson, the father of behavioral psychology. Although virtually no modern psychologist adopts Watson's classical behaviorism and its underlying reductionistic neuro-reflexology, his thinking was nevertheless enormously influential in its impact on psychology's empirical epistemology. Understanding how Watson justified behaviorism and how it impacted psychology is crucial for our task of understanding its meaning *within* the field of psychology and developing a proper conception of the relationship between behavior and mental processes.

Yet it is equally essential to understand how and why the concept of behavior seeped into usage across all the naturalistic scientific disciplines. That is, not only do we have the "behavioral sciences" and the various "sciences of human behavior" like anthropology, but the term moves *down* the scales of layered complexification into the biological and physical sciences. It is commonplace in the scientific literature to speak of cell or plant behavior; indeed, behavioral biology is a clearly identified subdiscipline that has its own professional journal, called *Behavioral Biology*. Even more striking, physics is frequently defined as the science of "the behavior of matter and energy" or the science that "explains how the universe behaves at every scale" (Henriques et al., 2019). It is perfectly natural to describe particle physicists as studying the behavior of very small entities (e.g., subatomic particles like electrons) and cosmologists as studying the behavior of very large entities (e.g., galaxies). Likewise, chemists, geologists, and electrical engineers readily frame their work in terms of the behavior of molecules, plate tectonics, and circuits, respectively.

This deeply curious phenomenon regarding the expansion of the boundaries of the word behavior has been largely overlooked. It can be stated in the form of a question: *How did a concept that originated to transform psychology into a "hard" science morph into a term that is regularly used across all the modern sciences, from physics to sociology?* By narrating this story, we will be able to clarify the role behavior plays in the Unified Theory's language system. This, in turn, will allow us to see why behavior is such a central concept in natural science.

Psychology as the Behaviorists Viewed It

The term behavior rarely appeared in science prior to the early 1900s, and it ultimately became part of the lexicon largely through Watson's efforts. He advanced behaviorism as a scientific approach to psychology that was directly opposed to the schools of thought that emphasized the structure and function of consciousness. To fully understand his arguments, we can turn to his classic 1913 paper *Psychology as the Behaviorist Views It.* Diving into Watson's manifesto allows us to peer into the justification systems that were operating in psychology just after the turn of the twentieth century. Doing so enables us to see that, consistent with UTOK's framing of the Enlightenment Gap, psychologists struggled enormously to effectively map the relationship between matter and mind via the methods of science. We also see the confusions about the term behavior, and the ambiguity regarding its epistemological meaning and ontological reference points.

The opening paragraph makes plain Watson's (1913) thinking regarding aligning psychology with natural science (p. 158):

> Psychology as the behaviorist views it is a purely objective experimental branch of natural science. Its theoretical goal is the prediction and control of behavior. Introspection forms no essential part of its methods, nor is the scientific value of its data dependent upon the readiness with which they

> lend themselves to interpretation in terms of consciousness. The behaviorist, in his efforts to get a unitary scheme of animal response, recognizes no dividing line between man and brute. The behavior of man, with all of its refinement and complexity, forms only a part of the behaviorist's total scheme of investigation.

Experimenting, predicting, and controlling behavior reflect epistemological concerns. In contrast, the claim that there is no dividing line between man and brute is an ontological assertion about the nature of behavioral kinds. As we will see, Watson clearly was operating from a reductive physicalist ontology, and by natural science he meant both an exterior empirical epistemology and the reductive materialist ontology of his day. However, Watson never clearly specified his philosophical framework in a way that allowed the epistemological and ontological commitments to be clearly differentiated and separately analyzed.

The manifesto was also a clear demonstration of the paradigm wars that characterized the early state of psychology. It reads as a direct attack against the two dominant paradigms of the day, structuralism and functionalism, with particularly sharp disagreements with the former. To understand the context, it is important to know that the primary frame for psychology at the time stemmed from Wilhelm Wundt, who is generally considered the father of the field. Although the concept of psychology has a long and complicated history, the science of psychology's official birthdate is 1879. This is when Wundt opened his research lab and shortly thereafter produced a professional journal. Wundt's approach was built from studies in psychophysics, which is the study of the relationship between changes in external stimulus levels and internal sensations. It remains an active domain of inquiry and has given us some of the most robust empirical findings that the field has achieved, such as Weber's Law of Just Noticeable Differences.

Wundt was more interested in inner perception than sensation per se, and he developed the methods of introspection to study perception and other forms of interior conscious states. This involved systematically training human participants on self-reporting methods, such that their observations about their interior mental states would serve as objective data. Just before the turn of the century, Wundt's student, Edward Titchener, had moved from Germany and established his version of Wundt's approach in the United States. He called it structuralism, and it was devoted to analyzing the elemental parts that made up human subjective experience and how they would form the full structure of human experiential consciousness. At one point, Titchener and his school had identified over 44,000 different mental states that purportedly served as the elemental structures of human consciousness (Schultz & Schultz, 2011).

In addition to structuralism, William James' functionalism was also influential at the time of Watson's writing. Unlike structuralism's focus on the elements that make up subjective consciousness in humans, functionalism emphasized evolution, purposeful action, and the processes of adaptation and adjustment in both humans and animals. James offered powerful analyses of how, in contrast to inanimate entities, the behavior of creatures like frogs and humans was apparently oriented toward end states, such as eating or safety. This meant that animals and people would demonstrate situational awareness and functionally adjust their actions based on feedback to achieve specific outcomes. James' frame for this kind of behavior was "mental life," which is how he characterized psychology's subject matter. James thought such functional or adaptive behavior patterns were clearly tied to conscious experience in some ways.

We can look back at behaviorism, structuralism, and functionalism via the three domains of mental processes we have identified. Watson's behaviorism corresponds to the overt functional responses of animals. That is, it is partly defined by Mind^1. Instead of a neurocognitive model, it simply posited a "physicalist" neurological model of the brain and nervous system. Placing Wundt's structuralism and its methods of introspection in the language of UTOK, we can say that such efforts represented attempts to explore the domain of Mind^2 in humans, via Mind^3. That is, Mind^3 is the intersubjective highway that allows humans to introspect on inner experience and narrate to others what they see in Mind^2. In his opposition, Watson was arguing that the only science that was possible was a science of Mind^1, framed as observable activities of organisms arising from complexes of neuro-mechanical reflexes. James' functionalist and pragmatic approach to "mental life" included each of the domains of mental processes, but he did not have UTOK's vocabulary or metatheoretical formulation for differentiating and integrating the domains.

Watson's ontology was a reductive physicalism that collapsed reality into a materialistic flatland that consisted only of physical causes. The vision logic of Watson's worldview could be represented as a single cone of Matter that stretches from particles to human behavior. In summarizing his analysis, he made his reductionistic position explicit, writing in a footnote (p. 170): "The elimination of states of consciousness as proper objects of investigation in themselves will remove the barrier from psychology which exists between it and the other sciences. The findings of psychology become the functional correlates of structure and lend themselves to explanation in physico-chemical terms." The flatland view is also apparent in that he saw no clear dividing line between humans and animals and, presumably, between animals and the behaviors of organisms and, ultimately, even of molecular objects.

For Watson, consciousness was a problem in science only in so far as it pertained to epistemology and the philosophy of how we can justify our observations and knowledge of events. This summary statement from Watson (1913) captures both his reductionism and his frame that consciousness plays a role in the philosophy of science rather than science (p. 176):

> The position is taken here that the behavior of man and the behavior of animals must be considered on the same plane; as being equally essential to a general understanding of behavior. It can dispense with consciousness in a psychological sense. The separate observation of 'states of consciousness', is, on this assumption, no more a part of the task of the psychologist than of the physicist. We might call this the return to a non-reflective and naive use of consciousness. In this sense consciousness may be said to be the instrument or tool with which all scientists work. Whether or not the tool is properly used at present by scientists is a problem for philosophy and not for psychology.

Here we see that Watson is using the concept of behavior in increasingly generalizable ways, moving from behavior of humans to animals and ultimately physics. Watson is also noting that consciousness is considered only to the extent of being the tool by which scientists make observations.

Watson (1913) was, of course, not completely blind to the fact that there are different kinds of behavioral patterns in nature. A final passage from Watson shows how he did differentiate different kinds of behaviors, although it also simultaneously shows how he was a reductive physicalist (p. 176).

> From the viewpoint here suggested the facts on the behavior of amoeba have value in and for themselves without reference to the behavior of man. In biology studies on race differentiation and inheritance in amoeba form a separate division of study which must be evaluated in terms of the laws found there. The conclusions so reached may not hold in any other form.

> Regardless of the possible lack of generality, such studies must be made if evolution as a whole is ever to be regulated and controlled. Similarly the laws of behavior in amoeba the range of responses, and the determination of effective stimuli, of habit formation, persistency of habits, interference and reinforcement of habits, must be determined and evaluated in and for themselves, regardless of their generality, or of their bearing upon such laws in other forms, if the phenomena of behavior are ever to be brought within the sphere of scientific control.

This passage is useful on several different accounts. First, it shows that Watson recognizes that there are reliably different patterns of behavioral activity for different species. We will come back to this as the ontological aspect of behavior (i.e., there are different kinds of behavioral patterns in nature). However, Watson's use of the "amoeba" as an example is telling, and it shows that his conception of behavioral complexity is fundamentally different than the separable planes of existence mapped by the ToK System. Consider that amoebas are single-celled organisms that lack a nervous system. As such, in the ToK taxonomy, they do not behave at the Mind-Animal plane of existence, but rather the Life-Organism plane. When this summary paragraph is considered alongside his claim that there is "no dividing line between man and brute" in his opening paragraph, we see Watson is operating from a single continuous and reductive form of physicalism that errantly collapses across the different dimensions of behavioral complexity.

Reviewing Watson's version of behaviorism allows us to clarify several key points. First, it highlights how difficult it was to define a science of psychology. This adds confirming evidence that there was no descriptive metaphysical system for sense-making that allowed for a coherent science of psychology to emerge based on a clear ontology of the mental (i.e., the Enlightenment Gap blocks the science of psychology from formulating its subject matter clearly). In contrast, consider how readily we can use the combination of the ToK System and the division of mental domains afforded by the Unified Theory to map the territory. Through this lens, we can see that Watson was primarily concerned with overt mental behavior mediated by the nervous system, which directly corresponds to Mind^1. We can also identify Wundt and Titchener as being concerned with human Mind^2 analyzed via Mind^3. James' functionalist pragmatic view of "mental life" overlaps with all three domains of mental process.

Of course, there was no clear, shared map of mental behavioral processes available during Watson's day, and the paradigm wars were the natural consequence of this lacuna. For Watson, consciousness and behavior are framed as an "either or but not both" choice for psychologists. His justification for why this is the case was ultimately grounded in the materialistic flatland view of both reality and science. Behavior fit epistemologically with biology and physics because it can be viewed from a third-person exterior perspective. For Watson, it fit ontologically because neuromuscular mechanical reflex-type mechanisms could be linked to matter-in-motion mechanics. The negative side of this argument is also apparent. That is, we can see that subjective phenomenology was an unworkable concept for Watson's science, both ontologically and epistemologically, and thus it had to be rejected.

It is important to note that Watson did acknowledge the existence of human consciousness, calling it a tool or instrument by which scientists make observations. However, he argued that the only proper place for it was philosophy, not psychology. This admission is useful on two accounts. First, it highlights how observation connects both to consciousness and scientific epistemology. Second, it reveals a significant performative contradiction in Watson's analysis. In acknowledging that consciousness is necessary for the philosophy of science because it pertains to the processes by which scientists make observations and generate intersubjective agreement that forms the empirical basis of scientific knowledge, we can flip this observation around and see that it is obviously a general point about human behavior. That is, it is simply an empirical fact that humans make subjective observations (i.e., perceptions and images appear in Mind^2) and proceed to generate narrative descriptions (i.e., Mind^3) about those observations to coordinate their actions. This is, after all, a straightforward description of how human persons build cultural systems of justification, from science to religions to laws. Watson's acknowledgment that subjective phenomenology is relevant for philosophy should have awakened him to the fact that intersubjective conscious evaluation is ubiquitous in human life and cannot be explained away by blind neuro-mechanical reflexes. This is a clear error in Watson's reductive version of behaviorism, and it is one of the many reasons why it becomes rejected by almost everyone, including the later versions of behaviorism promoted by scholars like B. F. Skinner.

Although we will not review Skinner's radical behaviorism in depth here, it is helpful to clarify that Skinner rejected Watson's formulation, as the two are often lumped together in nondiscriminatory ways. Skinner would regularly state that he was not a "stimulus-response" psychologist, which was his way of distancing himself from Watson's neuro-mechanical reductionism. In contrast, Skinner focused on behavior from an epistemological vantage point, and he paired these systematic empirical analyses with instrumental goals. That is, he emphasized the process of observing, predicting, and controlling behavior. He was far less concerned with making ontological claims about what behavior was or generating deep theoretical explanations for why animals behaved the way they did. This made him suspicious of the "intervening variables" that were posted by cognitive science. He argued that the role of the behavioral psychologist was to examine the environmental contingencies that shaped the complexity building process of animal behavioral repertoires and engineer systems that would afford selection of behavioral patterns in desired directions.

It is worth stating that Skinner did have a vague ontology of the world. He called it "three-tiered selection." He saw the behavior of organisms as being shaped by natural selection, the behavior of animals being shaped by operant selection, and the behavior of humans shaped by verbal selection. It is helpful to note that this three-tiered ontology closely aligns with the ToK System's Life, Mind, and Culture planes of existence. This makes good sense because both Skinner and the ToK System scientifically map the world in terms of unfolding behavioral processes. The difference between UTOK and Skinner is that he yoked science far too strongly to epistemology and saw science more in terms of a technological prediction and control system, rather than a justification system that functions to develop good descriptions of the world and explanations for why things unfold the way they do.

Despite its problems, Watson's version of behaviorism had a major impact on the science of psychology. The reason is that it set the stage for approaching psychology in a way that was fundamentally aligned with modern scientific epistemology. That is, to play by the rules of natural science, psychology had to be anchored to exteriorly observable behavioral responses. This epistemological and methodological connection to behaviorism remained even as Watson's ontologically reductive physicalism was overturned and was replaced by neurocognitive and phenomenological processes that play an intervening role. In his review of the history of psychology, Devonis (2014) summarized it as follows (p. 23):

> Behaviourism [sic] served an important historical function for Psychology in moving it methodologically into the realm of the natural sciences—really in

> a sense by insisting that psychologists actually behave like scientists. The theoretical price paid was an abandonment of concern with many genuinely profound philosophical and theoretical questions about Psychology's status and nature that had previously preoccupied the discipline, and commitment to a reductionism which many felt deeply unsatisfying and which seemed to discard the baby with the bath water.

This passage highlights how behaviorism's key contributions to Psychology were found in its epistemological and methodological critiques of prior mentalistic approaches, especially those that attempted to primarily rely on introspection or mysterious or untestable models of the subjective mind. That behaviorism served this function makes good sense given that an exterior epistemology is central to the language game of science in general. However, behaviorism failed at the level of "theory" because simply relying on an exterior epistemology does not resolve the problem of developing a workable ontology for mental behavioral processes.

The Concentric Circles of Behavior

Reviewing Watson's analysis enables us to understand the history of behavior and how it came to be prominent in psychology. Our task now shifts to disentangling the many meanings of the term. To begin this process, we can turn to basic definitions, of which the Merriam-Webster Dictionary (n.d.) gives four primary meanings. The first is the most common usage in everyday language, and it refers to the conduct or comportment of people, as in "the child's behavior was unruly and inappropriate." The second definition refers to the observable actions of organisms, which has its roots in Watson and is how most biologists and psychologists use the term. The third meaning refers to patterns and gives the example: "They tested the behavior of various metals under heat and pressure."

In his analysis of the concept of behavior in the academic literature, Lazzeri (2014, p. 66) identified four common meanings: "(i) behavior as the occurrence of an organism's action or reaction; (ii) behavior as a class or pattern; (iii) behavior as group behavior; and (iv) behavior as a change or movement of an object." Ultimately all these meanings are important, and we will be using the ToK System to untangle their interrelations. Doing so will enable us to see how behavior provides the conceptual linkages between an exterior view of reality and how the grammar of natural science maps that reality, via logic and exterior empirical analysis.

Missing from Lazzeri's scholarly analysis of the term is the original meaning of behavior, which pertains to comportment. It stemmed from the word *to have* and connected to an individual having appropriate manners or knowledge of the proper ways of acting. It is useful to note that behavior originally refers to an external assessment of people judged to be acting in good or bad ways, and this remains a common usage in everyday language. It was then adopted by the behaviorists to refer to observable responses in people, animals, and organisms. Because it was linked with observable change, it became linked with motion and action as these terms were used in physics, and over time became largely synonymous with these concepts in the material sciences. We can represent these various meanings of the term behavior as a set of nested relations. This is depicted in Fig. 9.2.

Moving from the outside in, first there is the general concept of behavior. We can consider this the metaphysical conception of behavior in that it refers to the key elements that go into the concept. Specifically, it involves entities or patterns changing on a field or grid. Such entities can be mathematically represented, either as point-like objects or as patterns of change, such as in groups or a wave. Next, behavior can be directly applied to objects in the real world, the most basic conception of this being an object in motion. In making this move, we have made a disciplinary jump from what some call the "formal sciences" (i.e., logic and mathematics) into the "natural empirical sciences" (i.e., physics).

The diagram moves from the behavior of inanimate material objects into subsets of different kinds of entities that exhibit unique kinds of behavioral patterns. In making this move, we are again shifting disciplines. That is, we are moving from the general physical sciences into what some call the "special sciences" of biology, psychology, and the social sciences. The concentric circles map the divisions between these broad domains of science. Organisms are material objects, but they also behave very differently from inanimate objects. They have unique, emergent behavioral properties that involve biological information processing and communication. In short, they exhibit *living* behaviors. Living behaviors exist above and beyond physical and chemical behaviors, such that we can consider them the higher-order patterns that exist after we factor out movements at the Matter plane.

Likewise, animals are organisms that exhibit unique behavioral patterns and processes, such as operant learning and decision making, which emerge from neuro-information processing and behavioral selection mapped by the basic psychological sciences. They exhibit minded

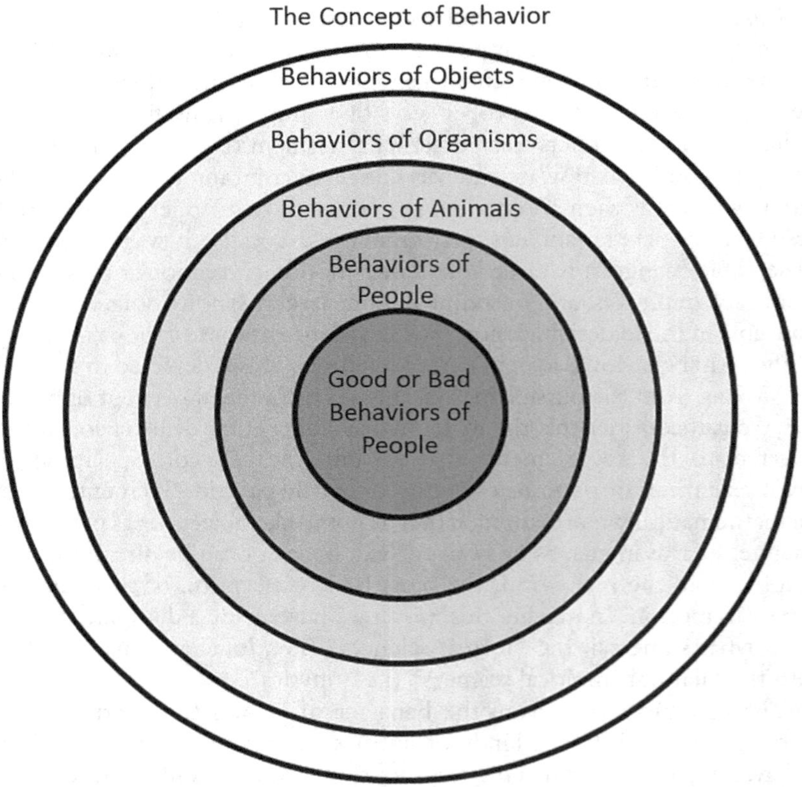

Fig. 9.2 The concentric circles of behavior

behaviors, which exist above and beyond living behaviors. And people are animals that exhibit novel elements of behavior, such as processes of justification and self-conscious narration, which resulted in the emergence of the Culture plane of existence, modeled by human psychology and the social sciences. Finally, a subset of the behavior of persons can be evaluated as good or bad. We can note that this final shift also involves a disciplinary shift. It moves us from the domain of descriptive or basic science into the domains of ethics and more humanistic considerations that explicitly consider value-based assertions regarding moral claims of what ought to be, which is why it was not included in Lazzeri's analysis.

Of course, the delineation of objects, organisms, animals, and people directly aligns with the four dimensions of behavioral complexification mapped by the ToK System. On the outer ring, we have added a link to the formal sciences via the idea that behavior can be represented mathematically. Moreover, after we move through the four domains of science, the center of the diagram enters the moral-ethical domain of the humanities. Framed this way, we can see that the concentric circles of behavior move from mathematics to physics to biology to psychology to the social sciences, and, finally, to the humanities. The point here is that are many layers associated with behavior, and they line up with key disciplinary endeavors in the academy.

Corresponding UTOK with Integral Theory on the Concept of Behavior

As reviewed in Chap. 3, a central feature of Ken Wilber's (1995) Integral Theory is the dividing of perspectives into epistemological quadrants on the interior versus exterior and individual versus collective axes. Directly aligned with the present argument, Wilber's exterior quadrants at both the individual-object and collective-systems level are framed in terms of behavior, namely the behavior of individuals and the behavior of systems. In addition, Wilber argues that the exterior behavioral view represents the epistemological position that is taken by modern science in its attempt to objectively describe and explain how the world operates and changes over time.

Wilber's analysis helps to elucidate some of the fundamental epistemological commitments of scientific justification systems and how they are distinct from humanistic justification systems. We can see this when we consider that he argued for a deep alignment between his four quadrants and the first-, second-, and third-person points of view. The first-person perspective is the "I" point of view from the upper left individual interior quadrant. The second-person perspective is the "We" point of view, which is framed as the collective interior position (lower left). Wilber argued that the interior perspective is what defines the humanities as humanistic in nature and is what separates them from scientific languages. He also aligned the first-person "I" perspective with the core concept of beauty, as well as the philosophy of aesthetics and the arts. He further corresponded the collective, intersubjective "We" quadrant with morality and the core concept of goodness.

In contrast, the third-person "It" perspective is aligned with science and the concept of objective or relatively knower-independent truth claims. Wilber argued, via his quadrants, that "It" should be divided into the singular (i.e., behaviors of specific entities) and plural or collective (i.e., behavioral systems). In *Integral Spirituality* (2007), Wilber described his position as follows (p. 33):

> The quadrants…suggest that any occasion possesses an inside and an outside, as well as an individual and collective, dimension. Taken together, this gives us the inside and outside of the individual and collective. These are often represented as I, you/we, it, and its (a variation on 1st-, 2nd-, and 3rd-person pronouns; another variation is in the Good, the True, and the Beautiful; or art, morals, and science, and so on—namely the objective truth of the exterior view of science, or it/its; the subjective truth of aesthetics, or I; and the [intersubjective] collective truth of ethics, or thou/we.

Figure 9.3 gives a visual representation of this analysis. It depicts the I, the We, and the It/Its in the four quadrants, and it then aligns them to core

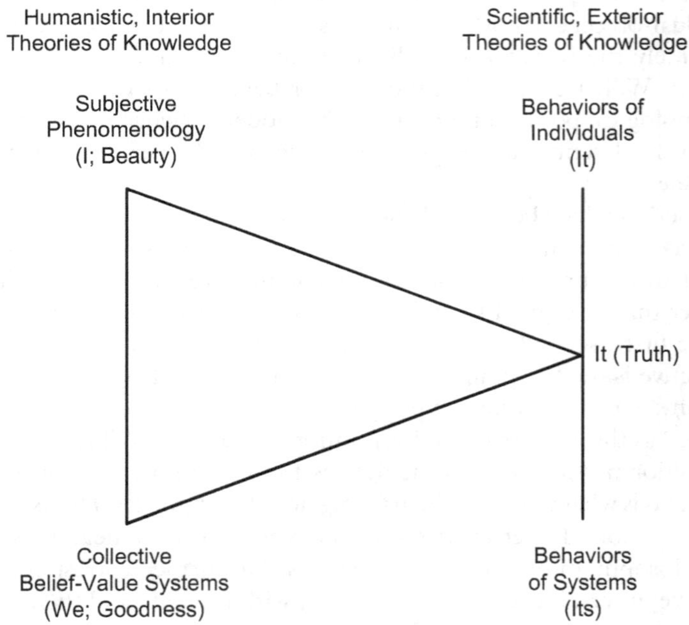

Fig. 9.3 Aligning Wilber's quadrants with scientific and humanistic perspectives

concerns of subjective beauty (interior individual), intersubjective ethical-morality (interior collective), and objective behavioral truth ascertained by scientific analysis at the individual and systems levels of analysis.

Although I developed the UTOK independently of Wilber, both our perspectives align regarding this epistemological distinction between the sciences and humanities. This can be seen when we look at how I described the relationship between scientific and humanistic forms of knowing in my article "Toward a Useful Mass Movement" (Henriques, 2005, p. 125):

> The two components, the scientific and the humanistic, reflect two different valuations of the [subjective experiences of the] knower. In attempting to construct general laws that objectively describe complexity and change, the scientist works to de-value the [position and] influence of the specific knower in the knower-known interaction. In other words, the task of the basic scientist is to describe 'reality' in as knower-independent terms as possible. Scientific methodology can be thought of as the tools by which this knower-independent knowledge is acquired. But...science is not the only way of knowing. And in the ToK System, science is seen as one particular type of justification system, which has particular strengths (accuracy) and limitations (amorality). Other justification systems (e.g., legal, religious, or political) are explicitly prescriptive, moral systems, [or are grounded in human subjectivity].

Like Wilber, UTOK thus characterizes modern science as a justification system concerned with rendering descriptions and explanations of entities and events as objectively as possible partly by "factoring out" both the unique subjective perspective and the moral value judgments of the social group.

Central to the UTOK is the idea that subjective phenomenology and objective behaviorism are two different epistemological perspectives on the world, and that science is anchored to the latter. The justification move that science makes from a first-person qualitative empiricism into a third-person quantitative empiricism is very much about factoring out the subjective, idiosyncratic, and moral values of individuals and groups to yield a more objective description of the unfolding wave of behavior. Figure 9.4 offers a different angle on the ToK System that depicts the argument articulated above. It captures how scientific knowledge involves human knowers attempting to map the knowable reality by factoring out the unique subjective, qualitative perspective.

The left side of the diagram represents the ontically real, whereas the right side represents the onto-epistemological position of the specific

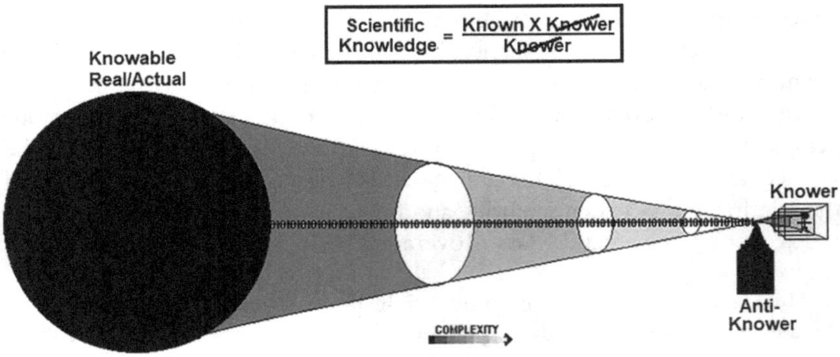

Fig. 9.4 The Tree of Knowledge System depicting how scientific knowledge factors out subjective knowledge

scientific knower attempting to map the real. The goal of science is an objective perspective. In the figure, the "anti-knower" represents the scientific method of removing the subjective biases and the unique positionality of specific knowers via third-person observation. The formulation in the box on the top represents how scientific knowledge attempts to generate maps of the objectively knowable (i.e., the known) via effectively canceling out the subjective and socially constructed elements of knowledge that are dependent on the individual or groups of people that generate them. The figure suggests that science achieves its generalizable knowledge by factoring out idiosyncratic subjective biases and unique positions and attempts to measure and quantitatively map complexity and change. The 0s and 1s that line the center symbolize the ideas of both measurement and quantification. They also symbolize the reducible complexification, or what Volk (2017) calls combogenesis, that science discovers in developing models of emergent behavior from Matter to Life to Mind to Culture.

To make the idea of the anti-knower more concrete, consider that the gold standard of experimental design in medicine is the double-blind randomized controlled clinical trial. In such a trial, the treating physicians and patients do not know which treatments are "real" and which are "placebo" (i.e., physiologically inert). This is what makes it "double-blind." The purpose of the design is to factor out the phenomenological and intersubjective biases and leave behind the causal mechanisms that are operative independent of human knowers. As such, it serves as an exemplar of how the scientific methodology can be thought of as a human "anti-subjective knower."

This analysis deepens the argument that there is a powerful connection with behavior, science, and the exterior epistemological position. Behavior can be seen from the outside and thus it is epistemologically available to a general observer trained in the methods of the field. This linkage is central in modern empirical natural science knowledge, both in what defines it and how it is separated from the humanistic lines of inquiry that embrace both the idiographic subjective "I" position and the moral-ethical intersubjective "We" position. We can now shift our focus to how the map of behavior provided by the concentric circle analysis aligns with Integral Theory's depiction of the irreducible layers of existence.

Behavior and the Ontological Alignment of Integral Theory and UTOK

Integral Theory is informed by a part-whole metaphysics that Wilber (2007) frames as "nested holarchies." The term holon was coined by Koestler (1967) in the book *The Ghost in the Machine* and elaborated upon in *Janus*. A holon is both a whole and a part. Koestler pointed out that the levels arrangement in nature (i.e., subatomic particles ↔ atoms ↔ molecules ↔ macromolecules ↔ organelles ↔ cells ↔ tissues ↔ organs ↔ organisms ↔ communities ↔ societies) was a holarchy. In his nested holarchy model, Wilber argued that the wholes at one dimension of analysis are nested as parts in another. That is, atoms are physical wholes, but they are nested inside cells and function as parts in that context. Cells are biological wholes, but they are nested in and function as parts of animals. Likewise, people are animals, but they are nested in and function as part of cultures.

Wilber contextualizes the part-whole nested holarchies in a variation of perennial philosophy, the Great Chain of Being. In *Marriage of Sense and Soul*, Wilber (1998) argued that each of the major links in the first four levels of the Great Chain (i.e., matter/mineral; cell/animate; animal; people) should be represented as a "Great Nest" of being. Wilber (1998) depicted the Great Nest as a set of overlapping circles that align in interesting ways with the concentric circles of behavior, and thus the ToK System. In his original diagram, he started with Matter/Physics and moved out to the higher levels. However, it is just as reasonable to invert the diagram.

As such, I am reproducing the diagram here to align more directly with the concentric circles.

Figure 9.5 shows how, for Wilber, matter, life, and mind are different layers of existence mapped by the physical, biological, and psychological sciences. In addition, and directly consistent with Hartmann's categorical ontology and Cahoone's different orders of nature, Wilber argued that each higher domain radically transcended the behaviors of the lower layers, such that the higher domains could not be reduced to or explained by the lower ones. In his book *Integral Spirituality* (2007, p. 214), he characterized the relationship as follows:

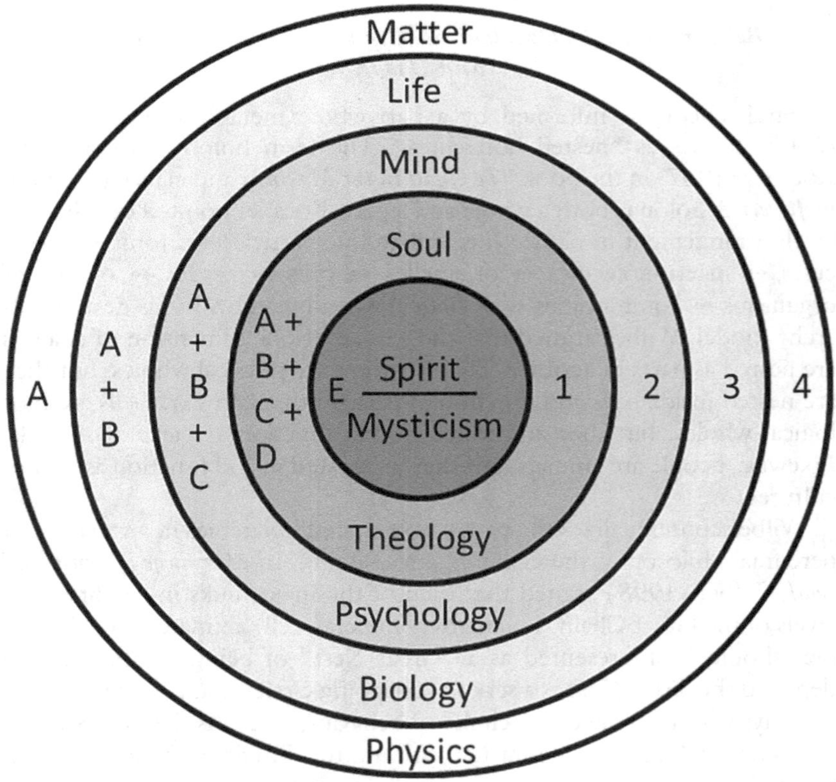

Fig. 9.5 Wilber's Great Nest of Being aligned with the concentric circles

> When body or life (A + B) emerges 'out of' matter ("A"), it contains certain qualities…that cannot be accounted for in strictly the material terms of 'A'. Likewise, when mind (A + B + C) emerges out of life, mind contains emergent characteristics ('C') that cannot be reduced to, nor explained by life and matter alone. When soul (A + B + C + D) emerges, it transcends mind and life and [matter]. Evolution, then is this 'unfolding' of Spirit from matter to body to mind to soul to Spirit itself.

As this quote elucidates, Wilber's Great Nest is very much in line with many elements of the ToK System. Indeed, with UTOK we can add much specificity to what is meant by "B," "C," and "D." That is, the ingredients of emergence that make the dimensions different include a complexity building feedback loop of variation, selection, and retention that gives rise to a novel complex plane of adaptive behavior that is connected via new forms of information processing and communication networks.

Although there is significant overlap, there are also some differences which should not be dismissed. One crucial difference is that Wilber ultimately embraces a spiritualist ontology that is different from the naturalistic scientific emergentist ontology espoused by UTOK. It is partly because of this that he represents the uniqueness of humans in terms of the dimensions of both soul and spirit on the diagram. Although this is an important ontological difference between the two systems, there remains a specifiable correspondence between UTOK's conception of behavioral patterns and Wilber's vision. This is achieved by making a connection between Wilber's distinction between soul and spirit and the behavior of persons in general and the good and bad behaviors that have an ethical-moral judgment.

To see this alignment, it is helpful to start by noting that there are many different meanings of soul and spirit. For Wilber, the soul represents the more everyday life concerns of relationships, personality, and emotions, whereas spirit represents concerns with the transcendent or ultimate nature of existence, which includes the foundations for the moral-ethical dimensions of being. If we set aside Wilber's ultimate spiritualist ontology, this distinction generally aligns with how I use these terms. That is, I interpret the soul as the functional form of a person's life, which is similar to how Aristotle used the term. Also aligned with Wilber, I see it corresponding more to the interior than exterior epistemological position. I consider the spiritual dimension of life to be each person's "transcendent calling" to their ultimate moral-ethical concern. Framed this way, there is a loose correspondence aligning the soul to the behavior of persons in general and the spirit referring to the moral-ethical aspects of human existence that are judged as good or bad. With this link in place, there becomes a general correspondence between Wilber's conception of the Great Nest and the concentric circles of behavior that emerges from the Unified Theory's analysis of the ontological domains of science. However, there is one more difference in ontology that is worth highlighting.

Behavior, Science, and the Ontology of the Upper Right Quadrant

Our review of Watson demonstrated that he made an important error when he conflated the epistemology of science with an ontology that is committed to physicalism. As we have seen, there is no necessary linkage between an epistemology that is framed by observing behavior and an ontology that commits to a reductive physicalism. I raise these issues here because Wilber makes a similar kind of error to Watson.

As we have reviewed, Wilber's overall ontology is that of layered emergence in a way that broadly aligns with the ToK System and larger UTOK metapsychology, minus the ultimate difference regarding a naturalistic versus spiritual ground of existence. However, there is another ontological difference that is worth highlighting. It pertains to the way Wilber frames the Upper Right quadrant. As those familiar with Integral Theory know, the Upper Right quadrant is typically framed by the following steps of complexification: (1) atoms; (2) molecules; (3) prokaryotes; (4) eukaryotes; (5) neuronal organisms; (6) neuronal chord; (7) reptilian brain stem; (8) limbic system; (9) neocortex (triune brain); (10) complex neocortex; (11) human "structure-function" level one; (12) human SF2; and (13) human SF3.

Wilber is making a connection between the exterior view of science that tracks behavior and the ontology of a materialistic worldview that interprets behavior as matter in motion. Thus, as the objects move up the scale, there is a shift in the behavior we observe. The description starts with the whole object (i.e., atoms, molecules) but there is a subtle shift between the fifth and sixth steps, such that there is a shift from behaviors to brain. The reason for this has much to do with the Enlightenment Gap and the confusion between matter, behavior, and mind. Wilber rightfully highlights that in the original Great Chain of Being concepts were based on inner experiences (i.e., first-person empiricism) rather than exterior positions and quantifiable entities (i.e., third-person empiricism). This creates some problems in alignment.

In the Appendix of *Integral Spirituality*, Wilber (2007) addresses this issue by considering the exterior view as being equivalent to a materialistic ontology. He writes (p. 219–220) that "matter is not the 'bottom' of all levels but the 'exterior' of all levels." This is interesting, and when we connect it to Watson's errors, we can make sense out of the claim by recognizing the strong historical connection between behavior and reductive materialism. This connection should be untangled, and UTOK gives plenty of justification for why. It shows clearly how we can both embrace the behavioral referent points and exteriorize Life, Mind, and Culture as behaving as complex adaptive systems. That is, in the descriptive metaphysics of UTOK, Life, Mind, and Culture are defined as planes of existence and are not some sort of force hidden in the interior, but there are directly observable patterns of complex adaptive behavior that are available from the outside-in as well as inside-out. When I am listening to someone tell me about their dreams, I have access to their justifications about those dreams on the Culture-Person plane of existence, as opposed to either the physical air molecules that are carrying the sound or their brain waves. The next chapter details how UTOK maps the various aggregate scales, levels and dimensions of behavior in nature via the Periodic Table of Behavior. It is the observable patterns of behavior in the world rather than a commitment to a materialistic ontology that should map the exterior view. The linkage highlights how important it is to separate the epistemological from the ontological considerations.

Behavior Frames the Metaphysics, Epistemology, and Ontology that Grounds Modern Science

The central thesis of this chapter is that the grammar of natural science is framed by the concept of behavior. Science is about observing, describing, and explaining patterns of behavior at various levels and dimensions of complexity. This final section delineates this argument in greater detail. Building from the previous section, it begins by breaking the concept of behavior down into its formal elements. This sets the stage for seeing how behavior can be quantified and mapped mathematically. The linkage is then made between mathematical representations of behavior and the way the human mind categorizes nature. These insights help deepen the claim that behavior frames the epistemology of science.

The core metaphysical concepts that go into behavior are: (1) entities; (2) fields, and (3) change. In addition, entities, fields, and change translate into how science maps the basic ontological structure of the world. This is apparent in the original matter in motion formulation in physics. It is a linkage that sets the stage for the next chapter's articulation of the Periodic Table of Behavior, which makes explicit the correspondence between the various domains of science and the levels and dimensions of behavioral change in nature. Together the analysis shows that science should not be thought of as grounded either in physicalism or materialism, but rather should be framed in terms of a naturalism and universal behaviorism across different levels and dimensions of analysis.

A Descriptive Metaphysical Frame

What are the essential elements that go into the concept of behavior? To understand how the UTOK frames behavior as the central metaphysical concept in science, we can return to a quotation from the first paper on the system, *The Tree of Knowledge System and the Theoretical Unification of Psychology* (Henriques, 2003, p. 157):

> The most general definition of behavior is change in an object–field relationship, which can be algorithmically represented as (X)(Xσ)t1—(X)(Xσ)t2, where X is the object, Xσ is the field (not X) and t is time. This is important because it highlights that all sciences are sciences of behavior. Physics is the science of the behavior of objects in general. Particle physicists study the behavior of very small objects (e.g., fermions) using quantum theory, and cosmologists study the behavior of very large objects (e.g., galaxies) using the theory of relativity. If it is agreed that physicists study the behavior of objects in general, then it logically follows that other scientists study the behavior of certain objects in particular. Chemists study the behavior of molecular objects; biologists study the behavior of living objects. This analysis highlights that there are obviously significant problems with defining psychology as 'the science of behavior.' It is not the fact that animals behave that makes them unique; it is that they behave so differently from other objects. The key then becomes defining the subset of behaviors that psychologists' study.

This description of the concept of behavior makes two points. The first, which we have already mentioned, is that psychology cannot be framed as the science of behavior, as it is far too general a concept. Second, behavior is defined via three interrelated aspects of objects, fields, and change. In this formulation, I framed difference in terms of time, as that is arguably the most generalizable frame that is used to track object-field change, and it works well at the macro- or classical mechanical level of analysis. However, at the fundamental levels of quantum physics, the standard meaning of time and its directional nature is more nuanced and ambiguous. In quantum field theory, there is the symmetry of relations that is not necessarily tied to the standard conceptions of linear time. Also, although I use the term "object" in the above paragraph, I currently prefer the broader term "entity," which can be framed as an object or pattern that can be extracted as a figure from the ground of its existence, which is the field. Waves are better framed as entities rather than objects, and they are patterns that present themselves and are extracted out of a field and analyzed in terms of change patterns.

The symbolic notion given in the quotation both allows us to specify the key concepts and categories (i.e., entities, fields, and differences or change over time) and enables us to make a bridge between our everyday language and how behavior is represented in mathematics. This is important because a major epistemological move that modern science made was the shift to more measurable and quantifiable aspects of behavioral change. With this background in place, we can now turn our attention to how this simple metaphysical description of behavior allows us to bridge into two of the important conceptual developments that would link mathematics to modern physics: the Cartesian coordinate system and the development of calculus.

Behavior as a Formal, Mathematical, and Quantifiable Concept

As noted in the previous chapter, René Descartes made several important contributions to natural science and the matter-in-motion worldview. One of the most important was analytical geometry. This emerged from his methods for locating entities on a Cartesian grid in three-dimensional space on a linear dimension of time. A Cartesian grid is the familiar depiction of the three dimensions of space, defined by the x, y, and z axes. By placing abstract entities on a Cartesian grid of three dimensions of space and one dimension of time, a quantifiable scheme emerges that can enable the use of much more precise mathematics to map the world. To see how this is the case, it is useful to tell the story of how Descartes came up with his coordinate system.

The legend is that as a precocious but physically frail teenager, Descartes lay sickly in his bed, reflecting on the nature of mathematics and natural philosophy. He watched as a fly buzz around the room, and he started thinking about how he could track it quantitatively. He wanted to develop a precise way to locate the fly in relationship to both other entities and how it moved across the room. Regardless of its veracity, it is a useful tale, because it helps make clear the key insights from analytic geometry. To link this to the symbolic representation of behavior given above, we can consider the fly as the "X" and the environment as "Xɵ." Descartes wanted a way of quantifying the fly's position and how its position changed over time. He saw that he could conceptualize the four walls of his room and the floor and ceiling as *x*, *y*, and *z* axes that would create a grid that could give the precise coordinates of the fly. In so doing, he created a coordinate plane, whereby any point could be identified on a three-dimensional grid that represented the three dimensions of space. He could then see that the fly (X) would move from one location on the grid to another. And this could be tracked via mathematical notations that came to be known as algebraic geometry.

Later, Newton and Leibnitz would build from algebraic geometry and create a new form of mathematics called calculus. Calculus was the mathematization change and the rate of change, and it allowed Newton to develop the mathematical tools for a general theory of gravity. Prior to calculus, mathematics was not well suited to describing and mapping acceleration or other features of behavior when the rate of change was changing. Of course, very little—if anything—in the universe is completely still. Rather, at its core, the universe is constantly oscillating across a multiplicity of frequencies of behavioral patterns. Calculus could be used to map how change changes, and this became deeply relevant in many fields of scientific inquiry.

Behavior, Mathematics, and the Structure of Human Perception

In addition to the obvious claim that behavioral processes can be mathematically represented by advances such as algebraic geometry and calculus, the Unified Theory also provides a way to understand how the human mind perceives reality in behavioral terms. This is one of the advantages of having a theory of the human knower as part of a comprehensive system of understanding. This theory of the human knower will become apparent as we progress into understanding how UTOK describes the fundamental architecture of the human mind. For example, we will see how the human mind operates to perceive objects in fields, which is called perspectival knowing. Humans can also generate propositions about those objects and justify claims via argument and logic. This is called propositional knowing. Humans can also develop procedural knowledge that functions like a recipe or algorithm to produce reliable effects. The procedural knowing becomes crucial in science in terms of experiments.

For the purposes of the current argument, we simply want to note that there is a deep connection between human phenomenology and behavior. To understand why the linkage between how humans observe behavior and human phenomenology is a crucial piece of the puzzle, consider this quote from Gribbin (2019) recounting the advances made by Galileo, who is generally considered to be the father of modern natural scientific justification systems (p. 101):

> By analyzing mathematically subjects which had previously been the prerogative of philosophers, [Galileo's Discourses and Mathematical Demonstrations Concerning] *Two New Sciences* was the first modern scientific textbook, spelling out that the Universe is governed by laws which can be ***understood by the human mind*** [italics added] and is driven by forces whose effects can be calculated using mathematics.

This quote highlights that a full philosophy of science needs to include a way to understand the relationship between behavior, mathematics, and the way the human mind perceives reality subjectively (i.e., via $Mind^2$) and analytically justifies it intersubjectively (i.e., via $Mind^3$) at the individual and collective levels.

As Lakoff and Núñez (2000) explore in their book *Where Mathematics Comes From: How the Embodied Mind Brings Mathematics into Being*, there is indeed a fascinating relationship between mathematics, reality, and human thought, both phenomenologically and intersubjectively, as well as behaviorally. Albert Einstein captured the profound nature of the mystery when he asked: "How can it be that mathematics, being after all a product of human thought independent of experience, is so admirably adapted to the objects of reality?" With its holistic vision, UTOK helps us to weave together the picture of the relationship between mathematics, reality, and the human mind into a more coherent tapestry. It achieves this in part by showing the deep commensurate connection between the human mind and behavior writ large. In addition, via the ToK, it locates the place of the logical relations between measurement, observation, and justification.

What will become apparent in the next three chapters is that the human mind is a neurocognitive behavioral investment system that symbolically tags objects and changes in the form of nouns and verbs. We will see that, via Behavioral Investment Theory, the neurocognitive architecture of the preverbal human mind (i.e., both Mind[1] and Mind[2]) is structured to: (a) attend to relevant objects in the environment; (b) identify changes that are happening in predictable and unpredictable ways; (c) determine what are desirable and undesirable outcomes; and (d) direct subsequent work effort (i.e., behavioral investment) in an attempt to realize desirable outcomes and avoid undesired ones based on probabilistic expectations. In short, evolution built our primate phenomenology to see the world in behavioral terms. In addition, nature added a language acquisition device (Pinker, 1994) that allowed us humans to generate a symbolic-syntactical representation of our phenomenological experience.

The Unified Theory goes further than neurocognitive functionalism and subjective perceptual phenomenology in helping make the linkages of the human mind and the concept of scientific knowledge and its intersubjective propositional networks. Via Justification Systems Theory, we can see how the essence of symbolic language is anchored to the grammar of perceiving behavioral change via the senses. Recall that behavior is fundamentally about change in entity–field relations. Now consider that nouns, verbs, and adjectives are the three most basic kinds of words. Nouns correspond to objects, verbs correspond to changes, and adjectives correspond to differences within kinds of objects or change (i.e., an object is big or small; a change is fast or slow).

For those familiar with Kantian philosophy, this analysis will be reminiscent of the Kantian notion that there are a priori synthetic categories of mind that allow humans to develop phenomenological concepts of the noumena (i.e., the things of themselves, independent of human perception). One way to interpret Kant's categories of mind is to consider the way that human perception is structured to pull out figures that focus our attention relative to a background and track how they change, which we experience as time and cause-effect relations. The UTOK metapsychology advances the Kantian formulation because it provides a cosmic evolutionary picture of how the human mind evolved and the requisite structural and functional relations therein. It also clarifies with more precision the nature of perception and its relation to justification and how justification systems evolved in Culture to generate science, and its epistemological structure that affords us a transcendent realism that is somewhat detached from phenomenological experience. That it does the latter is perhaps most obvious when we consider quantum mechanics. Quantum mechanics is a science based on mathematics and experimentation. It essentially defies our commonsense categories of perception and phenomenology. We were able to construct quantum theory precisely because science is a propositional network grounded in the exterior empirical position, rather than inner experience. This becomes clearer when we consider how behavior frames key aspects of scientific methodology.

Observable Behavior as the Dependent Variable in Science

In his *Principles of Topological Psychology*, the pioneering social psychologist Kurt Lewin (1936/2015) offered what became a classic formulation, namely $B = f(P, E)$, where B is behavior, P is person, and E is environment. The idea represents Lewin's claim that human (psychological) behavior emerges as a function of both the person and the environment. Lewin's work was influential precisely because it effectively integrated several different domains of psychological thought, including gestalt, developmental, and social psychology. It is useful to draw attention to this idea here because we can show how it corresponds to, but also is somewhat different from, the general conception of behavior we have offered above.

Consistent with the general definition of behavior, we can see that the Person in Lewin's equation is the object, whereas the E is the environment which is the field, and the change over time aligns with the function. The difference between the two conceptions is found primarily in the generality of the formal representation of behavior offered above, which gives the general categories as entities and fields rather than Persons and Environments. The second difference is that Lewin is following in the tradition of experimental psychology and is using the term behavior to capture change as a dependent variable. This is because the grammar of science is such that one needs to be able to formulate cause–effect relations to determine outcomes that can be observed or measured. This means that, for psychologists in particular, behavior becomes framed as a dependent variable.

In Chap. 3, we referenced how Charles Stangor came to see behavior and empiricism as the fundamental concepts that defined scientific psychology. Behavior was the dependent variable that psychologists were using to predict and control via their empirical methods. Aspects of this way of thinking are present in Lewin's formulation and in many conceptions of behavior in psychology. But it is different than how we are using behavior here. Behavior is not a dependent variable produced by the functional relation between animal and the environment. Rather, we should think about behavior as capturing the entire ontological field, as opposed to what the methodologist is observing and measuring. The difference becomes clearer as we turn to the ontological meaning of behavior.

Matter in Motion: The Behavior of Objects in Space and Time

It was Galileo's work on matter in motion and Descartes' work on analytic geometry that set the stage for the full flowering of the modern scientific justification system. That flowering was realized via Isaac Newton, whose classical mechanics laid the foundation for our scientific understanding of the material universe for over 200 years. Newton's laws of motion clearly involve describing changes in object–field relationships. The first law pertains to constancy or inertia and refers to the fact that an object either remains at rest or continues to move at a constant velocity unless acted upon by a force. The second law is the idea that changes in inertia are a function of force and force is equal to the mass of the object times the acceleration of the object, which gives rise to the famous equation, $F = m \cdot a$. The third law is that the exertion of any force is met by a force equal in magnitude and opposite in direction.

Newton's classical mechanics revolutionized Aristotle's metaphysical system of understanding, as it eschewed any reference to formal or final causes that took place at the material dimension of existence. Newton also shifted the fundamental conception of inertia. For Aristotle, inertia meant being at rest. That is, absent a force acting upon it, an entity will eventually come to rest. In contrast, Newton argued that, absent any external force, inertia is the current state of movement, such that an object in motion will remain in motion unless impacted by a force. This resulted in a third change, whereby Newton introduced the concept of gravity as a ubiquitous attracting force, based on the mass of an entity. Curiously, the attractive nature of the force of gravity was found to be identical to the nature of inertia, which remains a fascinating physical and metaphysical question that scholars continue to ponder.

With the concept of inertia and gravity, we can see that metaphysical considerations remain. Of course, the metaphysics of classical mechanics and its frame for changes in object field relations would be upended at the turn of the twentieth century. The idea of a standard "Euclidian" relationship between the three dimensions of space and one dimension of time would be blown up by both Einstein's general relativity and quantum mechanics. The nature of space and time and the relationship between the system and the nature of measurement or observation would be thrown into disarray, such that our commonsense frame of understanding was shattered. Despite this, it is worth noting here that the concept of behavior remains central. Planck's famous constant, *h*, is a quantum of action, a mathematical-physical concept that is behavioral in nature. Indeed, with its key ingredients pertaining to objects, fields, and the complex and confusing relationship between knowers/observers/measures and known/observed/measured events, we can be assured that the metaphysics of behavior and its observation or measurement resides at the heart of the most central questions regarding foundational physics.

Ontology and the Different Kinds of Behavioral Patterns in Nature

Our final task is to untangle the various kinds of behavior patterns in nature. According to the Unified Theory, the natural sciences employ a descriptive metaphysical grammar of entities, fields, and change and are anchored to an exterior epistemology. As such, the metaphysics and epistemology of behavior connects the modern empirical natural sciences into a shared system of justification. These aspects of behavior unite physicists, chemists, and biologists, and, with the ToK System and UTOK metapsychology can now align psychologists and social scientists who embrace the onto-epistemology of a universal natural behaviorism. In addition to uniting these fields, the concept of behavior also differentiates them. It is precisely because there are various kinds of entities in the world that exhibit different kinds of behavioral patterns in different contextual fields that we have different sciences with different methods and vocabularies. Simply put, rocks behave differently than organisms, which behave differently from animals, which behavior differently from people, and the different domains of science are interested in different kinds of behavior. This ontological analysis corresponding the behavior patterns in nature with the different domains of science is deepened in the next chapter via the Periodic Table of Behavior.

Conclusion

This chapter has made the argument that behavior is a central concept in science, and that objects or entities, fields, and change provide the conceptual ingredients that allow natural scientists to map the world. This can be considered either a conventional or radical argument, depending on the angle it is approached from. It is conventional in that it proclaims that modern science is a particular kind of system of human thought that eschewed Scholastic metaphysics and emphasized empirical observation that systematically and analytically maps both the structure of entities and how they are influenced by forces and change over time. We can shorten this to say that modern empirical science systematically maps the natural behavioral patterns in the world. Although it is rarely framed exactly this way, this is nevertheless a good description of the conventional understanding of what constitutes the emergence of modern science as a particular kind of inquiry.

The UTOK metapsychology arrives at this rather conventional framing via an unusual path that has not been well charted. It follows the trail of behavior through Watson's behaviorism and shows how and why the concept emerged and why it was flawed. Specifically, Watson (and many others) has erroneously yoked behavior to physicalism. The ToK System shows why this is flawed, mapping behaviors to changes across levels and dimensions of complexification. This insight allowed us to connect the concept of behavior to both the ontic reality and the onto-epistemology of science. The epistemological framing of behavior connects to the process by which science justifies objective knowledge via systematic observation and measurement. The ToK System captures the ontological meaning of behavior via its depiction of different kinds of behavioral patterns in nature. The correspondence between the dimensions of behavioral complexity on the left side of the ToK represents the ontic reality of Matter, Life, Mind, and Culture, whereas the major domains of science emerging out of Culture on the right side represent the scientific onto-epistemological knowledge systems that function to model and map the ontic reality.

This chapter also corresponded UTOK's framing of behavioral patterns in nature with Integral Theory and clarified overlap and some key differences in ontological commitments. It also proceeded to show how the concept of behavior stretches from formal abstractions that lend themselves to mathematical logic to the behaviors of objects, organisms, animals, and people, and even moral-ethical considerations of good and bad behavior. This concentric map of the hidden layers of behavior creates a powerful alignment with how many conceive of the proper relations between mathematics, physics, and the special sciences of biology, psychology, and sociology, and finally philosophical and humanistic considerations pertaining to moral action.

In addition to these connections, UTOK's analysis of behavior can be further specified because of the problem of psychology. Specifically, the prediction that arises is that the structure of scientific knowledge should already reflect a correspondence between the domains of ontic reality and "hard" sciences that move from physics into chemistry into biology and finally neuroscience. However, as the transition moves into the domain of psychology (i.e., animal behavior, mindedness, and mental processes), the correspondence should become murky and lacking in clarity. This is because psychology lacks a clear ontological referent, in part because of the knotty conceptual issues associated with science, behavior, and mind, especially subjective conscious experience (i.e., Mind^2). The next chapter deepens the argument by adding clarity to the nature of aggregates, levels, and dimensions, which gives rise to the Periodic Table of Behavior and an even clearer picture of the relationship between science and behavioral patterns in nature as mapped by the ToK System.

References

Devonis, D. C. (2014). *History of psychology* (Vol. 101). Springer.

Gribbin, J. (2019). *The scientists: A history of science told through the lives of its greatest inventors*. Random House.

Henriques, G. R. (2003). The tree of knowledge system and the theoretical unification of psychology. *Review of General Psychology, 7*(2), 150–182.

Henriques, G. R. (2005). Toward a useful mass movement. *Journal of Clinical Psychology, 61*(1), 121–139.

Henriques, G., Michalski, J., Quackenbush, S., & Schmidt, W. (2019). The tree of knowledge system: A new map of Big History. *Journal of Big History, 3*(4), 1–18.

Koestler, A. (1967). *The ghost in the machine*. Macmillan.

Lakoff, G., & Núñez, R. (2000). *Where mathematics comes from*. Basic Books.

Lazzeri, F. (2014). On defining behavior: Some notes. *Behavior and Philosophy, 42*, 65–82.

Lewin, K. (1936/2015). *Principles of topological psychology*. Martino Fine Books.

Merleau-Ponty, M. (1963). *The structure of behavior* (A. Fisher, Trans.). Beacon Press.

Merriam-Webster. (n.d.). Behavior. In *Merriam-Webster.com dictionary*. Retrieved August 5, 2022, from https://www.merriam-webster.com/dictionary/behavior

Pinker, S. (1994). *The language instinct: How the mind creates language*. Harper.

Ryle, G. (1949). *The concept of mind*. Routledge.

Schultz, D., & Schultz, S. (2011). *A history of modern psychology* (10th ed.). Wadsworth.

Skinner, B. F. (1988). What are the scopes and limits of radical behaviorist theory: Reply to Harnard. In A. C. Catania & S. Harnard (Eds.), *The selection of behavior: The operant behaviorism of B. F. Skinner: Comments and consequences* (pp. 465–473). Cambridge University Press.

Volk, T. (2017). *Quarks to culture: How we came to be.* Columbia University Press.

Watson, J. B. (1913). Psychology as the behaviorist views it. *Psychological Review, 20*(2), 158–177.

Wilber, K. (1995). *Sex, ecology, spirituality: The spirit of evolution.* Shambhala Publications.

Wilber, K. (1998). *The marriage of sense and soul: Integrating science and religion.* Broadway Books.

Wilber, K. (2007). *Integral spirituality: A startling new role for religion in the modern and postmodern world.* Shambhala Publications.

CHAPTER 10

The Periodic Table of Behavior: Mapping the Levels and Dimensions in Nature

The scientific worldview given by UTOK is starting to come into focus. Modern empirical natural science is a particular kind of justification system that gives rise to a naturalistic, behavioral view of the world and the place of humans in it. Metaphysically, science is framed by entities, fields, and patterns of change. Epistemologically, modern science is defined by the exterior empirical position and methodological practices such as systematic observation, measurement, and experimentation, and institutions such as research programs, professional journals, and organizations. Ontologically, it is about mapping patterns of behavior in nature at various scales, levels and dimensions of complexification. Because the ToK System functions as a map or taxonomy of behavioral kinds in nature it is crucial that we understand what its boundaries are and how to apply them.

To do so, let us consider seven entities that, at first glance, seem to potentially defy the basic categories of the ToK taxonomy. The examples are: (1) virtual particles; (2) viruses; (3) jellyfish; (4) great apes trained to use language; (5) fossilized bones; (6) spider webs; and (7) iPhones. These examples are entities that, at first glance, do not seem to be easily placed in the categories given by the ToK System. However, as we will see, once both the boundaries of the ToK are considered and we are explicit about what the system is mapping, the placement of these entities will become apparent.

G. Henriques, *A New Synthesis for Solving the Problem of Psychology*, Palgrave Studies in the Theory and History of Psychology, https://doi.org/10.1007/978-3-031-18493-2_10

Virtual particles are entities that arise as a function of the weird features of the quantum world. They pop in and out of potential existence in the perturbations of quantum fields. They can only be indirectly measured, and their "true" ontic nature remains a topic of significant debate (Jaeger, 2019). Nonetheless, they are established phenomena in at least some senses of the word. As such, we can ask: Where do virtual particles reside in the ToK System taxonomy? Consistent with a critical realist view of science, the ToK depicts the unfolding wave of the grid of energy-matter-space-time, such that there are real generative mechanisms that operate on a wave of probability that transforms the real potential into the actual occurrence, which, to become the data of science, must be quantitatively measured and observed. Virtual particles reside at the base of the matter cone, in the weird quantum space between the real generative mechanisms that create the potential into actual events that appear in what, following Bhaskar, might be called the "empirical present." Although this formulation does not immediately clarify the thorny issues that emerge when examining this question of the reality of virtual particles in rich detail, it nonetheless provides a broad framework for addressing the debate and thus properly locates the conceptual coordinates of the issues.

The next three entities are examples that can be thought of as existing in the spaces between the major dimensions of complexification. Starting with viruses, most biologists do not consider them as living creatures (Villarreal, 2008). At the same time, viruses must be understood via the lens of biological evolution. Moreover, they interface with living creatures at the level of living processes. That is, they become intricately interwoven into the way cells function and they essentially function to "re-engineer" the complex adaptive behaviors the cell engages in. Via the descriptive metaphysics of the ToK System, we can say that viruses are biotic entities that reside between Matter and Life. Careful attention to the diagram reveals an overlapping circle that captures both continuity and "in-betweenness" regarding spaces of adjacency that reside across the dimensions. This is where we can locate viruses. And it is one of the key conceptual features that the ToK taxonomy has that Aristotle's scales of nature lacked.

Jellyfish are animals that move and have nervous systems. Thus, at first glance, it would seem that they should be included in the Mind dimension. This is problematic on two accounts. First, it is hard to argue that jellyfish are "mental" in any meaningful sense of the word. Second, in terms of complexification, surely a network of trees is vastly more intricate and complicated than a simple jellyfish. The response here is twofold. First, jellyfish do not meet the full criteria for Mind, because, as we will see, for something to be considered an exemplar of mental behavior it requires both a centralized brain and a complex active body, of which the jellyfish has neither. Second, the ToK maps the layers of emergent and evolving complexification, rather than the absolute amount of complexity across scales. Thus, a collection of trees is indeed far more complex than simple neuronal networks or even flies, which do exhibit mental behaviors. The ToK is a taxonomy of complexification rather than complexity per se, which will become clearer when we introduce the Periodic Table of Behavior below. The bottom line regarding jellyfish is that, as was the case with viruses residing in between Matter and Life, jellyfish reside in the conceptual space in between Life and Mind.

Many bonobos and chimpanzees have been taught the basics of language. For example, Washoe, a chimpanzee trained in approximately 350 signs, arguably combined the signs for "water" and "bird" in a novel way to refer to a duck. This raises the question about whether these animals should be thought of as existing on the Culture-Person plane of existence. Although a minority of primatologists answer this in the affirmative, there are good reasons to dispute this claim. The fact of the matter is that no other primate has been found to engage in processes of justification or question-and-answer dialogue. Indeed, in all the research in animal language, there is no clear example of any other animal ever asking a question or giving a reason for why they did what they did (Jordania, 2006). As such, they are not exemplars of fully functioning persons. Thus, paralleling viruses and jellyfish, linguistically trained apes can be placed in the space in between Mind and Culture.

The fifth example is fossilized bones. Although they have been shaped by life processes and their structural organization carries remnants of that history, they are well characterized as existing at the Matter dimension. Spider webs are interesting as they give rise to the question of where the boundaries are between "the mind" and the Mind-Animal plane of existence. In the vocabulary of the ToK, the web is clearly part of the spider's mental behavior, and thus the whole of the spider's web-spinning activity resides on the Mind plane, which is defined as the set of mental behaviors. However, the web itself would not be formally a part of "the mind" of the spider in the technical sense. As will become clear when we lay out the Map of Mind[1,2,3] in the next chapter, "the mind" can be technically defined as the neurocognitive processes that take place within the nervous system.

The web is thus part of the material world that plays an active interfacing role between the spider's mind and its overt mental behavior. Put differently, in the descriptive metaphysics given by UTOK, the web is part of the mental adaptive behavioral landscape (i.e., Mind), but is not explicitly part of the spider's mind (i.e., neurocognitive processes within the spider's nervous system).

Finally, iPhones are characterized as "material culture." All human technological innovations are put under the hybrid category of material culture, which is terminology for artifacts and tools that aligns with the vocabulary of archeologists. As we have noted, technology is one of the key aspects that differentiates Culture from the more general concept of society, as the latter is clearly constituted in part by technologies. The idea of material culture makes clear that hybrid categories are possible, which we also saw with the concept of fossils (i.e., their form is in part a function of their history as living entities).

It is also worth noting that iPhones and other kinds of digital information processing tools are seen through the lens of the ToK as unique developments in human technological evolution. The reason is grounded in the logic of the original ToK diagram. Each dimension of complexification is framed as making a qualitative jump as a function of information processing systems and communication networks that give rise to a new complex adaptive plane. That pattern suggests that what happened in the twentieth century was the emergence of the ground of a new dimension of complexification that has radical implications for what the twenty-first century will look like. That is, we can think of the digital technology in the twentieth century as being akin to jellyfish with distributed neural networks, but no real centralized pathway that created a coherent, interconnected digital virtual world. The suggestion is that we will see such an emergence in the twenty-first century, which will thus spark the generation of a whole new complex adaptive landscape. In UTOK, this theoretical phase transition is known as the fifth joint point (Henriques, 2011).

With these clarifications in hand, we are now positioned to extend the analysis linking science and behavior patterns in nature by introducing the Periodic Table of Behavior (PTB). The PTB is an extension of the ToK System that advances both the descriptive metaphysical systemic and ontological analyses that align our scientific knowledge with ontic behavioral patterns in nature. The PTB expands and elaborates on the crucial difference between emergence that happens within the dimensions (e.g., the levels of emergence from particles to atoms to molecules within Matter) and emergence processes that give rise to new planes (e.g., the jump from

Matter to Life). This division adds substantial specificity to the correspondence between the kinds of behavior in nature and the domains of scientific inquiry. It also results in the claim that for each plane of existence there is a primary object level of analysis. As we will see, this insight also affords greater clarity regarding the naturalistic picture being generated by UTOK because it shows why we need an ontology grounded in natural behavior patterns.

The Periodic Table of Behavior: A Taxonomy of the Patterns of Behavior in Nature as Mapped by Science

The Periodic Table of the Elements was a wonderful advance in how scientists thought of the atomic elements because it shifted the categorization system into a formulation that included the now familiar groups (the columns) and periods (the rows). It was Dimitri Mendeleev who had the insight that allowed the empirical data gathered about the various elements to be systematically categorized this way. The Periodic Table of the Elements can be considered a descriptive metaphysical system because it presents the concepts and categories that are used to map the atomic elements. Of course, it is also empirical in the sense that it was developed by formal processes of data gathering, prediction, and experimentation. Although it is not common to refer to the Periodic Table as a descriptive metaphysical empirical system, it is nonetheless perfectly acceptable to do so. Indeed, there are ongoing philosophical investigations about the metaphysics regarding the essence of atomic elements and the arrangement of the Periodic Table (see, e.g., Scerri, 2005). The Periodic Table of Behavior (PTB; Henriques & Michalski, 2019) draws its name from this inspiration.

As we have seen, most big picture systems (e.g., Big History) depict the evolution of complexity along a single axis, from particles to atoms to molecules to cells to multi-celled organisms and ultimately to human societies. The ToK System, in contrast, characterizes nature as consisting of four different planes of existence. In addition, the ToK System depicts the process of complexification as also taking place within each plane. The PTB builds on this insight by highlighting that, for each plane of existence, there is a primary unit of organization. For the Matter dimension, that primary unit is the atom. For the Life dimension, the primary unit is the cell. For the Mind dimension, it is the animal. And for the Culture dimension, it is the person. These primary units can then be analyzed both in terms of the behaviors of the primary parts at the level of analysis below the primary whole, and in terms of behaviors of groups of entities. In addition, the behaviors of patterns across aggregate scales is framed by the general object-field relations.

The result of this set of insights is a dual-axis vision of the evolution of complexification. Splitting what has traditionally been conceived as a single dimension of emergence into two separate axes provides a new way to organize and classify behavior patterns in nature and align them with various domains in scientific inquiry. One axis (the columns) consists of the four dimensions of behavioral complexification depicted by the ToK System (i.e., Matter, Life, Mind, and Culture). The other axis (the rows) consists of the primary level of object-change analysis that is being considered (i.e., part, whole, or group). In addition, there is a general column that clarifies how aggregates at the part, whole, or group might be analyzed (i.e., how atoms might aggregate to form planets or stars). What emerges from this combination is a table that separates aggregates from primary levels, and then provides three rows representing the primary levels of emergence by four columns representing the different dimensions of emergence. The result is a zoomed-out view of a general descriptive theory of behavioral patterns in nature classified by different domains of scientific inquiry writ large (Fig. 10.1).

The Periodic Table of Behavior

		Dimensions of Behavioral Complexity			
Domains of Reality and Science	Planes of Existence	MATTER	LIFE	MIND	CULTURE
	Major Class of Science	Physical	Biological	Psychological	Social
General Object-Field Relations	Context of Behavior	Field	Ecology	Environment	Society
	Behavioral Entity	Object	Organism	Animal	Person
Primary Levels of Object Analysis (Part, Whole, Group)	Groups of Wholes	Molecule	Multicell/Colony	Family-Group	Family-Community-Nation
	Primary Whole	Atom	Cell	Mind^1	Mind^3
	Primary Part	Particle	Gene	Neural Network	Symbolic Justification

Fig. 10.1 The Periodic Table of Behavior in nature

An Overview of the Periodic Table of Behavior

The PTB is a taxonomy of behavioral kinds in nature that builds upon the ToK System's theory of reality and science and gives a more detailed picture. It achieves this because it adds the insight that there are primary levels of analysis within each dimension that can be separated from aggregates across scale, and then uses that insight to develop a three primary levels by four major dimensions table that maps the patterns of behavioral complexification in nature and the domains of scientific inquiry that correspond to those patterns. The first row on the PTB is labeled "Domains of reality and science." This represents the most general correspondence shown on the ToK System between the ontic planes of existence and the domains of scientific knowledge about those planes (i.e., the Matter/Physical, Life/Biological, Mind/Psychological, and Culture/Social relations between the dimensions and major classes of science).

The second row is labeled "General Object–Field Relations." This refers to the various kinds of entities associated with that dimension of behavioral complexity (i.e., object, organism, animal, and person), and the fields in which such objects reside (i.e., field, ecology, environment, society). Importantly, it also represents the range of various aggregate scales of analysis that might be undertaken at any level of analysis within that dimension. For example, a physicist might track the behavior of either a raindrop or a cannonball traveling through the air. Both raindrops and cannonballs are "general objects" that exist at the material dimension. A biologist might explore a specific cell or organ system or species, or might examine a population of genes or cells or organisms. A basic psychologist might examine the reinforcement schedule of a rat compared to a pigeon and then connect with an ethologist who compares these findings to the behavioral repertoires of these animals in their natural environments. An anthropologist might study the roles, traditions, and identities of specific community members in a particular culture.

In considering the general relationships, it is important to consider the size or scale across the dimensions of space and time, which range from the smallest to the largest entities that can be studied in the relevant category. Indeed, a primary reason for the presence of this "general relations" row is to represent the fact that entities being examined in the various dimensions of complexity can be framed by scale. Consider, for example, the truly stupendous scale differentiations at the Matter-Physical Sciences domain. Entities on this dimension can range from the absolute smallest measurements that are theoretically possible to the behavior of the largest objects imaginable. When we go to the smallest scales in physics, we find things like the "quantum of action," which is framed by Planck's constant, which is 6.626×10^{-34} joules per second, and the Planck time, which is 10^{-43} seconds, and the Planck length, which is 10^{-33} centimeters. These are unfathomably small units of scale and change.

When we go the other direction, we find similarly incomprehensible scales of hugeness relative to our "normal" human frame. For example, the Hercules–Corona Borealis Great Wall is a massive superstructure of galaxies that is spread across approximately 10 billion light years (Horvath et al., 2020). One could even consider the Universe itself in all its entirety as the entity under consideration. Indeed, the Energy Information singularity at the time just prior to the Big Bang can arguably be thought of as the absolute largest entity, such that out of it extends all matter, space, and time across the eons that make up our observable universe.

The same basic logic of scale across aggregates applies to the other dimensions. For example, the general behavioral unit in the biological-Life dimension is the organism-ecological relation. Starting at this level, the scale in biology would range in scope from the smallest units of biological information or biochemical structures that make up organisms, such as a gene or an organelle like a ribosome, and would stretch into the largest groups of organisms and broadest ecologies. Consider, for example, the aspen tree, which makes clones of itself and grows into colonies. One such colony of aspen trees, which many biologists classify as a single organism, covers over 106 acres (Rogers & McAvoy, 2018). Ecologies refer to the context in which the system is embedded. This can extend from specific contexts and niches, like a pod or a lake, into biomes, which are regions like tropical rainforests, temperate forests, tundra, and deserts. As we have seen with climate change, many environmental scientists are tracking the ecology of the Earth as a whole system. Indeed, the Gaia hypothesis advanced by Lovelock (2003) frames the Earth as a kind of superorganism, and even if the idea of the Earth as a superorganism is problematic, nonetheless it can be taken as the largest scale of biological-ecological systemic analysis.

For the Mind dimension, the general unit of analysis is the animal-environment behavioral relation. In terms of scaling, this would run down into animal parts consisting of specific neuronal networks into neuronal reflexes into brain areas into specific mental behavioral patterns of individual animals all the way up into colonies of animals and finally interacting super-colonies. The largest documented entity of this kind consists of

Argentine ants. A super-colony of Argentine Ants has been found that extends over 6000 km, from Italy to the Spanish Atlantic coast. It consists of billions of individuals and functions to form the largest cooperative unit ever recorded in the animal world (Giraud et al., 2002).

Moving to the Cultural dimension, the primary unit of analysis is the human person. Later chapters will specify exactly how we can define a person, but the basic ingredients are framed by Justification Systems Theory, Mind^3, and the Culture-Person plane of existence. The smallest units here are symbols of shared meaning. We then move up into propositional sentences and then ideas or languages and then into human persons behaving in sociolinguistic contexts that can range in scale from dyads into families or small groups and then communities, states, nations, and then global patterns of interaction. Indeed, many wonder about an emerging digital geo-political world order that would be the largest Culture-Person societal system of organization (Volk, 2017).

It is useful to remember that the ToK System and PTB can be thought of in both descriptive and prescriptive-normative terms. That is, they function as a map both for how the sciences *are* currently organized as far as the physical and biological sciences are arranged and how the sciences *ought to be* organized regarding psychology's core definition and its relationship to neuroscience (from below), ethology (horizontally), and the social sciences (from above). More explicitly, grounded in the UTOK metapsychology, the PTB carries with it the idea that the institution of Psychology should be explicitly divided into basic (or animal) psychology, which aligns closely with ethology and human psychology, which functions as the base of the social sciences. As has been noted, making this dividing line is necessary because the "behavior and mental processes" of humans are radically different from those of animals. Of course, this is not how the institution of Psychology is currently defined, but it represents UTOK's prescription for how it ought to be defined, given an effective descriptive metaphysical system that maps the ontological dimensions of existence in right relation.

The Primary Part, Whole, and Group Levels of Analysis

The third set of rows captures the key contribution of the PTB over and above the ToK System. This is the recognition that there are "primary units of behavioral analysis" in each dimension that then have part and group relations that can be effectively placed below and above the primary whole, respectively. To start, we have atoms making up the primary whole units in the material dimension. Atoms are not single entities, but themselves are made up of parts, which are entities like electrons, which are aptly known as particles. These particles and the force interactions that connect them represent the very bottom of the Matter dimension on the ToK. If we move to the group level above the atom, we find the world of molecules, which have many novel emergent features studied by chemists.

When we shift to the Life dimension, we find that cells are the primary units. Of course, there is much complex behavioral space in between chemistry and cells. Framed by scientific disciplines, we can follow the trail from organic chemistry into biochemistry into molecular biology. When we do, we find the four major classes of biological macromolecules, which are carbohydrates, nucleic acids, proteins, and lipids. There are also many different parts of the cell that perform specific functions, such as the cell membrane, ribosomes, Golgi apparatus, and mitochondria. These are called organelles. These and many other domains of biochemistry and molecular biology are legitimate "biological parts." Nonetheless, because the ToK System highlights how a fundamental shift in behavioral organization pertains to information processing and communication patterns, an obvious primary part is the gene. The gene can be framed as representing a fundamental unit of biological information.

When we shift to the cell and upward, we notice a major division in cells between those that have a nucleus and those that do not. Those that have a nucleus are called eukaryotic, and they had emerged by 2.7 billion years ago. Eukaryotic cells may well have emerged from engulfing and merging with simpler cells. This is an interesting jump in that it represents a kind of aggregate shift in the structure of the cell. In addition, it is crucial in that they are the kind of cells that make up the major multi-celled kingdoms of plants, fungi, and animals. Thus, the jump into eukaryotic cells resulted in the capacities for cells to much more systematically self-organize into functional groups.

The primary unit of analysis for the Animal-Mental dimension is the animal behaving-as-a-whole, mediated by neurocognitive processes. This is the definition of Mind^1, which can be thought of as the (neurocognitive functional) mind, brain, behavioral patterns of animals. This is why Mind^1 is placed here, although we could have simply put animal as the primary object of analysis. As we will see, Mind^2 (i.e., subjective conscious experience of being) arises out of Mind^1 processes but represents a part of the expansion of the Mind dimension rather than a specific "cleave" in patterns of animal behavior. Moreover, Mind^2 is not included because the epistemological frame of the PTB is an exterior rather than interior perspective. Indeed, as we will see, scientific investigations of subjective conscious experience gain their access via functional behavioral patterns of awareness and responsivity, because that is the epistemology of scientific language games. This is relevant because it offers a contrast to ontological frameworks that focus on subjective conscious experience as representing a new order of existence (e.g., Hartman and Cahoone).

As with Matter and Life, there are many possible parts that can be referenced in the Mind dimension. Because the ToK homes in on the crucial role of information processing, the primary part that emerges is the neuronal network, and the role it plays in the instantiation and processing of information in the nervous system. As those who work in artificial intelligence and machine learning know, neural networks often refer to computational systems that model learning and complex adaptive behavior. However, neural networks can also refer to the "real thing," whereby the most basic arrangement consists of an input neuron that is tied to an intermediate neuron which is connected to an output neuron.

The exemplar case of Mind^1 is of an animal with a brain and complex active body acting in a way that demonstrates functional awareness and responsivity. There are a series of videos on the internet of hydras, which are animals that have long tentacles like the arms of an octopus, eating sandfleas. I mention them because they provide a good visual example of the kind of behavioral patterns that reside at the base of Mind^1. Indeed, even though the hydra exhibits functional awareness and responsivity in the way it latches onto the sandflea and proceeds to eat it, this feeding behavior is not a full exemplar of Mind^1. The reason is that the hydra does not have a brain and lacks a complex, active segmented body with an integrated sensory motor loop. The sandflea, on the other hand, does have a brain and complex, segmented active body and thus an integrated sensory motor loop. As such, the sandflea has the core ingredients that go into exemplars of mental behavior. The videos can be framed as residing at the very base of Mind on the ToK System and showing the overlap and crossover and fuzziness of creatures in this territory of complexification.

Shifting up from the mental behavior of specific animals into the behavior of animal groups in the form of families, herds, colonies, and troupes is captured by the move from the Mind^1 box into the animal "family-group" box. A good example of animal group behavioral patterns was provided by a large-scale study of elephants that was reported on by the

New York Times (Angier, June 4, 2021). These elephants provide us an excellent example of the mental behavioral patterns on the top of the Mind dimension in terms of scale and complexification.

The primary unit at the Culture-Person plane of existence is a self-conscious person who can justify their actions on the social stage. Mind^3 is the center of this process of self-conscious justification. Of course, Mind^3 requires the individual to have Mind^1 and Mind^2 processes operating in support of the self-conscious justifying activities. As suggested by this comment, there are again many possible parts that could be placed as the center of the tenth floor. However, consistent with the focus on units of information, the primary part here is the symbolic justification, which is the fundamental unit of propositional meaning-making that serves as the "bricks" that make up the systems of justification that network and coordinate human activity. Above the human person is the group or sociocultural system that is coordinated by large-scale systems of justification. We will elaborate further on these distinctions below and in later chapters.

Implications and Predictions from the Periodic Table of Behavior

The PTB carries many implications. First, as a proposed map of the natural behavioral kinds mapped by science, a prediction emerges that the distribution of the sciences and the major divisions between them should be effectively classified by the Periodic Table of Behavior. That is, each of the 12 floors of the PTB should clearly align with a domain of scientific inquiry. In addition, because of the problem of psychology, a second prediction emerges regarding how scientific knowledge is currently organized. Specifically, it predicts that there will be relative consensus that science is effective at classifying behaviors from quantum mechanics into chemistry into biology and finally neuroscience. However, the clarity of scientific disciplines and how they map the levels and dimensions in nature that are above neuroscience (i.e., psychology) should become more nebulous and ambiguous.

A second set of predictions is that the PTB affords a novel taxonomy that should be effective at classifying various types of behavior in general. Indeed, the general argument being made is that (a) there has been much confusion about the meaning of behavior and (b) individuals often attempt to consider whether something is or is not an example of a behavior, whereas the PTB suggests that the issue should be framed by asking what kind of behavior is being offered. Consider, for example, the difference between a blink and a wink. Both are behaviors. However, they are different kinds of behaviors. A blink is a neurological reflex and thus resides somewhere between the neuronal and Mind[1] levels of analysis. A self-conscious wink that functions as a symbol that is part of a shared system of justification (i.e., whereby the person could be called to give an account of the wink and offer a narrative for it) is better considered a Mind[3] behavior.

Finally, the combination of the ToK System and PTB provides a new way to frame the ontology of the mental from the vantage point of science. Specifically, the argument is that animal-minded behavior is a coherent dimension that can be separated from the dimension of living behaviors, and that cultural-personal behaviors can be separated from animal-minded behaviors. This separation is crucial to understanding the ontology of the mental in terms of natural behavioral science. This will then set up the argument for how to define mental processes, including subjective conscious experiences in animals. Turning to the first prediction, the next section delineates how the PTB leads to the idea that there are "12 floors" in science.

The 12 Floors of Science Mapped by the PTB

The prediction from the ToK and PTB is that if these are the primary levels and dimensions of behavior in nature, then this taxonomy *should* broadly align with how the sciences are arranged, at least up until neuroscience. An additional prediction can be made because of the problem of psychology. The correspondence should be much clearer at the level of the physical and biological sciences. However, there should be confusion as to where, exactly, the biological sciences transition into the psychological sciences and where the psychological sciences transition into the social sciences. In what follows, the PTB is applied to demarcate four domains and "12 floors" of science, breaking each dimension into the primary parts, wholes, and groups.

The First Six Floors of the Periodic Table of Behavior

We can start by envisioning a "basement" for the 12 floors of the PTB. Based on the logic of the ToK System and modern physics, we can label this the "Energy Information Field" that grounds the Matter dimension of complexification. This Energy Information Field is represented at two points on the standard ToK System. The first is at the Big Bang/Big

Beginning of the observable universe. This is when the universe was in a super hot, super dense state of pure radiation. Moreover, the four fundamental forces (electromagnetic, gravitational, and strong and weak nuclear) were collapsed into a "singular superforce" field (Henriques, 2021, October). It is from this Energy Information ground state that a chain reaction takes place and gives birth to the Matter dimension approximately 13.8 billion years ago. In addition, at the bottom of the ToK cones, there is a circle that is listed as the "present." This small circle represents the quantum fields out of which the particles of the Matter dimension emerge. As Eastman (2020) skillfully notes, these quantum fields can be framed in terms of input-output contexts, such that the quantum super-positioned potentials decohere into actual events in the macroscopic world.

If we consider the ground of the PTB to be "energy information fields" framed by the four fundamental forces (and the Higgs field), then we are situated to move into the first floor. Thus, in the descriptive metaphysics given by the ToK and PTB, Floor 1 represents the jump from Energy into Matter, as mapped by particle physics. This is where we get the emergence of particles. According to quantum field theory, particles can be thought of as emergent field fluctuations. Here is the astrophysicist and science educator Ethan Siegel's (2019, January) explanation for how quantum fields create particles:

> The field exists everywhere in space, even when there are no particles present. The field is quantum in nature, which means it has a lowest-energy state that we call the zero-point energy, whose value may or may not be zero. Across different locations in space and time, the value of the field fluctuates, just like all quantum fields do. The quantum Universe, to the best of our understanding, has rules governing its fundamental indeterminism.
>
> So if everything is fields, then what is a particle? You may have heard a phrase before: that particles are excitations of quantum fields. In other words, these are quantum fields not in their lowest-energy—or zero-point—state, but in some higher-energy state… In our pre-quantum picture of the Universe, particles are simply points and nothing more: individual entities with a set of properties assigned to them. But we know that in the quantum Universe, we have to replace particles with wavefunctions, which are a probabilistic set of parameters that replace classical quantities like 'position' or 'momentum.'

Our focus here is on the basic outline that gives us an optimal grip on how to put the pieces together in a coherent way without getting lost in the details. The point here is that with particles we have moved from the fundamental ground of substance (i.e., the Energy Information Field) into the first floor of Matter. This gives us the first floor of the PTB, which is mapped by one of the greatest scientific achievements of all, the Standard Theory (or Model) of Elementary Particle Physics. As Siegel's (2019, January) essay notes, "the Standard Model gives you fermion fields, which correspond to the matter particles (quarks and leptons), as well as boson fields, which correspond to the force-carrying particles (gluons, weak bosons, and photon), as well as the Higgs."

With the particles mapped by the Standard Model, we can now take a step up to Floor 2 on the PTB. We can do so by noting the alignment with primary part (i.e., particles are parts of atoms) and primary whole. Of course, Floor 2 on the PTB is the floor of atoms. This is the atomic theory of Matter, and the logical organization of atoms is mapped by the Periodic Table of the Elements. This brings us to a happy relationship between the PTB and the Periodic Table of the Elements, which is that the latter fits into the former, as it is the map of the second floor of science. As the primary unit of Matter, atoms are also the primary unit of scale, especially for entities that we humans regularly encounter. That is, as we move from the subatomic into atomic levels and then move up in scale (i.e., size and number), we move from the quantum world into the world that is generally well described by Newton's classical mechanics. This is the world of rocks and cannonballs, as well as the moon and stars.

Floor 3 moves us into entities made up of "groups of atoms," more commonly known as molecules and their interactions. This is the world of chemistry. Although chemistry has not figured prominently in our analyses up until this point, its role in the natural sciences is central. Indeed, some individuals consider chemistry the "central science." The popular textbook *Chemistry: The Central Science* (Brown et al., 2018) defines chemistry as the scientific study of the properties and behavior of matter. Note that physics extends "down" into energy and the world of the quantum, and thus can be said to be the broader domain of inquiry, such that everything that is chemical is also physical, but not everything that is physical is chemical. In short, in the language of UTOK, the subject matter of chemistry emerges as we move from the basement and first floors, which are the foundation of physics, into atoms and molecules across scale. If we return to the original ToK diagram reviewed in Chap. 6, we can see that chemistry is listed as the primary science that studies Matter dimension, and that correspondence holds here.

Particles, atoms, and molecules make up the three primary floors of analysis for the Matter dimension, and these floors correspond to the heart of the physical and chemical sciences. Of course, there are other material sciences and other kinds of entities in the material world. However, these entities arise as a function of scale and represent material aggregates. For example, cosmology and Earth science are material sciences, but the entities they study emerge as a function of scale and aggregates. It is important to keep in mind why this is the case. The PTB is a taxonomy that functions to highlight the primary levels of analysis and then uses that to frame other sciences that focus on behavior patterns across different aggregates. This is because the PTB maps onto what Tyler Volk (2017) calls combogenesis, rather than complexity in general. Volk defines combogenesis as "the combination and integration of things from a prior level to make a new level of things."

With the jump to Floor 4, we shift into a new dimension, Life. When jumps between dimensions happen, the prediction is that we will see what might be called "hybrid sciences" (Henriques, 2004), in that they necessarily involve analyses from both the dimension beneath and the dimension above. As was noted earlier, Floor 4 is the domain of "biological parts" and would include everything from genes and proteins to organelles (i.e., all the parts that go into making up cells). The sciences concerned with this floor represent the jump from chemistry into biology and the interface between them. Thus, we find biochemistry as being primarily located in chemistry and moving from there into the biological parts, and we have molecular biology and genetics, which start more in biology and move down into chemistry.

Floor 5 is the world of cells, which are the primary entities or units of living behavior. The science of cells and their behavior is called cytology. Disciplines that study multi-celled creatures, such as mycology (i.e., fungi) and botany (i.e., plants) map the behaviors on Floor 6. Zoology is another discipline in this domain. It is the biological study of animals, both living and extinct. It studies their structure, species classification, embryology, evolutionary history, and genetic lineages, as well as habits and behavior. As the focus turns to behavioral patterns, zoology transitions into ethology, which we will describe more below, because this means that, in the frame of understanding given by UTOK, we have transitioned into the dimension of Mind. We should also note, when considering the biological sciences, that ecology and environmental sciences that study ecosystems are part of biology but operate across larger scales and systems.

The UTOK posits that the first six floors of the PTB correspond well with the physical and biological sciences. Specifically, we can see that the natural sciences demonstrate clear divisions that correspond to the parts, whole, and groups of the primary levels of analysis at both the Matter and Life dimensions. We can also see that there are disciplines like geology and ecology that deal with various aggregate scales in the physical and life sciences. This is all consistent with the claim that UTOK provides a useful description for how the physical and biological sciences are currently structured. When we shift into the next six floors, the argument is that there is much confusion, especially pertaining to the Mind into Culture dimensions. This makes perfect sense given the problem of psychological science, and UTOK provides a clear prescription for how the disciplines should be arranged.

Floors 7 Through 12 on the PTB

Floor 7 on the PTB is the domain of neuroscience. Just as molecular biology and genetics bridge the level of chemistry with the level of the cell, neuroscience bridges the biological/living and mental/psychological domains. As such, although neuroscience is primarily a biological discipline, it is also something of a hybrid, as it must be considered via the lens of both biology from below and psychology from above. As Tryon (2016) explains in detail, computational neural networks serve as a good example of the neurological-psychological interface, and they can be thought of as residing at the fundamental transition point between the biological/life and psychological/mental sciences.

Floor 8 on the PTB is where things get rearranged. Or, more prescriptively, the Unified Theory proposes a rearrangement to generate a coherent language system and arrangement between the domains of science and the proper account of the dimensions of existence. As predicted by the ToK System and the problem of psychology, there is enormous ambiguity about the domain of science that maps animal behavior. At the most basic level of categorization, we can ask whether animal behavior science is a biological or a psychological discipline. The answer is that there is confusion about the nature of animal behavior and its classification. The confusion is predicted by UTOK because, as we have seen, most accounts of science fail to clarify where and how the domain of Life/biology transitions into the domain of Mind/psychology. This means there is massive ambiguity in the relationship between the animal behavioral sciences, the neurosciences, and psychology in such frameworks.

The ambiguity about the place and nature of the animal behavioral sciences can be observed empirically. For example, Bollen et al. (2009) examined the "clickstream data" of scholars using web portals and developed a network representation of journals that showed the interconnectedness of various disciplines. The clusters reveal a similar pattern to what we have seen, such that the physical/material sciences (e.g., in this case represented by journals covering topics such as applied physics and analytic chemistry) formed one large cluster, the biological or life sciences (represented by journals covering topics such as plant biology, physiology, and genetics) formed a second cluster, and the social sciences (which overlapped quite a bit with the humanities) formed a third cluster (represented by journals that covered sociology, anthropology, and economics). When the authors divided their scheme, two broad divisions of the natural sciences and social sciences became clear.

Interestingly, and consistent with the map provided by the ToK and PTB, there were journals and topics at the intersection and overlap between biology and psychology that were not readily classified in one or the other direction. As the authors noted, the clickstream data revealed "highly connected clusters corresponding to biology and psychology [that] contain a mix of journals classified in either the social or natural sciences." Animal behavior was one such domain.

The history of science helps us clearly see the nature of this conundrum, and we can start by noting that the first systematic scientific approaches to dealing with animal behavioral patterns were undertaken by psychologists. In the nineteenth century, comparative psychologists were among the first scientists to study "behavior and mental processes" in animals, comparing and contrasting different patterns in different species. And as we saw in our review of behaviorism, early in the twentieth century the scientific analysis of animal behavior in the laboratory was done largely by psychologists. Indeed, for those who argue that psychology is the science of behavior, the primary referent is the behavior of the animal as a whole.

However, by the latter part of the twentieth century, most naturalists studying animal behavior in the world were biologists. Specifically, we see the rise of ethology, which is defined as the science of animal behavior, and is considered a branch of biology. So, historically, we have a situation where scientists studying the exact same phenomena (i.e., animal behavioral processes) carried different professional and institutional identities (i.e., psychologists were doing experiments on animal behavior in the lab, whereas ethologists studied animal behavior patterns in nature).

With its clear frames for behavior and mental processes, the ToK, PTB, and different domains of mental behavior allow us to clarify the metaphysical and ontological issues, and from there we can make proposals about the institutional identity of the various sciences involved. Of course, this is not how things work in the real world, where traditions get established and practical matters of convenience and accidents of history set down institutional arrangements. This is especially so in the absence of a shared, coherent framework of understanding. But nonetheless, we can proceed to articulate the logic of UTOK for why basic psychology corresponds to the animal-mental behavioral dimension.

First, the behavior of animals with brains and complex active bodies (e.g., insects, vertebrates) clearly operates on the eighth floor in the levels by dimensions arrangement given by UTOK. As such, there should be no confusion regarding *where* such patterns of behavior take place in the ontological layers in nature. Thus, if we refrain from any institutional claims, we can simply note that the cognitive and behavioral neurosciences, along with ethology and comparative psychology, all clearly work to analyze behavior of animals in experimental and naturalistic conditions. This is Floor 8 on the PTB, and it is the primary unit that is framed by the third dimension of complexification on the ToK System, which corresponds to the behavior of minded animals. It is confused in science precisely because of the Enlightenment Gap's confusion regarding the proper relation of matter and mind.

In numerous prior writings (e.g., Henriques, 2004, 2008, 2011), I have argued that the science corresponding to Floor 8 should be referred to as "basic" or "formal" psychology. However, this argument is about the institution and is thus very much open to debate. Institutions need not be defined by naturalistic ontology but emerge and operate as a function of the social construction of justification systems. Indeed, because the twentieth century saw Psychology become so tilted toward the human, a strong argument can be made that we should label the sciences corresponding to the Mind dimension the "mind, brain, and behavior sciences" or something of that ilk, and explicitly separate it from the institution of Psychology. The point is that because modern science has been confused about the meaning of behavior and mental processes, the proper institutional alignment and way of conceptualizing animal behavioral science has been massively convoluted. This requires attention, and new institutional considerations ought to be generated so that the boundaries can be more effectively arranged and aligned to correspond to the actual levels and dimensions of behavioral complexity found in nature and mapped by the scientific enterprise.

When making these considerations, we should be clear about how UTOK conceptualizes Mind with a capital M and Mind^1 and Mind^2. As has been noted, when capitalized Mind refers to the set of mental behavior. This can also be defined as mindedness, which is the state of being minded or operating on the Mind dimension of complexification. Being a minded creature in UTOK refers to the sensory-motor looping functions that enable animals with brains and complex active bodies to behave so differently than plants or rocks. Regardless of how we define the institution of Psychology, Mind or mindedness is a concept that both science and modern society need to cultivate awareness about. That is, to have a clear map of natural behaviors in the world, we need to adopt the concept of mindedness. It should be second nature for people to take a walk through the woods and see the plants and fungi as living organisms and see the squirrels, bees, and birds as minded animals. Mindedness is the domain of Mind^1, which refers both to the overt activity of animals and the way they demonstrate sensory-motor functional awareness and responsivity, as well as the within nervous system neuro-information processing that constitutes the specific definition of "the mind" in UTOK.

We can also consider Mind^2 in relationship to the PTB. The structure of the PTB helps make clear that science is about observing behavior patterns in nature, and thus because subjective conscious experience is not directly epistemologically available via an exterior lens, it is one of the great challenges for a science of psychology. Indeed, the combination of this challenge and the ontological difficulties subjective conscious experience brings with it, and the role of subjective conscious experience in the intersubjective construction of science points to the difficulty a science of subjective experience might encounter when it is framed via the traditional modern, empirical, natural science language game. Things get more convoluted when we consider that, for many, the primary referent point for psychology is Mind^2. As we will see in the next section of the book, UTOK provides the necessary metaphysics, onto-epistemological mapping, and metatheoretical architecture to address this issue and provide a clear conceptual framework for approaching the domain of Mind^2 in animals.

The behavior of animal groups, which corresponds to Floor 9 on the PTB, is generally analyzed by sociobiologists and behavioral ecologists. Here again, we currently have mostly biological disciplines where UTOK argues that there should be basic psychological disciplines. The perspective afforded by UTOK and the PTB results in an interesting prediction about the conceptual relations between sociobiology and traditional, experimental animal behavioral science. Specifically, this analysis from UTOK asserts that the experimental analysis of animal behavior represents a "lower" floor of analysis than does sociobiology. Given that B. F. Skinner and E. O. Wilson are exemplar theorists of these different domains, an interesting set of predictions arises regarding their ideas. Namely, even though Wilson is a biologist by training and Skinner a psychologist, according to the ToK System and PTB, the latter's analyses of operant behavior should be more fundamental in nature's patterns of behavior than Wilson's analyses of altruism and the behavior of animal groups.

This is a prediction that has been confirmed. It can be seen when one examines the correspondence between Wilson and Skinner as analyzed by Naour (2009). He obtained a set of interviews and exchanges between the two men. And, consistent with this formulation, over the course of several exchanges, it becomes clear to Wilson that Skinner's operant theory is in many ways more foundational in terms of hierarchical levels of analysis in accounting for groups of animal behavior than his sociobiological investigations. Although this is interesting, we should be aware that it is hardly surprising that the lab-based experimental analysis of behavior provides the conceptual ground for understanding the behavior of animal groups in nature.

The next set of floors operate in the Culture-Person plane of existence. Like molecular biology and genetics and neuroscience, Floor 10 can be thought of as the hybrid space between basic psychology/ethology and the social sciences. In terms of the cluster of sciences that operate in this domain, we find the interdisciplinary intersection called cognitive science. This includes (human) cognitive psychology, linguistics, and cognitive anthropology. Artificial intelligence, neuroscience, and the philosophy of mind are also important contributors to cognitive science. However, according to the Unified Theory, and its classification of modern scientific knowledge, these latter domains should be considered supportive rather than primary science domains of Floor 10. The reasons are that philosophy is not really an empirical science, neuroscience is already accounted for, and artificial intelligence is about technology, which is not formally part of the natural-into-human sciences. The next chapter explores the cognitive revolution and recent developments in cognitive science.

Floor 11 is most clearly made up of human developmental, personality, and social psychology, which involve the analysis of human mental behavioral patterns over time in relational and cultural contexts. With the Influence Matrix and JUST extending from BIT, this is the domain where the Unified Theory does its most specific metatheoretical work in generating a coherent and comprehensive frame for understanding human persons in their biological, developmental, and social contexts. Key ideas from the Unified Approach, such as Character Adaptation Systems Theory (Henriques, 2017), make this especially clear. This is also the base of anthropology, at least to the extent that it is focused on human individual and small group behavior. The final section of the book makes this explicit.

Along with social and cultural psychology, anthropology transitions into human groups in context, with cultural anthropology clearly moving into the 12th floor. This domain would then extend to reach the rest of the social sciences, such as political science, economics, and, of course, sociology. These sciences take as their unit of analysis the behaviors of large human groups, cultures, or societies. It is worth noting, however, that proper theorizing at these final levels of social science must bridge from the natural-into-human sciences mapped by the ToK to the material cultural technological interventions that become so central as humans move into agrarian organizations and then cities. That is, to make sense of societal behavior, one must include not only shared, learned behavioral investment repertoires and the large-scale justification systems that constitute the Culture-Person plane of existence, but also technology and its evolution and influence. However, as was noted at the beginning of the chapter, the ToK System and Periodic Table of Behavior are concerned with natural entities, and so one must step outside it to bridge to technology.

Applying the Periodic Table of Behavior to Disentangle Behavioral Kinds

Central to UTOK's critique of mainstream academic psychology is that the concept of behavior is deeply muddled. It is a claim that has been empirically demonstrated. The behavioral biologists Levitis et al. (2009) explicitly examined this issue and found many different frames and definitions of behavior that carried different implications and led to different classification systems. Levitis et al. (2009) concluded that much work was needed to clarify the meaning of behavior. They ended their article with the following definition: "The internally coordinated responses of whole living organisms (individuals or groups) to internal or external stimuli, excluding responses more easily understood as developmental changes."

This is a potentially useful definition, but things remain complicated and muddled. Comporting with the language and perspective of UTOK, the Levitis definition means that bacterial cells behave. However, it is a definition that presents a problem for Psychology. Indeed, the very notion of *behavioral biology*, which is the professional identity of Levitis and colleagues, raises a problem for those who argue that psychology is *the* science of behavior. The behavior of bacteria does not fall under the purview of even the broadest conception of the subject matter of psychology. But a problem with the Levitis et al. definition is that molecules, planets, and atoms also behave, as do genes and organelles. Although the behaviors of particles, atoms, planets, and galaxies are not what Levitis, Lidicker, and Freund mean by biological behavior, it does *not* follow that these are *not* behaviors.

Especially relevant for the PTB is the fact that Levitis and colleagues asked experts to characterize different kinds of events and rate the extent to which they were classified as good examples of behavior, or not. The list they generated was as follows: (a) a person decides not to go to the movies if it is raining; (b) a beetle is swept away by the current in a river; (c) a spider spins a web; (d) a plant bends toward the sun; (e) geese fly in a V formation; (f) a person's heartbeat speeds up following a nightmare; (g) algae swim toward food; and (h) a rabbit's fur grows over the summer season. The results from the study demonstrated that these examples generated significant confusion and disagreement among the experts, and confirmed that there is no general agreement about what is meant by behavior.

The primary framing offered by Levitis et al. was whether the example listed was an example of behavior or not. Using the PTB taxonomy, we can see why this is a problematic frame. The PTB posits that if there is an entity defined relative to a field and there is an identified pattern of change over time, then, by definition, behavior has occurred. With this broad definition of behavior in place, the PTB makes clear that each of the examples given by Levitis et al. is a behavior of some kind. This means that the PTB lens shifts the frame from "Is this a behavior?" to "What kinds of behaviors are these events?" Put another way, Levitis and colleagues did not recognize that to be successful they needed to first define behavior in general and then secondarily specify the kind of behavior in which they were interested (which in their case was the behavior of the living organism, as a whole).

A clear prediction emerges for the PTB as a taxonomy of behavior, which is that it should lead to a much more reliable way to classify the descriptions offered by Levitis et al. (2009). To foster clarity in the classification scheme, we can follow the 12 floors of science that we have just reviewed and see if we can classify the behavioral examples from the study accordingly. Their first example was "a person decides not to go to the movies if it is raining," which is a deliberate, reflective act, justified by a human person. This example can be categorized as taking place on the fourth dimension, the Cultural-Person plane, and it is readily placed at the level of the individual human person. Thus, we can place it as an example of human mental behavior that resides on Floor 11 of the PTB.

The second example is "a beetle is swept away by the current in a river." The movement being framed refers to physical forces, and the fact that the entity is a beetle as opposed to a dust particle is not relevant for describing the pattern of change being referenced. This example shows why it is important to be clear about the frame of reference. This makes the example a case of a general object-level, material behavior. This means it is not an example that fits at a primary level of analysis, but rather is an aggregate of matter, which in this case is in the form of a beetle. The third example, "a spider spins a web," offers a straightforward case of a general animal-mental behavior, thus falling in the animal-Mind[1] category, and fits into a behavior patterns that takes place on Floor 8.

The fourth example is "a plant bends toward the sun." This describes the behavior of an organism-as-a-whole, which represents an example of the second dimension of complexity (biological-organic) behavior, at the level of the whole organism. The example could be considered a general organism–environment relation (Life/Biology-General). If one were inclined to correspond it to a floor on the PTB, it would go into the multi-celled domain and thus be placed on Floor 6, which fits with a primary science associated with that floor, botany. The next example is "geese fly in a V formation," which offers another clear example of mental behavior. However, unlike the spider example, the reference is at the animal group level, and thus is placed on Floor 9.

The sixth example is "a person's heartbeat speeds up following a nightmare." This example is a bit less clear because it potentially involves several different kinds of behaviors as mapped by the PTB. The dream itself can only be directly experienced by the subject and is thus a Mind[2] event. Mind[2] events are tricky because of the epistemological gap and are not directly accessible from the exterior PTB lens. Note that if it were a person describing how they experienced the nightmare, this would be different (i.e., it would be a Mind^2 event reported on via Mind^3). It is not fully clear from the description if the referent is the nightmare (Mind^2), the verbal description of the nightmare (Mind^3), the neurocognitive activity associated with the nightmare (Mind^1) as measured by a brain scan, or the increasing heartbeat rate, which would be classified as the domain of biophysiological processes (i.e., Life). Thus, the PTB analysis on this example suggests it is confused in that it has several different referents that are intertwined, and that more specification is required.

The seventh example is "algae swim toward food," which is a biological-organic life behavior. This could be interpreted to be at either the individual or group level, depending on whether the referent is the coordinated group as a whole or a bunch of separate individual cells. Thus, it could be a Life-Biology/General referent, or Floor 5 (e.g., individual cell) or Floor 6 (e.g., groups of organisms behaving in a coordinated fashion). Finally, "a rabbit's fur grows over the summer season" is a Life-Biological/General behavior (not mental, because it is not mediated by the brain/nervous system). This also raises an interesting question regarding how a "generic" part of a multi-celled organism (i.e., fur) rather than a primary part might be framed.

The summary point is that by viewing behaviors through the lens of the PTB we can shift the analysis from the dichotomized question of whether something is a behavior to the question of what kind of behavior is being examined. Moreover, this can be done by specifying the level-dimension frequency on which the behavior in question is operating as mapped by the PTB taxonomy. It is worth noting that this claim regarding the effectiveness of this classification scheme can be empirically tested. For example, researchers could essentially repeat the Levitis et al. (2009) study and then introduce and educate participants about how to place different categories, and then give a similar set of examples and see if much greater rates of reliable classification are achieved. I have informally done this in my classes and the results seem unambiguously clear that the introduction and application of the PTB, with minimal training, affords a much clearer and more reliable classification scheme than the basic and often vague frames the students brought to the task regarding the concept of behavior.

Seeing the World Through the PTB: An Everyday Example

The PTB affords us a new epistemological lens that creates an ontological taxonomy of entity–field relations across the levels and dimensions of analysis that make up the stratification of behavioral complexification in nature. The previous section showed how that taxonomy can classify different behaviors. Here we apply the taxonomy by depicting a straightforward, everyday situation and decoding behavioral patterns operating at different level-dimensional frequencies. To do so, we can imagine an individual at a furniture store considering whether to buy a table. The salesman has just made the pitch to buy the table, and the shopper responds that he can "see the table."

Our goal is to clarify how one can classify the various behaviors associated with this everyday observation through the lens of the PTB. The example aligns with a famous analysis by Arthur Eddington (1928) and his "Two Tables" paradox. The paradox highlights how there are two dramatically different ways modern humans can frame, describe, and experience a table. The first of Eddington's tables is the everyday commonsense experience (i.e., as experienced by the person via Mind[2]). The table is of a particular size, structure, and stability and can be used for eating on or writing or any other function. The second table is the table described by physics. It posits that the table is made up of quantum fields out of which particles and then atoms and molecules emerge. The table according to physics is mostly empty space. It is crucial to note that Eddington was not operating from a coherent naturalistic ontology that can effectively weave these two tables together. The ToK and PTB, however, give us a frame for doing just that.

Figure 10.2 depicts the basic components that are likely to be salient to a human observer. It is important to be aware that these elements would not be what is salient to other creatures, like dogs or woodlice. This is a reminder that the human mind categorizes the world in particular ways and that knowledge is always an interaction between knowers and that which is known. Thus, we are not situated from a "god's eye view," but rather are situated in this example as a "generalized human knower" interpreting a series of events that could be seen via a video camera (i.e., an exterior epistemological position).

The given situation is such that the individual, the statement, and the table are the figures that stand out from the environmental background.

Fig. 10.2 Seeing a table through everyday and scientific behavioral lenses

The intersubjective knowledge necessary for the tasks of observation become more obvious if we focus on the language. Consider, for example, that if the individual had said "wǒ kàn zhuōzi," I would not have been able to observe those words in any meaningful sense because I do not speak Chinese. This point highlights the fact that what is observable in part depends on the schema or knowledge of the observer. Professional chess players *literally* see different objects in front of them when they look at a chessboard relative to a novice.

With this point made, we can look at the scene anew, this time through the lens of the Periodic Table of Behavior (Fig. 10.3), and we can begin to label the various domains in the diagram.

A scientist who is informed by the ToK System and PTB divides the world into the various dimensions of complexity. She sees that some behaviors are physical-material, others are bio-organic, others are mental-psychological, and others are cultural-sociolinguistic. The behaviors of the atoms (Floor 2) in the table are taking place at the Matter dimension of complexity. We know that the light coming from the lightbulb consists of electromagnetic radiation that is being released in the form of photons (Floor 1). Those photons are bouncing off the table and entering the individual's pupils. Technically, photons are not matter in the way that physicists use the term because they have no mass; they are instead a bundle of energy that carries the electromagnetic force. However, they are still

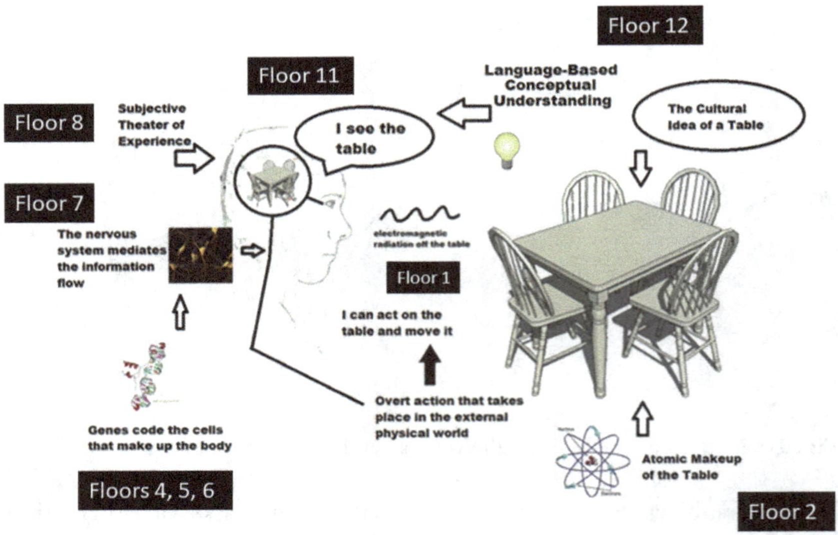

Fig. 10.3 Seeing human behavioral patterns through the lens of the PTB

very much a part of the energy-matter-space-time grid that makes up the Matter dimension.

The ToK scientist understands that the person is an organism and is made up of organ systems that are constantly communicating, which in turn are made up of collections of cells. This is the Life-Organism dimension of behavioral complexity. Inside those cells are genes (Floor 4) that provide informational content that has guided and shaped his physiological development. Although it used to be believed that the genes-to-physiology causal direction was only one way, we now know that there is much more of a bi-directional relationship between genes and epigenetic development. Of course, cells organize into organ systems, and the heart, liver, and many other organ systems self-organize to make up the whole human organism.

We know that the human being is an animal, and that the individual has a nervous system and brain. More specifically, we know he is part of an evolutionary lineage that flows through mammals and primates and into the great apes. The nervous system is a special kind of bio-organic system that connects the cells together in a "connectdome" (e.g., Sporns et al.,

2005) and coordinates the individual as a whole. The stream of information in the man's nervous system plays a coordinating and mediating role in his observable actions. The overt actions he performs, such as walking around the table, are all part of Mind^1 and framed in terms of neurocognitive functionalism and operate on Floor 8 on the PTB.

We know that the flow of neuronal information is crucial in generating his subjective, interior experience of the table, which we label Mind^2. This is the "everyday table" that Eddington referenced, as it appears in the man's subjective conscious experience of being. From an exterior, intersubjective perspective, we can be confident that the man has conscious access to this experience because he is reporting on it verbally. Both his actions and his perceptual experience of the table reside at the "mental order of behavior," labeled Mind in the ToK's descriptive metaphysical system.

We know that linguistics is a specific kind of symbolic information processing system that connects human minds, and that English is a particular variant of human language. His comment that he can see the table suggests that perhaps he is thinking about his options, and he does not want to commit to a purchase under the pressure of the salesperson. The self-conscious narration (Mind^3) and shared, intersubjective dialogue takes place on the Culture-Person plane of existence (Floor 11) on the PTB. Finally, the sociocultural context that allows for furniture stores and market exchanges would be Floor 12, the macro-level analysis.

We also need to note the table itself. Although the table's behavior is material, the form and function of the table did not emerge via physical-material causation alone. Rather, it emerged as a cultural idea that was enacted and shaped into a technology by the behavioral investments of people. It is an example of technology. On the ToK System, technology can be thought of as a hybrid between the material and the cultural dimensions. The PTB is a map of modern natural empirical sciences. It does not depict material culture or technology, and a science of technology or related domains like computer science are not included.

Natural Science, Behavior, and Mental Behavior

The PTB completes the UTOK bridge between science and behavior. It shows how modern natural science is about empirically mapping behavior patterns at various scales and frequencies that correspond to the levels by dimensions taxonomy. The argument developed in these two last chapters shows conclusively why science should not be framed in terms of a structural physicalism or materialism. These frames were tied to physically reductionistic frames for cause and effect. The emergence of information theory, information science, chaos and complexity sciences, and philosophical analyses of living and mental creatures (Deacon, 2012) all show why a strong physical reductionism or eliminative materialism is a fallacious frame for scientific knowledge.

The ToK and PTB show how the natural world is singular and monistic in the sense that the foundational ground of existence is an Energy Information Field, out of which the dimensions of behavioral complexification called Matter, Life, Mind, and Culture emerge. Novel properties emerge via things such as the linking of parts together to make new wholes (e.g., electrons merge with protons to make hydrogen), and via aggregate groups across scales (e.g., the way groups of water molecules give rise to fluidity and rivers and lakes). In addition, there have been four great dimensional emergences of Matter, Life, Mind, and Culture. The latter three all give rise to complex adaptive dynamic systems, whose patterns of behavior are not just matter and energy on a field of space and time but include dynamically self-organizing complex adaptive systems that engage in information processing and communication behaviors that require different metaphysical concepts above and beyond efficient causation.

The natural behavioral view of scientific ontology grounded in UTOK also shows why any special framing of behavior by either behavioral biologists or psychologists must be clarified with an appropriate adjective that places it in the proper behavioral box. Behavioral biologists are interested in behaviors that take place in the Life-Organism dimension of complexification. Similarly, basic psychologists, along with ethologists and animal behavioral ecologists, are interested in mental behaviors that emerge in the Mind dimension. And human psychologists and other social scientists are interested in human mental behaviors that involve Mind^3 and the Culture-Person plane of existence.

Conclusion

The Enlightenment Gap left us with the BM^3 problem at the center of psychology's ontological confusions. The Unified Theory untangles this knot by first separating behavior from mental processes and then analyzing the epistemological and ontological aspects of both. It then gives us the ToK System, which allows us to clearly see that the universe is an unfolding wave of behavioral complexity that operates on four different planes of existence. The Periodic Table of Behavior deepens this analysis via the addition of the primary levels of analysis that result in the 3 × 4 grid. With it, the PTB clarifies the levels and dimensions of the stratification of nature and how they align with the disciplines in the natural sciences.

The PTB shows a generally clear correspondence between the ontic levels and dimensions and our scientific disciplines as we move from physics into chemistry into biology into neuroscience. However, ontological clarity breaks down when one gets to the Mind dimension, which is what would be expected given the problem of psychology. The Unified Theory argues that the reason for confusion found in the transitions from Floors 7 to 11 is clear. The Enlightenment Gap does not provide modern science the proper grammar to sort out the proper relation between matter and mind. The UTOK metapsychology, in contrast, provides a clear descriptive metaphysics to map the ontology. And with modern natural science mapped by the concept of behavior, which is then properly framed by the ToK and PTB, we can now move to mapping the domains of mental processes and specifying their metatheoretical interrelations in greater detail.

References

Bollen, J., Van de Sompel, H., Hagberg, A., Bettencourt, L., Chute, R., Rodriguez, M. A., & Balakireva, L. (2009). Clickstream data yields high-resolution maps of science. *PLoS One, 4*(3), 1–11. https://doi.org/10.1371/journal.pone.0004803

Brown, T. L., LeMay, H. E., Jr., Bursten, B. E., Murphey, C. J., Woodward, P. M., Stoltzfus, M. W., & Lufaso, M. W. (2018). Introduction: Matter, energy, and measurement. In *Chemistry: The central science* (14th ed.). Pearson.

Deacon, T. (2012). *Incomplete nature: How mind emerged from matter.* W. W. Norton.

Eastman, T. E. (2020). *Untying the Gordian Knot: Process, reality, and context.* Lexington.

Eddington, A. S. (1928). *The nature of the physical world.* Dent.

Giraud, T., Pedersen, J. S., & Keller, L. (2002). Evolution of supercolonies: The Argentine ants of southern Europe. *Proceedings of the National Academy of Sciences of the United States of America, 99*(9), 6079.

Henriques, G. R. (2004). Psychology defined. *Journal of Clinical Psychology, 60*(12), 1207–1221.

Henriques, G. (2008). The problem of psychology and the integration of human knowledge: Contrasting Wilson's consilience with the tree of knowledge system. *Theory and Psychology, 18*, 731–755.

Henriques, G. R. (2011). *A new unified theory of psychology*. Springer.

Henriques, G. R. (2017). Character adaptation systems theory: A new big five for personality and psychotherapy. *Review of General Psychology, 21*(1), 9–22. https://doi.org/10.1037/gpr0000097

Henriques, G. R., & Michalski, J. H. (2019). Defining behavior and its relationship to the science of psychology. *Integrative Psychological and Behavioral Science, 54*, 328–353.

Henriques, G. (2021, October). *The "singular energy-information superforce field" that started it all*. [Blog] Medium, unified theory of knowledge. Retrieved August 1, 2022, from https://medium.com/unified-theory-of-knowledge/the-singular-energy-information-superforce-field-that-started-it-all-2330bfc331d7

Horvath, I., Szécsi, D., Hakkila, J., Szabó, Á., Racz, I. I., Tóth, L. V., ... & Bagoly, Z. (2020). The clustering of gamma-ray bursts in the Hercules–Corona Borealis Great Wall: the largest structure in the Universe?. *Monthly Notices of the Royal Astronomical Society, 498*(2), 2544–2553..

Jaeger, G. (2019). Are virtual particles less real? *Entropy, 21*(2), 141.

Jordania, J. (2006). *Who asked the first question?: The origins of human choral singing, intelligence, language and speech*. Logos.

Levitis, D. A., Lidicker, W. Z., Jr., & Freund, G. (2009). Behavioural biologists do not agree on what constitutes behaviour. *Animal Behaviour, 78*(1), 103–110.

Lovelock, J. (2003). Gaia: The living Earth. *Nature, 426*(6968), 769–770.

Naour, P. (2009). *E.O. Wilson and BF Skinner: A dialogue between sociobiology and radical behaviorism*. Springer.

Rogers, P. C., & McAvoy, D. J. (2018). Mule deer impede Pando's recovery: Implications for aspen resilience from a single-genotype forest. *PLoS One, 13*(10), 1–19.

Scerri, E. (2005). On the formalization of the periodic table. *Poznan Studies in the Philosophy of the Sciences and the Humanities, 84*, 191–210.

Sporns, O., Tononi, G., & Kötter, R. (2005). The human connectome: A structural description of the human brain. *PLoS Computational Biology, 1*(4), e42.

Siegal, E. (2019, January). Ask Ethan: How do quantum fields create particles? *Forbes*. Retrieved August 7, 2022, from https://www.forbes.com/sites/startswithabang/2019/01/13/ask-ethan-how-do-quantum-fields-create-particles/?sh=479bda16330a

Tryon, W. W. (2016). Transtheoretic transdiagnostic psychotherapy. *Journal of Psychotherapy Integration, 26*(3), 273.

Volk, T. (2017). *Quarks to culture: How we came to be*. Columbia University Press.

Villarreal, L. P. (2008, August). Are viruses alive? *Scientific American* [Online]. Retrieved August 5, 2022, from https://www.scientificamerican.com/article/are-viruses-alive-2004/

PART V

Defining Mental Processes and Grounding the Domains in Metatheory

CHAPTER 11

Mental Behaviors and the Map of Mind[1,2,3]

In the first half of the twentieth century, behaviorism in America evolved from Watson's neuro-reflexive model into a much more sophisticated drive-reduction model. One of the most influential leaders in these developments was Clark Hull. At the time, many thought his work was on the verge of a genuine breakthrough in terms of the mathematization of animal behavior. For example, one of Hull's most successful students, Sigmund Koch, argued strongly for the drive-reduction model in several prominent journals in the early 1940s. However, as Koch would document in his later writings (e.g., Koch, 1961), the paradigm would come crashing down in a surprisingly fast manner, such that by the late 1960s Hull's work was already drifting into the dustbins of history.

From the vantage point of the Unified Theory, a major reason for the fall was that Hull's system was grounded in a problematic conception of behavior, which, as we have documented, is tangled up with the problem of psychology more generally. If Hull had been grounded in a conception of mental behavior and had seen the potential of an information processing view of the nervous system, the story might well have unfolded in a different way. The following quote from Hull's (1943) influential book, *Principles of Behavior*, illustrates his justification system. Consistent with themes we have been emphasizing, Hull is explicitly grounded in natural science, and, not surprisingly, insists on an exterior epistemology. In a

G. Henriques, *A New Synthesis for Solving the Problem of Psychology*, Palgrave Studies in the Theory and History of Psychology,
https://doi.org/10.1007/978-3-031-18493-2_11

section titled "Suggested prophylaxis against anthropomorphic subjectivism," he wrote (pp. 27–28):

> One of the greatest obstacles to the attainment of a genuine theory of behavior is anthropomorphic subjectivism.…One aid to the attainment of behavioral objectivity is to think in terms of the behavior of subhuman organisms, such as chimpanzees, monkeys, dogs, cats, and albino rats. Unfortunately, this form of prophylaxis against subjectivism all too often breaks down when the theorist begins thinking what he would do if he were a rat, a cat, or a chimpanzee; when that happens, all his knowledge of his own behavior, born of years of self-observation, at once begins to function in place of the objectively stated general rules or principles which are the proper substance of science.
>
> A device much employed by the author has proved itself to be a far more effective prophylaxis. This is to regard, from time to time, the behaving organism as a completely self-maintaining robot, constructed of materials as unlike ourselves as may be.…It is a wholesome and revealing exercise, however, to consider the various general problems in behavior dynamics which must be solved in the design of a truly self-maintaining robot…
>
> The temptation to introduce an entelechy, soul, spirit, or daemon into a robot is slight; it is relatively easy to realize that the introduction of an entelechy would not really solve the problem of design of a robot because there would still remain the problem of designing the entelechy itself, which is the core of the original problem all over again. The robot approach thus aids us in avoiding the very natural but childish tendency to choose easy though false solutions to our problems, by removing all excuses for not facing them squarely and without evasion.

This passage highlights the deep concerns that Hull had regarding imputing a subject into his animal subjects. Our specifying the different domains of mental processes allows us to see that there are two different issues that Hull is wrestling with. One is the epistemological problem of the impossibility of seeing Mind^2 from the exterior position. The other is the ontological problem of how mentation in general and consciousness in particular work and generate causal consequences in the world. Consistent with UTOK's argument that psychologists have long been blind to the fact that these are two separable issues, there is no indication that Hull was aware of the importance of disentangling these two aspects of the problem.

When Hull was writing his *Principles of Behavior*, he was also unaware that the decade that would follow would launch the cybernetics-into-cognitive science revolution. His comment about robots highlights why at least a part of him would have likely been positively inclined. Indeed, in his analysis, we can see why many scholars left behaviorism and started investing in the emerging cognitive approaches and related areas like cybernetics and artificial intelligence. If animals are like robots, then it follows that it would be advantageous to build robots and see how they work. It is worth noting here that Hull is using the concept of a robot primarily to emphasize the epistemological blindness associated with the problem of Mind[2]. However, it also points to the ontological issue regarding the nature of mental processes that would become very real in the decade that followed. In the language of UTOK, scientists were developing models of Mind[1] as information processing that would ground a new ontological picture of the mental.

The Mind as a Neuro-information Processing System

Cognitive science is an interdisciplinary endeavor that includes cognitive psychology, linguistics, anthropology, neuroscience, philosophy of mind, and artificial intelligence. As we noted in the last chapter, it corresponds to the tenth floor of the PTB, in that it deals with the parts and processes that enable human mental behavioral patterns. The central organizing idea in cognitive science is the claim that the nervous system in general—and brain in particular—is a kind of information processing system. The most general version of this claim is that the brain works by translating changes in the body and the environment into a "language" of neural impulses that communicates messages that in some ways model or represent or symbolize the animal–environment relationship. Computational operations can then be performed on these representations, which in turn can be used to guide outputs (i.e., actions).

The broad and general structure of the computational theory of the mind was a breakthrough because, for the first time, it allowed scientists to conceptually separate the mind from the biophysical brain and the rest of the body. Scientists could now conceive of "the mind" as the flow of information through the nervous system. As we have noted, this is essentially how the Unified Theory defines the mind. Whereas Mind is the third plane of existence on the ToK and references mindedness and the set of animal-mental behavior, "the mind" is defined as the information instantiated within and processed by the nervous system. The information content, storage, and processing are also what the word "cognitive" broadly refers to in the neurocognitive functionalist meaning of Mind[1]. As we will see, the difference between Mind and the mind is important and sits at the heart of some of the confusions and debates in modern cognitive science between traditional cognitive neuroscientists who focus on "the mind" and 4E cognitive scientists who focus on Mind and mindedness.

For an analogy about how information can be conceptually separated from the biophysical matter that makes up the brain, think of a book. The book's mass, temperature, and size can be considered as being akin to the physical brain. The story the book tells or claims it makes is the semantic information content. The analogy is far from perfect because the book is not an information processing system. This raises the question of exactly what is meant by "information." As is the case with all fundamental concepts, the concept of information has many layers. In the context of the Unified Theory, we can start by noting the ubiquity of information. That is, the concept is everywhere, and it can seem to apply to almost everything. For example, in UTOK's language game, everything can be framed as being in and spawning from a web of energy and information, which resides at the base of the ToK System. This meaning of information can be thought of as a web of interconnected data points and their differentiation. However, when we move to information in the context of neuro-information processing, it has different referents.

Three Meanings of Information in Cognitive Science: Processing, Semantic, and Theoretic

To get clear about what information means in the context of cognitive psychology, we can start by dividing it into three different referents as follows: (1) information processing; (2) information as semantics, schematics, or epistemic meaning-making; and (3) information theory and its relation to prediction and the reduction of uncertainty. Starting with the first concept, the concept of information processing at a minimum involves: (1) an input/translation system; (2) a rule system that can perform recursive operations on the input; and (3) some form of output mechanism that generates a response or work effort. Consider, for example, a calculator. When you push on the numbers and operations, say 5 × 6, those inputs are translated into an algorithm and then electronic signals are decoded such that 30 shows up on the screen.

This simple description allows us to see why the nervous system can be readily described as an information processing system. It has afferent neurons that are triggered to respond to some change in the environment or the body. This is the input/translation function, which is technically called transduction. These signals are then transmitted to intermediate neurons that are arranged in hierarchical networks that are in constant communication with each other. Finally, there are efferent neurons that connect to muscles and other systems and regulate motor outputs. There is much more to be said, of course, but the point here is straightforward and basic. The informational input-computation-output processing sequence is apparent in the structural and functional arrangement of the nervous system.

The second meaning of information pertains to, well, meaning. To start to understand meaning-making, we can turn to the field of semiotics, which focuses on the semantic meaning of information, and how signs and symbols have specific referents. Earlier in this book, we introduced the semiotic triangle, and showed how the symbol of "dog" was different from but connected to both the idea or mental representation of dogs and actual dogs in the world. The idea of a dog refers to the semantic or schematic aspect of information. This brings us to semantics, which involves meaning-making and the way a subject perceives, models, or represents the outside world.

Understanding information as semantic meaning is one of the more complicated aspects of information—and in science and philosophy more generally. This is because it rather quickly translates into the problem of meaning in general. That is, it requires an understanding of how the information is processed or interpreted or epistemically organized or experienced by different entities. There are deep questions to be asked about the meaning aspects of information, and the extent to which that meaning exists independently of the knower or is dependent on the entity doing the processing. The more general point is that there are some kinds of interpretive schema or model-making or representational processes that organize the inputs and allow for a determination of what the incoming forms are and their implications for the agent that is construing the situation.

To see the relevance of this, consider, for example, the odor of feces. Is there such a thing as "the odor of feces" independent of individuals with various schematics for detecting it? When we humans think of the odor of feces, it is usually tied to an aversive evaluation of disgust and the sense of lacking in cleanliness. However, dung beetles interpret such stimuli very differently than humans do. One way to conceive of cognition or even consciousness in functional terms is that it is the semantic meaning the neurocognitive system generates to frame the animal–environment relation and identify what is relevant in the landscape of affordances that might guide its path of behavioral investment. We will come back to this issue on several occasions in the chapters that follow. For now, we need to be aware that information in cognitive science can refer to semantic meaning-making or the subjective epistemic construal of how the animal is making sense of the inputs to guide its actions.

The third key idea pertains to information theory, which involves the logic of probability, predictability, variation, and the ways in which data can be used to reduce uncertainty. This referent connects to a set of ideas developed most notably by Claude Shannon in the 1930s and 1940s. As a set of concepts, information theory connects most directly to mathematics, statistics, physics, and computer science. One way to think about information theory pertains to how much uncertainty there is in a particular situation and how data about that might allow you to reduce uncertainty. Consider, for example, that I have a coin and will flip it and you must guess whether heads or tails will come up. Since there are only two states, the uncertainty regarding the prediction is low and one bit of information will give you certainty about the message (i.e., if you know one side, you know the other). However, if we were talking about the outcome of a lottery that has over a billion possible combinations, you would need more than a single bit of information to predict the outcome.

Karl Friston is a psychiatrist who has made substantial contributions to cognitive neuroscience because he has worked on developing a predictive processing model of the brain and nervous system that draws on each of these conceptions of information. His work shows that we can usefully consider the neurocognitive system as attempting to model and predict the future in a way that is deeply consistent with the information theoretic meaning of the term (see, e.g., Friston, 2010). In *The Mind: Consciousness, Prediction, and the Brain*, the cognitive scientist Bruce Goldstein (2020) similarly emphasizes the idea that one of the key functions of neuro-information processing is the reduction of uncertainty and anticipation of the future.

It is important to note that when, in the context of UTOK, we consider the mind as consisting of the information instantiated within and processed by the nervous system, we are invoking all three aspects of these meanings. First, as we have noted, the structural and functional organization of the nervous system can be framed as an input-computation-output processing system. Second, the nervous system is structured as a semantic system that determines the forms of inputs and references them against a sense- or meaning-making system of some sort that enables the animal to generate a functional response. The last element links to the information theoretic notion and is connected to recent work in active inference and predictive processing which frames the neurocognitive system as attempting to reduce uncertainty between its model of the world and the inputs it receives. With this brief description of these different meanings of information, we can turn to the history of cognitive science, and why it generated a revolution in how we think about the ontology of mental processes.

Cybernetics and Cognitive Science

The decade that spanned 1943–1953 saw the emergence of "cybernetics," which the founders believed was a pathway to the new science of mind. The Unified Theory shares much affinity with the cybernetic approach. Indeed, the neurocognitive functionalism that frames our understanding of both the mind and Mind can be appropriately considered a neurocybernetic approach. In *The Embodied Mind: Cognitive Science and Human Experience*, Varela et al. (1991) review how the developments made by the cyberneticists produced a remarkable array of lasting insights, which included: (1) the use of information theory to understand signal and communication channels; (2) mathematical modeling of the connections and communications inside the nervous system; (3) the establishment of a general discipline of systems theory that would connect to many fields, including laying some key groundwork for modern complexity science; (4) the invention of information processing machines; and (5) the modeling and examples of self-organizing systems via feedback and control processes.

Although these developments are foundational to modern cognitive science, for a host of complicated reasons, cybernetics became separated from the mainstream. One reason for this was that cybernetics was connected to many fields and had many different branches and applications, such that it did not become structured in a way that set the stage for a centralized science. What did emerge in the subsequent decades (i.e., the 1960s through the 1980s) as a primary focus was the idea that the mind functioned as a logical, symbolic, algorithmic, computational processor. The famous physicist Richard Feynman once remarked that if you want to show that you understand something, you should proceed to build it. Consistent with this mindset, much attention was focused on the construction of artificial intelligence systems. Artificial intelligence refers to machines that can "think" or, more accurately, at least simulate important aspects of human thinking; for example, computer programs like the chess-playing Deep Blue, which, in 1997 became the first computer program to beat the best human player in the world (Newborn & Newborn, 2003).

Although the cognitive revolution was an important move forward relative to behaviorism, significant problems emerged. Some of these problems stemmed from the fact that framing the "mind as information" allowed it to be readily separated from the biophysical brain. Researchers became fascinated with models of algorithmic processors (i.e., computers) that had little connection with the other elements of mental phenomena as they appear in nature, such as conscious experience, overt actions, evolution, culture, or even the brain itself. In a word, the artificial intelligence and symbolic information processing models were "disembodied." Given that animal and human minds are fully embodied, it is perhaps not surprising that the idea that mind-brain systems work like disembodied electronic computers led to many dead ends. This is in large part because the way the nervous system processes information is very different from a computer.

The Emergence of 4E Cognitive Science

Valera, Thompson, and Rosch's work on the embodied mind was an influential turning point in the field. They highlighted how science tends to focus on an exterior epistemology and how relatively few thinkers grappled deeply with the connections between science, subjective experience, and the world. The authors drew significantly on the work of the philosophers Edmund Husserl and Maurice Merleau-Ponty and encouraged a scientific perspective that could bridge to human phenomenology to engender a more embodied conception of mentation. They also made many connections between Buddhist traditions and cognitive science. Their work played a key role in the emergence of what is now known as "4E" cognitive science, which emphasizes how cognitive processes are embodied, embedded, extended, and enacted in the agent–arena relationship. The authors characterize embodied cognition in terms of highlighting the fact that "cognition depends upon the kinds of experience that come from having a body with various sensorimotor capacities, and second, that these individual sensorimotor capacities are themselves embedded in a more encompassing biological, psychological and cultural context." Put differently, unlike computers that are stationary and algorithmically programmed, animal and human mentation is characterized by an agent–arena relationship, and the information processes must be situated in that context and its development, both evolutionarily and ontogenetically, to be properly framed.

What emerges in the 4E cognitive science tradition is a view of mentation that shifts from the abstracted symbolic algorithmic information processing models prominent in fields like artificial intelligence and computer science and returns these processes back in the lived experience and (mental) behavioral contexts of both animals and humans. It argues that we must deeply consider phenomenology as part of the human mental field and recognize that the embodied context and interface with the environment is crucial to understand how mental activity works. A few examples can help clarify the 4E cognitive science viewpoint. Consider, for example, how so much of our language relates to our embodied place in the world. When we say we "look up to someone" or things are "over our heads" or she is "close to him" we are using metaphors that are drawn from our embodied relationship to the environment. Other examples come from how we can gain understanding by acting things out. For example, researchers have shown that when scientists were attempting to guide a rover on Mars, they worked to internalize the structure of the rover so that they imagined themselves becoming the rover's eyes and arms, as well as imagining the rover as becoming more human-like (Chiappe & Vervaeke, 2018).

Other considerations involve how thoughts can be extended into the environment. Consider the following: Mr Jones is headed to the grocery store and needs to remember seven items. To do so he imagines the list in his head, and he repeats them over and over, and when he gets to the store, he retrieves the memory. This is a clear example of processing that seems to take place fully in his head. Now consider Mrs Jones, who needs to go to the hardware store and get seven items. She sometimes has trouble remembering, so instead of reciting the items she takes out a piece of paper and makes an actual list. Then, upon arriving at the hardware store, she retrieves the list from her pocket. The point here is that by writing the list down, a strong argument can be made that Mrs Jones has extended her cognition into the environment, such that now there is a cognitive system that is no longer confined to events within her nervous system. Similar arguments have been made with animals, such that some scientists consider a spider's web to be an extension of its mind in that it uses the web to detect various kinds of movements and vibrations in the larger environment (Sokol, 2017).

The move to 4E cognition is particularly relevant from the vantage point of UTOK. One key reason, although one that is often overlooked, is that it results in the cognitive perspective having much closer connections with behavioral science. We can see this in the history of behaviorism. As it moved from reflexology into Hullian drive theory, clear parallels can be drawn between Hull's work and the control theory feedback loops that are at the heart of cybernetic thinking. It is also the case that there were a group of psychologists who identified as "neo-behaviorists." These scientists were focused on understanding the mediating and intervening elements that coordinated such overt behaviors. For example, the neo-behaviorist Tolman argued that rats who spent time in mazes made "cognitive maps" of the territory. This could be empirically demonstrated because they would learn to run the maze much faster if they were exposed to the maze, even if there were no rewards or punishments included.

The point here is that as 4E cognition places mental processes in an embodied, embedded, enacted, and extended context, it becomes increasingly aligned with viewing the functional behavior of the animal in its environment. Once we link overt animal behavior with an exterior epistemology and note that cognitivism is about tying overt actions and physiological responses together with the concept of the mind as an information processing system, then the idea that neurocognitive functionalism is one side of the coin and overt mental behavior is the other side becomes increasingly apparent. This is the set of mental behaviors and the dimension of Mind or mindedness in nature. The result is a powerful bridge that gives rise to a clear frame of understanding for the first domain of mental processes, Mind[1].

Mind, Mental Behavior, the Mind, and the Domain of Mind[1]

The UTOK's descriptive metaphysics for the domain of the mental is now coming into view. It is aligned with developments in the evolution of thought from behaviorism into the cognitive revolution and, more recently, 4E cognitive science. Because it affords a new map of Big History, the ToK gives UTOK a unique angle upon which to frame this complicated set of metaphysical and ontological issues. When combined with the

PTB lens, UTOK allows us to see clearly that mental behaviors are a particular kind of behavioral pattern in nature. And we can frame the set of these behaviors as Mind, which is as readily definable as the concept of Life. Moreover, we can align with the cognitive revolution in defining "the mind" as the information instantiated within and processed by the nervous system. This aligns strongly with the traditional cognitive perspective. However, UTOK also situates neuro-information processing in a natural context and an animal behavioral context, which brings us back to a more embodied conception, aligned with 4E cognitive science.

This dynamic regarding the difference and relation between Mind and the mind is apparent in recent debates in the cognitive science literature. Gallagher's (2017) *Enactivist Interventions: Rethinking the Mind* throws down the gauntlet and challenges mainstream cognitive psychology to shift from a traditional neuro-information processing model, where representations mediate inputs and outputs, to seeing the mind as an entire complex adaptive system. In this view, the grocery list that Mrs Jones generates is part of the mind. Of course, to traditional cognitive psychology, this seems odd because it does not mediate between the stimuli and responses, but rather is simply a stimulus outside the neuro-information processing system. The metaphysics and ontology of UTOK allows us to see that traditional cognitive psychology is referring to the neurocognitive information patterns within the nervous system, whereas modern 4E cognitive scientists are framing the mind in terms of the set of mental behaviors identified by Mind on the ToK System.

This background means we are now ready to more clearly specify Mind[1], the first domain of mental processes. Mind[1] refers to the neurocognitive functional processes that enable animals to move as coordinated singularities in their environment (Henriques, 2003), whereas Mind refers to the system of complex adaptive mental behavioral patterns. To enrich the picture of how UTOK frames mental behaviors and the mind, consider the following narrative:

> The hunter carefully positions himself, well camouflaged against the backdrop of a tree. He blends in naturally. An ambush predator, he waits patiently, for hours, even days at a time, sitting motionless. He is good at estimating distance, using binocular disparity to contrast visual images to see where exactly he needs to strike. Although his distance detection is excellent, there is no color.

> The stimulus in front of him is referenced against templates for prey and nonprey. When an unsuspecting prey emerges in his visual field, he tracks it by repositioning his head and forelimbs ever so slightly. His eyes are large and forward-looking. He can detect, fixate, and track visual objects. His delicately coordinated head movements keep the object in an acute zone of highest spatial resolution. He identifies his prey via a combination of visual cues including overall size, contrast to background, location in the visual field, and estimated speed. He will only strike when the prey is at the appropriate distance. His largest and most ambitious target is a hummingbird. When he strikes and captures one of them, the struggle is intense. If he can, he will go straight for the brain of the hummingbird to subdue it as quickly and efficiently as possible. In addition to tracking prey, he is attuned to dangers from above. Any sudden movement can signal his own life to be at risk, and he will take evasive action, perhaps launching himself out of his hiding place, and spiraling downward to avoid danger.
>
> This coordinated pattern of hunting behavior is tied together via neuro-information interface, processing, and transfer. Photons from the sun illuminate the environment around him. They enter his eyes and the patterns in the light are sorted and processed by the sensory neurons in retina. That information, encoded in neural impulses and shared between specific parts of the brain, flows first through the optic nerve and then into the optic lobe. The signals are processed in three distinct neuropils in the optic lobe: the distal lamina, the medulla, and the proximal lobula complex. The former portion of his brain is particularly well-differentiated, which is what allows for him to process distance so accurately.

The "he" in this narrative is a praying mantis and the description is based on a chapter titled "In the Mind of a Hunter: The Visual World of Praying Mantis" from a book that explored the complex activities of insects called *Complex Worlds from Simpler Nervous Systems* (Prete, 2004) When the authors talk about "the mind of the hunter," they are referring to the neurocognitive functional information processing that is playing a key role in coordinating and regulating the mantis' overt mental behavior. This is the interior part of Mind[1].

Mental behavior is functional in that it is organized based on goals or paths of behavioral investment, which can be framed via the language of complex adaptive systems. The key features of this description include the fact that the mantis coordinates his actions in response to relevant cues in the adaptive landscape and in ways that have functional outcomes toward attractor states. We can say that he hunts, seeks mates, and works to avoid being eaten precisely because there is a pattern of functional awareness and responsivity that can be explained via behavioral landscapes that are oriented to approach certain states and avoid others. The mantis has a network of brain systems that are constantly processing information and sending signals to regulate the expenditure of overt action, depending on present circumstances (i.e., determining if certain movements are from a prey or predator). All of these are mental behaviors of the Mind[1] variety. It is important to be clear that we are not directly addressing the issue of whether the mantis has any subjective conscious experience of being in the world. That is, no claims are being made about whether the mantis has Mind[2] processes at this point.

Consistent with the work in 4E cognition and complexity science, we can map this description of the praying mantis onto a general map of complex adaptive systems. Figure 11.1 provides a diagram of the key components that go into such systems.

The diagram captures how complex adaptive systems take in energy, matter, and information to maintain self-organization. In the case of the praying mantis, it breathes, eats, and scans the environment for relevant signals so it can realize its goals. The "rule system" in the middle of the diagram represents the way the mantis processes incoming information via detectors, computes that information based on rules, and then generates output via effectors. This corresponds to the neurocognitive functional meaning of the mind in the sense that it refers to the information instantiated within and processed by the nervous system. All of this is nested inside the mantis' body, which refers to the biophysiological or dimension of living processes. We can correspond that to the "metabolic system" that surrounds the rule system. Of course, the mantis lives in the material environment. And, unlike stationary plants, the mantis moves as a unit that mates, hunts, and avoids predators. This is labeled the "adaptive learning" feedback loop on the diagram. Directly aligned with 4E cognitive science, this "agent-arena" movement is foundational to Mind, the set of mental behavior.

Mental Behaviorism

The mantis example allows us to readily divide Mind[1] into two distinct domains. On the one hand, there are overt actions that can be observed via the third-person perspective. On the other are the neurocognitive processes that take place within the nervous system. This book opened with

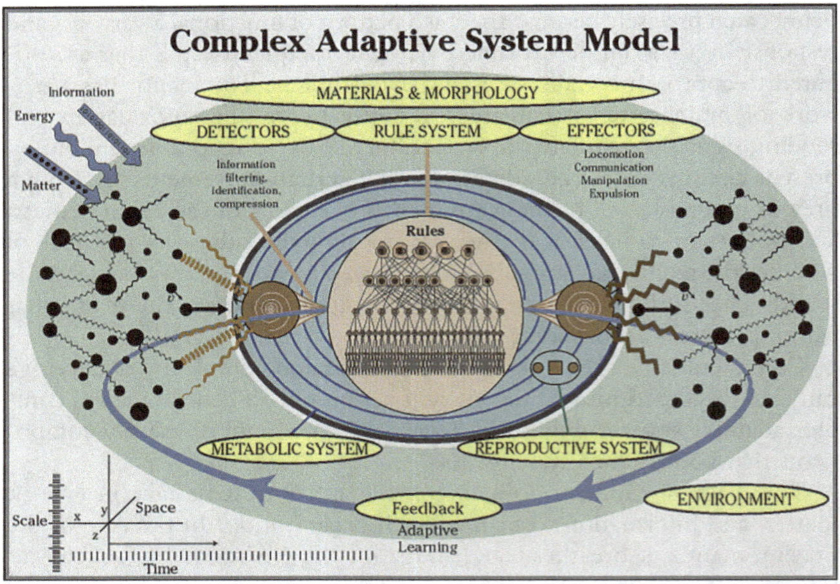

Fig. 11.1 A generic representation of a complex adaptive system https://en.m.wikipedia.org/wiki/File:Complex_adaptive_system.gif

an analysis of cognitive behavioral approaches to psychotherapy and how they represented a blend of techniques and perspectives on thinking and actions that pragmatically made sense but that theoretically were suspect. The reason is that the CBT paradigm carries with it clear lineages from both behavioral and mentalist traditions, but there never was a deep dive into the descriptive metaphysics and ontology of mind that enabled practitioners to clearly address the long-standing divide that exists in the two traditions. The fact that some academic textbooks still define psychology as the science of the mind sans behavior whereas other textbooks define psychology as the science of behavior sans mind is good evidence that the mentalist versus behaviorist problem in psychology has never been successfully addressed (i.e., imagine if some textbooks defined biology as the science of life, but not organism behavior, whereas others did the reverse).

As was detailed in Henriques (2011), the Unified Theory resolves the mental versus behavioral divide via the frame of mental behaviorism. Mental behaviors are the sensory-motor looping processes that produce a functional effect on the animal–environment relationship. Mental behavioral patterns are what is being described in the description of the praying mantis. To more precisely explicate the patterns of behavior that are captured by the concept of mental behavior, imagine three cats being dropped from a tree: one dead, one anesthetized, and one awake and aware. As they fall, all three cats behave. However, they behave very differently. The first falls through the air, lands on the ground, and bounces. Applying the lens of the PTB, we can frame these behaviors as purely physical or material behaviors. The falling is a function of gravity, the mass and shape of the cat, and things like air resistance. The second cat also falls and bounces. However, if you look inside the second cat, you see behavior of a different kind. The heart pumping, the lungs breathing, and cellular activity that maintains the cat's biological integrity are physiological behaviors. These are behaviors taking place on the Life dimension and are the subject matter of biology.

The final cat, awake and aware, lands on its feet and takes off. This is behavior of a different sort. Mental behavior is defined as the behavior of the animal as a whole mediated by the brain and nervous system that produces a functional effect on the animal–environment relationship. If the cat took a swipe at its handler because it was angry that it had been dropped, that aggressive behavior would also be mental behavior or, equivalently, minded behavior. However, when we ask the question of what it is like for the cat to be angry "from the inside," we find this question to be much more difficult for the language game of modern science to answer. That is, we can see that the domain of Mind^1, defined in terms of mental behaviors, is readily accessible to be analyzed from the vantage point of science. However, as we saw in the PTB chapter, Mind^2 is qualitatively different when it comes to observing it from the outside.

Framing Subjective Conscious Experience Via Mind^2

We have already defined Mind^2 as subjective conscious experience, which will be hereafter abbreviated as SCE. Defining Mind^2 in terms of SCE allows us to emphasize two key aspects of Mind^2. The "subjective" captures the epistemological aspect, whereas "conscious experience" characterizes the ontological referent. To deepen our understanding of this framing, we can correspond it to everyday definitions of consciousness. Merriam-Webster (n.d.) defines consciousness in five primary ways: (1) the explicit awareness of inner experience, objects, or the world; (2) a state of mind or experience characterized by sensation, emotion, volition, and thought; (3) the totality of conscious states; (4) the normal state of conscious life (framed in terms of arousal and activity, as opposed to being asleep or in a coma; i.e., unconscious); and (5) the "upper level" of mental life, as contrasted with unconscious processes.

The first definition describes both an inner experience of the world and a secondary, reflective awareness of that experience. Using our taxonomy of mental processes, we can say that, applied to humans, it describes a combination of and interface between $Mind^2$ and $Mind^3$. The second definition corresponds directly to the domain of $Mind^2$. The third definition regarding the totality of states reminds us that there are many facets to consciousness (e.g., states of arousal or breakdown into psychotic states or spiritual insights achieved in "higher" transcendent states). The fourth definition of consciousness corresponds more closely to $Mind^1$ and aligns with what we have described as functional awareness and responsivity, which can be perceived from an exterior, behavioral vantage point. Finally, the fifth definition can be interpreted as corresponding to a combination of the domains of $Mind^2$ and $Mind^3$, as differentiated from the domain of $Mind^1$. With this survey of basic definitions framed by our three domains of mental processes, we can now shift our focus to more philosophical issues associated with consciousness.

The Easy and Hard Problems of Consciousness and Their Alignment with $Mind^1$ and $Mind^2$

One of the most influential philosophical approaches to consciousness has been provided by David Chalmers and his analysis of the hard problem. Chalmers' argument traces its lineage to a famous essay by Thomas Nagel (1974) appropriately titled "What is it like to be a bat?" In that essay, Nagel argued that the core of the problem of consciousness was the subjective experience of being. He argued that the problem was so confounding that it would never be solved by science. He wrote (p. 437):

> The fact that an organism has an experience at means, basically, that there is something that it is like to be that organism…[Moreover] every subjective phenomenon is essentially connected with a single point of view, and it seems inevitable that an objective physical theory will abandon that point of view.

Chalmers (2007) built off Nagel's analysis and added clarity by separating out what he called the (relatively) easy problems of the mind from the hard problem of consciousness. He argued that the easy problems consist of neurocognitive functional analyses that provide insights on consciousness. For example, we can surmise that if you are reading this paragraph and thinking about it, the following is happening: (1) light patterns are coming off the screen and (2) flowing into your retina where they are (3) translated into the "language" of neurobiological information. The (4) incoming information is sorted and tracked back into the occipital lobe, where it is sorted further, integrated with higher-order processes, and connected (5) to your semantic-linguistic processing system. We can track all this via the activity of your nervous system, your overt actions, and your self-report. A psychologist could assess your "functional awareness and response patterns" by asking questions to assess how well you processed the information.

For Chalmers, these are easy problems for science to tackle. However, although such investigations provide information about the nature and function of consciousness, they leave the hard problem essentially untouched. Like Nagel, Chalmers argued that science may never be able to crack the nut of subjectivity. Here he is recently explaining the hard problem of consciousness to Sam Harris on the "Making Sense" podcast (Harris, 2020, pp. 4–5):

> It is useful to start by distinguishing the easy problems—which are basically about performance functions—from the hard problem which is about experience. [Some] easy problems are: How do we discriminate information in our environment and respond appropriately? How does the brain integrate information from different sources and bring it together to make a judgment and control our behavior? How do we voluntarily control our behavior to respond in a controlled way to the environment?...
>
> The easier problems fall within the standard methods of neuroscience and cognitive science What makes the hard problem of experience hard? Because it doesn't seem to be about behavior or about functions. You can in principle imagine explaining all my behavioral responses to a given stimulus and how my brain discriminates and integrates and monitors itself and controls my behavior. You can explain all that with, say a neural mechanism, but you won't have touched the central question, which is, 'Why does it feel like something from the first-person point of view?'

Of course, we have already encountered a variant of this analysis. Indeed, it started our framing of the BM^3 problem, which began with a review of Ray Jackendoff's splitting neuro-information processing off from phenomenology and calling it the mind-mind problem. We then added "mind as behavior" and "mind as self-conscious justification" to bookend the analysis and specify it as the BM^3 problem. This suggests we should be able to align Chalmers' analysis of the easy and hard problems of consciousness with the domains of Mind^1 and Mind^2, and this turns out to be the case. Mind^1 refers to the mental behaviors that are analyzable by the standard epistemological frames given by science, whereas the domain of Mind^2 is qualitatively different.

The Epistemological and Ontological Aspects of the Hard Problem

Chalmers is right to call the subjectivity of experience a hard problem in science and philosophy. However, as has been suggested by our analyses through this point, he is wrong to consider it a *single* problem. In fact, there are at least two different aspects that need to be separated (indeed, later in this work I will introduce a third aspect of the problem relating to historicity and uniqueness). More specifically, it is essential that we identify and differentiate the epistemological from the ontological aspects that make the hard problem hard. We have already encountered this issue in the context of exploring the science of psychology and the difficulty in bridging the epistemological gap, which is the fundamental distinction between the interior, subjective, phenomenological, first-person perspective and the exterior, objective, behavioral, third-person point of view.

We can add to this by thinking of Mind^2 as representing each sentient creature's unique experienced perspective on the world. We can call this one's subjective "epistemological portal" to knowing about the world. The combination of the unique epistemological portal and the gap between first and third positions will help us maintain our footing as we proceed. It will help us keep in mind that individual subjects are always anchored to their particular epistemological portal, and another can never directly bridge the gap and observe one's experience. This is true even if we consider or imagine another's perspective. It is even true for things like telepathy. Consider, for example, the science fiction notion of the "Vulcan mind meld" from Star Trek, which refers to the capacity of Vulcans like Mr. Spock to meld with the minds of others. At first, it looks like this might be an example whereby the epistemological gap is transcended, as the mind reader is seeing the thoughts of another. However, closer reflection reveals this is not exactly true. The mind reader is still seeing the thoughts of the other through their own epistemological portal.

Real life is sometimes even more remarkable than science fiction, and there is one real life case that shows that maybe the epistemological gap–portal relation may not be forever unbridgeable. As depicted in this documentary (https://www.cbc.ca/cbcdocspov/features/the-hogan-twins-share-a-brain-and-see-out-of-each-others-eyes), the Hogan twins are conjoined at the head. They are also joined at the brain, such that both twins share a "thalamic bridge." This means that their brains are intimately intertwined at the physiological level. The thalamus is often called the master relay center of the brain, as it is a central hub that picks up and distributes sensory and motor input. Although the girls have distinct selves and distinct subjective experiences of being, it is also the case that their minds "meld" together. To quote them, they can "see through each other's eyes" and "read each other's thoughts." These twins are perhaps the closest example of bridging the epistemological gap that I have seen. However, despite this remarkable example of interconnectedness between minds, it nevertheless remains the case that, even here, each of the twins has a separate observing self, and their awareness of being comes through each of their unique epistemological portals. They describe the connection by saying that "we talk in our heads." Put that way, there is still a first-person portal and third-person gap between them.

The epistemological portal–gap distinction is key, but we need to press further to clearly see the nature of the difficulty in applying the language game of science to SCE. This is because the epistemological portal of subjective knowing is almost directly antithetical to the epistemological grounding of "objective" science. That is, modern science is very much about factoring out that which is specific, idiographic, qualitative, and private to generate knowledge that is objective, general, nomothetic, quantitative, and public. For example, consider my first-person experience of sitting in this chair and working out the right way to describe SCE from the UTOK perspective. What do the laws of science say about this unique, idiographic SCE at this point in time? They give broad and general frames, but they do not do much to explain the unique, contingent specifics.

Another way of describing my unique experiences at this moment is that they are subjective, qualitative, unverifiable, and unreliable. These are exactly opposite to the kind of justification rules the science language game plays by. This is not to say that these claims are not true. It is just that they are first-person empirical claims that are unique and specific rather than third person and generalizable. The point here is that the grammar of modern science is fundamentally different than specific idiographic subjective experiences. This is thus another issue that must be grappled with. Subjective conscious experiences are real, specific, idiographic, particular, unique events. This nomothetic/general/theory versus idiographic/specific/real becomes another crucial aspect of the problem of consciousness and how it relates to the onto-epistemology of natural science.

With the epistemological issues framed in terms of both the point of view and the differences in the language game of science from the historical experiences of the particular individual, we can now consider the ontological aspect of the hard problem. This is variously referred to as the "explanatory gap" or the "neurobiological binding problem" or the "engineering problem." The nature of the ontological problem is first found in understanding how the "stuff of matter," especially the "stuff of the brain," turns into the "stuff of consciousness." This is indeed a difficult problem at the level of specific mechanisms. Our current understanding gives us a reasonable grasp on the Matter and Life dimensions. By this I mean that the fact that the chair I am sitting on is made up of atoms that are held together by covalent bonds makes conceptual sense. And I can understand that, at the level of an organism, I am made up of billions of cells, each of which has a nucleus and metabolizes energy and communicates with others. I can also understand that my nerves work something like wires, and that I control my fingers as I type, such that if you cut my spinal cord, the message would be lost, and I would not be able to move them. All that fits with my basic naturalistic worldview in a relatively complete way.

However, when I reflect on exactly how of my SCE comes into existence, I experience deep awe and wonder. It remains difficult to conceive of exactly how or where the "water" of material, mechanical brain processes give rise to the magical "wine" of first-person experience and my broader sense of self. Some of this mystery has to do with epistemology; we are used to thinking of science as our primary explanatory system, but the nature of science is such that it does not speak the language game of the subjective, unique, and qualitative. Nevertheless, some deep ontological problems remain. This becomes clearer when we consider that, as our analysis of Mind[1] demonstrates, most neurocognitive functional processes operate without any consciousness at all. For example, if I were to hop on a bike right now, I could ride it; however, I would really have no idea how this works. I am only conscious *that* I can ride the bike; I am not conscious of how I can do that. Those operations, like most neurocognitive processes, produce functional outcomes without any SCE attached to them.

We can frame the ontological aspect of the hard problem as follows: *What exactly are the mechanisms of brain activity that function to generate SCE?* Although there is progress being made in this area, the neurobiological mechanisms that play a specifiable causal role in generating SCE remain largely opaque. That is, in contrast to other biological processes like digestion, we do not have good models that explain in mechanistic detail why specific brain activities result in specific subjective experiences. Thanks to brain imagining technologies, we have made significant and substantial advances in exploring the pattern of neurocognitive correlates associated with subjective experience. That is, we know that visual experience is associated with activity in the occipital lobe and other aspects of the visual cortex. However, the specific causal relations are not well elucidated. That is, science does not really have a clue as to why cortical arrangements in the occipital lobe give rise to the experience of vision, whereas other neurological activity generates the subjective experiences of smell or sound. Similarly, engineers have no idea how to construct a robot that *feels* things subjectively. All of this summarizes the ontological aspect of the hard problem, and it remains hard.

Even though there is a hard problem, it is also the case that the size of the hard problem has been exaggerated. This is because of the Enlightenment Gap and the failure of science to generate a clear picture of behavior and mental processes, and related concepts like functional awareness and responsivity, sentience, and the self. With its descriptive metaphysical mapping of science and reality in a way that delineates the ontology of the mental, UTOK sets the stage for clarifying our language and advancing our understanding considerably. This allows us to make the nature of the hard problem clear and place it in its proper context.

Mapping Human Mental Behavior

With the domain of mental behaviors specified by the ToK and PTB, and the linkages made clear between behaviorism in psychology and modern 4E cognitive science approaches, we obtain a clear picture of Mind[1]. By linking that to philosophical analyses of the easy and hard problems of consciousness, we then achieve a more fine-grained analysis of the differences and relationship between Mind[1] and Mind[2]. With this grounding, we can proceed to map the domains of human mental processes in greater detail. We can start by refuting John Watson's fallacious claim that consciousness is only something for philosophers to deal with. We know, scientifically and in our lived experiences, that the domain of Mind^2 is, of course, central in the world of the human, one obvious reason being that processes and systems of justification (i.e., Mind^3) are built on this.

To see this clearly, consider a clinical interview I did years ago when I worked at a psychiatric hospital. In the middle of the interview, the woman started looking all around and then started slapping at her wrists frantically. Had I not been informed of her condition in advance of the interview, I would have been completely confused about what she was doing. "Are you experiencing them now?" I asked. "Yes," she said. "There they are, can't you see them?" The woman I was interviewing had been diagnosed with paranoid schizophrenia. She had periodic bouts of haptic and visual hallucinations that would involve her experiencing gnat-like bugs crawling out from under her skin and then flying around her head.

From my third-person perspective, her overt mental behavior looked completely chaotic. That is, there was no functional awareness/response or input–output relations that I could track that made sense of her actions. However, my knowledge of schizophrenia and hallucinations, along with her capacity to introspect and self-report on her inner subjective experiences, allowed us both to make sense of her actions. Specifically, her Mind^2 system was registering gnat-like bugs. And given that she was willing to share her experiences via self-report, I had direct access to her verbal narrative via Mind^3 processes. She was considered to have a serious mental illness in large part because there was a profound disconnect between her perceptual experiences and what I and others deemed to be true about the external world. Indeed, the definition of psychosis is a break between a person's subjective perspective and the "real" world, at least as it is intersubjectively agreed upon by the rest of the community.

This example highlights how, with the addition of Mind^3, we can scientifically "box in" Mind^2 and set the stage for achieving a clear vocabulary of Mind, the mind, mental behaviors, mindedness, and the three distinct domains of mental processes. To enable the process of mapping human mental behavior, let us return to the scene from the last chapter of the shopper in the furniture store looking to buy a table and divide it up based on our updated model of behavior and mental processes. We start with Mind^1, which is the neurocognitive functional analysis of mental behavior. This includes both the man's overt actions as he walks around the table, demonstrating functional awareness and responsivity from the outside, and the covert neurocognitive processes that function to regulate the inputs and outputs and guide actions.

The next two domains of mental processes move us into the two primary domains of human consciousness (subjective conscious experience and self-conscious narration). As the individual is considering purchasing the table, we can identify his emotional reaction to it and how it relates to his perceptual experience of other tables. We can also consider how he feels in his body, that is, the sensations of hunger or perhaps pain in one of his knees and his sense of balance and body position. Then there is the domain of Mind^3, which is the "talking" mind (Fig. 11.2).

Only human persons have fully developed talking minds, in part because this requires socialization via language to develop. Language-based thoughts are different from subjective experiences both because of the informational medium they travel in and because they can be shared directly with others without losing their informational form. Although the individual could not directly share his subjective experience of the table with the salesman, he could explain directly why he did or did not want to purchase it.

The three different domains of mental processes are crucial in getting an effective solution to the BM^3 puzzle. Figure 11.3 maps this terrain as a group of nested circles of mentation. The outer-most circle represents the environmental or situational context (e.g., purchasing a new table in the furniture story). Overt activity refers to the "doings" of the individual as he effects some change in the environment (e.g., walking around the table). Here we will refer to the externally observable overt activity of Mind^1 as Mind^{1b}.

We can then move into the body. As Cahoone makes clear in *The Orders of Nature*, when thinking about Mind or mental behavior, we need to first place it in a biological or living context. That is, Mind does not emerge out of Matter, but rather out of Life. And with living organisms, we already see much dynamic complexity and functional, goal-oriented activity. We can further enrich this by considering the description of the praying mantis and viewing it through the complex adaptive behavioral lens described earlier in the chapter.

Then we have Mind^2 as subjective witnessing and experiencing. The Unified Theory considers this a special kind of neurocognition, which has been called "neurophenomenology" by some scholars associated with 4E cognitive science. As noted, the jump from Mind^1 to Mind^2 is marked by

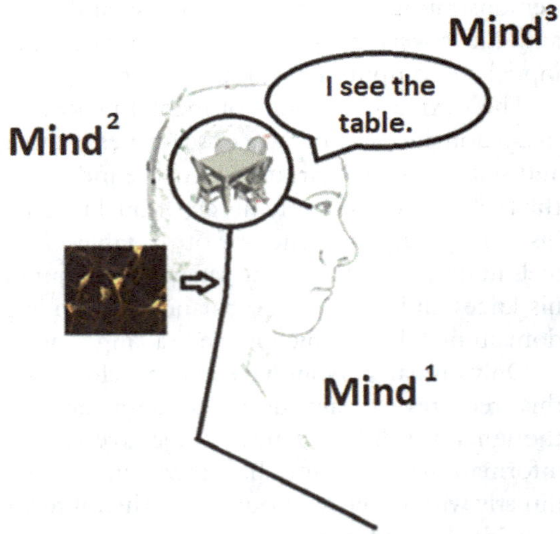

Fig. 11.2 The three domains of human mental processes

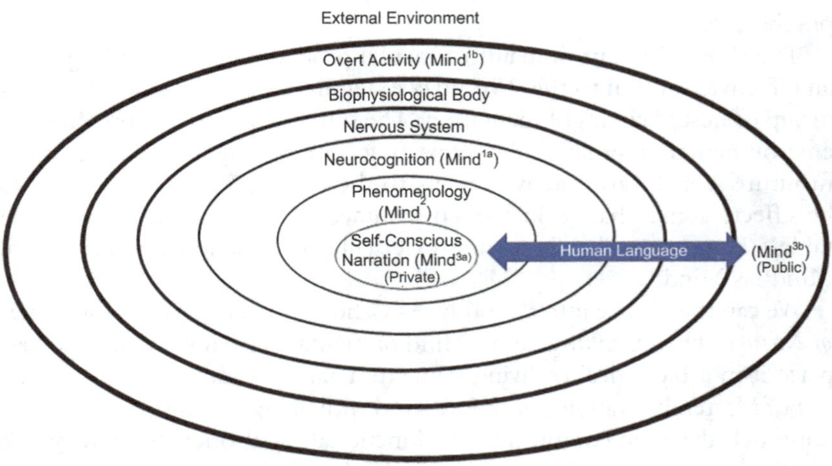

Fig. 11.3 The nested domains of human mental behavior

an "explanatory gap," which is what most mean by the hard problem of consciousness. We have noted that, in addition, there is an epistemological portal/gap problem, and so we remain unclear if a praying mantis has a

Mind^{2}, as there is no way to directly observe another creature's experience of being in the world. Chapter 14 explores current knowledge about subjective conscious experience and argues that it is plausible that a praying mantis does have some minimal conscious experience, but currently our knowledge is tentative.

Of course, we know humans have subjective conscious experiences. This is obvious because of Mind^{3}, which has an interesting position in the diagram. In one sense, it is a subset of experiential conscious awareness, as it emerges out Mind^{2}. However, it is a unique subset in that it can explicitly self-reference, and it is deeply connected to language. The arrow in the diagram shows that Mind^{3} can also connect directly to the outside world via the information highway that is human language. This is the reason that the "I see the table" statement in the first depiction is framed as existing between the individual and the external world.

The Map of Mind^{1,2,3}

We are finally ready to lay out the full map of human mental behaviors. The Map of Mind^{1,2,3} (Fig. 11.4) provides a descriptive metaphysics of the three domains of mental process in a way that also effectively delineates the epistemological and ontological considerations. First, we have the distinction between the interior and exterior epistemological perspective. The domains of Mind^{1b} and Mind^{3b} are readily available from the exterior epistemological position. Mind^{1a} represents informational constructs embedded in the nervous system. Information theory, semiotics, and the information processing sciences provide a conceptual frame for scientifically understanding the various meanings of information. These elements can then be bridged to the structure and behaviors of both the nervous system and the functional awareness and response sets of animals acting in and on the environment.

Mind^{2} is the witnessing, experiencing portion of mind and the view from within. It is directly available only from the interior epistemological perspective. This epistemological portal/gap relation is part of what makes understanding the subjectivity of other animals such a complicated and difficult process. Despite this fact, later in the book we will qualify this and argue that there is an implicit intersubjective field between people that warrants consideration as "Mind^{2b}." It emerges as we develop relations with others and track them and engage in activities. Mind^{3a} refers to the private narrator in humans, and the inner dialogue that characterizes

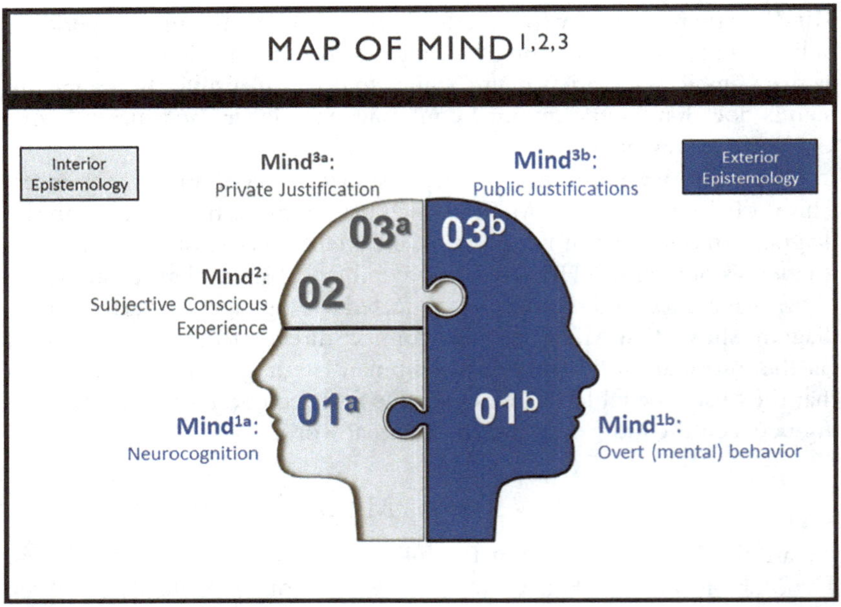

Fig. 11.4 The Map of Mind[1,2,3]

self-talk and self-conscious egoic functioning. Mind[3b] refers to "verbal behaviors," which include both talking and other forms of language-based communication that are available publicly, such as writing.

This book is about resolving the problem of psychology and bridging the Enlightenment Gap in a manner that allows for the development of both science relative to subjective and social knowledge and mind relative to matter. To provide some forecasting of where we are headed in the upcoming chapters, it is useful to consider the Map of Mind[1,2,3] in relationship to the Updated Tripartite Model (UTM) of human consciousness, which we covered in Chap. 5. Mind[1a] is not represented on the UTM, but Mind[1b] is, labeled as "overt actions." Mind[2] corresponds to the "Experiential Self" which includes both conscious awareness and the core primate self, whereas the Ego aligns with Mind[3a], and the Persona corresponds to Mind[3b]. Because the diagram provides the metatheoretical framework to understand the dynamic interrelations between the domains, we are now explicitly bridging a metaphysical description of mental processes given by the Map of Mind[1,2,3] to the metatheoretical formulation given by JUST. This is the kind of knitting together of descriptive metaphysical systems that clarifies the ontological referents with metatheoretical causal explanatory frameworks that make UTOK's mental behaviorism a fundamentally new approach to the science of psychology.

The dynamic causal explanatory linkage becomes clearer when we consider the filters. First, there is the relationship between Mind^{1a} and Mind^{2}. This is captured by "the attentional filter," which, as we will see more clearly going forward, refers to how the consciousness system detects salient aspects of the information landscape in the perceptual field and brings them into experiential awareness. The Freudian filter captures the relations between Mind^{2} (i.e., experiential self) and Mind^{3a} (i.e., ego). This is where the psychodynamic defense mechanisms of suppression and repression and rationalization and intellectualization operate to maintain a legitimizing internal narrative of self and the world. Finally, there is the Private-to-Public Rogerian filter, which refers to how people manage impressions to regulate their social influence and place in the relational matrix, as they engage in public speech acts (Mind^{3b}) and explicit dialogue with other persons (i.e., the domain of the persona). To further set the stage for this linkage, we can introduce the concept of informational interface to start to frame the relations between these domains of mental processes.

Linking the Mental Domains Via the Concept of Information Interface

Informational interface refers to the transfer of information between systems and includes feedback loops and communication between different kinds of domains or through various mediums. Examples of informational interface are all around us. Consider what happens when you are speaking on a cell phone. Information is being translated through various media throughout the process. In a phone conversation, an explicit, self-conscious linguistically mediated thought is translated into motor speech, and the linguistic information flows as sound waves into the receiver in the phone. Those sound waves are translated into electrical signals which are then beamed as radio waves to the nearest cellphone tower, where they are then often sent to a satellite in space and beamed back into another cellphone tower, and then into the person's cell phone. They are then retranslated into the speaker, which projects the information through the air via sound waves. These sound waves are then translated into liquid waves via the ear drum and three small bones (i.e., hammer, anvil, and stirrup), which cause vibrations in the cochlea, which results in a pattern of fluid that is picked up by auditory receptors and then translated into words that can be processed and pulled together to form meanings. And the person responds, "OMG, I can't believe she did that!"

We can start to understand human mental processes by observing the informational interface occurring between the domains of human mental behavior. Specifically, for Mind[1] we can consider how different parts of the human nervous system communicate messages and store information, and how information flows from the nervous system into the muscles (and back again) to give rise to controlled, functional movements and purposeful actions that can be observed by others (overt mental behaviors). We can see the centrality of neuronal communication networks in Goldstein's (2020) closing comments of his book *The Mind: Consciousness, Prediction, and the Brain* (p. 181):

> In closing, it is important to acknowledge that cognitive scientists have created many stories about the mind…The stories told in those books have in common the idea that behind every cognitive function, the brain is working silently behind the scenes: neurons are arranged in patterns, connecting [and communicating] with each other in changeable ways.

We can also consider how brain-based information processing gives rise to and is influenced by subjective conscious experience. This is the Mind[1]–Mind[2] interface, and the next two chapters lay out the metatheoretical architecture that allows us to outline the interface between these two domains as framed by UTOK. Finally, we can consider how both neurocognitive processes and phenomenology interface with language, inside the subjective, private domain of the internal narrator (Mind[3a]) and those verbal thoughts that are expressed publicly (Mind[3b]). All of this suggests that we can think of the human mind in its totality as a neurocognitive–behavioral–experiential–linguistic informational interface system.

With this Map of Mind[1,2,3] delineating more clearly the different domains of the mental, we can consider what happens when communications become disconnected or disrupted between the domains. For example, consider the way amyotrophic lateral sclerosis (ALS) impacts mental behavior. Sometimes called Lou Gehrig's disease, ALS is a motor neuron disease that causes the death of neurons controlling voluntary muscles. It sparked the well-known "ice bucket" challenge that went viral and raised millions of dollars to help research the condition. The famous physicist

Stephen Hawking had the disease. In him, we can see the dramatic consequences of severely disrupting the information interface between Mind[1b] and the rest of his mind. We can also see how, via technology, information interface was able to be retained with his Mind[3], so that contact with his mental world could be maintained via language even as his body became completely paralyzed.

The phenomenon of blindsight, as described by Weiskrantz (1997), is another example of disrupted informational interface that demonstrates the utility of the Map of Mind[1,2,3]. Blindsight involves damage to the occipital lobe which results in the loss of all visual experience. That is, it knocks out the Mind[2] experience of seeing. However, what Weiskrantz noticed in working with these patients is that they would sometimes behave as if they could see some things. For example, they might adjust their walking in a room to avoid a table. This Mind[1b] act of avoidance is an example of a scientist seeing "functional awareness" from the exterior vantage point. However, if asked, his subjects would report that they were completely blind, that is, they had no Mind[2] visual experiences. Weiskrantz tested this empirically by setting up an empty room with a large object on one side. He then asked his patients to point to which side of the room they thought the object was on. At first, they would understandably object, commenting that they were completely blind. Yet, when asked to guess, they would, with remarkable consistency, point to the correct side of the room. There are similar examples of "deaf hearing".

Another example of the disconnect between domains of informational interface comes in the form of split-brain patients and one of its most notable consequences, alien hand syndrome. Beginning in the 1950s and 1960s, neurosurgeons began to cut the corpus callosum in the brains of some patients with severe seizures to minimize the spreading of the out-of-control neural firing. The corpus callosum is the set of neural fibers that connects the two hemispheres of the brain, and thus when it is cut, communication between the two hemispheres is broken. These patients—who came to be called split-brains—generally lived normal lives. However, careful research revealed some striking findings.

If simple commands were flashed to the right hemisphere, such as "walk around" or "laugh," the patients would follow these commands (the right hemisphere does have rudimentary linguistic capacities). However, when asked to explain why they were performing these behaviors (e.g., walking or laughing), patients would make up a reason and say, "I am going to get a drink" or "Because you guys are so funny." In other words, their Mind^{3} justified the behavior in the absence of necessary information. Sometimes even more dramatic examples emerged. For example, some patients would experience alien hand syndrome, in which the left hand (guided by the right hemisphere) would seemingly act as if it were controlled by a mind of its own. A patient with alien hand syndrome might go into the closet to get a blouse and find that both her hands would reach for separate garments, and a literal struggle between the two hands would ensue.

A final example involves Mind^{3b} and the way we regulate the flow of information to different people in different contexts. I am on many different email list serves, and I am often engaged both in conversations on the list serve and "back-channel" exchanges off list. As the term "back channel" suggests, we humans set up distinct lines of communication with different parties. Justification Systems Theory makes clear why this is so, although it is so much a part of our folk experience that it hardly needs stating. It is patently obvious that we do not want to share everything with everyone. This fact was brought home to me in a dramatic fashion one day when, on a list, a friend of mine reached out via back channel to ask about a new participant who was starting to make frequent posts.

In replying to the query, I shared that I knew the person from other contexts, and that I thought that many of his arguments were empty and that he could be annoying in his combativeness. In replying to my friend via back channel, I failed to notice that I started to reply to the full list serve prior to shifting my focus. However, I did not change the email address, and so my intended back-channel communication went to the entire list. Upon pushing send, the email almost immediately popped up on the general list. Seeing it, I literally let out a yelp of pain and reached toward the screen, as if I could somehow tunnel into the email server and delete the message. This is an example of a failure of information interface in the opposite direction. That is, rather than a breakdown, there was a failure in expected constraints and filters that resulted in unwanted communication.

The point of these various examples is to show how the domains of mind can be separated and disruptions of the information interface between them can emerge, often with dramatic consequences. Such examples are strongly suggestive of the importance and utility of dividing mental processes in these domains and linking them via the concept of information interface. The dynamics regarding the relations between the domains of mental processes will be a central focus of the remainder of this work. Specifically, we are now positioned to bring in UTOK's metatheoretical frames of Behavioral Investment Theory, the Influence Matrix, and Justification Systems Theory to assimilate and integrate key insights from various perspectives and weave this picture together in a more rigorous and much richer, complex, dynamic way. The result will be the capacity to interpret human mental behavior as a function of justification, investment, and influence dynamics.

Conclusion

The BM³ problem is at the center of the problem of psychology, and psychologists have failed to develop a clear descriptive metaphysical map that specifies the ontological referents of behavior and mental processes. In contrast, UTOK is explicit regarding how it defines natural science, behavior, and the domains of mental processes. As such, it enables us to develop a scientific psychology based on the ontology of the mental rather than the epistemological methods of science.

Mental behavior specifies the kinds of behaviors that psychological scientists are interested in, and the set of mental behaviors can be characterized by the Mind dimension on the ToK System. Mental behaviors can also be characterized as mindedness and framed in terms of the functional awareness and response patterns of animals. We traced the transition from behaviorism into cybernetics and the cognitive revolution into cognitive science and embodied cognition to ground the analyses and set the stage for the Map of Mind[1,2,3]. It does this by clarifying the behavioral and cognitive divide and the necessary epistemological and ontological issues to generate the vocabulary needed to talk about mental behavioral processes with the appropriate specification.

The Map of Mind[1,2,3] extends the analysis of the differences that we have identified throughout this book between neurocognitive processes associated with overt mental behaviors, subjective conscious experience, and self-conscious justification processes. Specifically, by delineating these domains, differentiating the ontological and epistemological considerations, and linking them via the process of informational interface, we now can clearly specify five different areas of mental activity (i.e., overt mental behavioral activity, neurocognitive processes, subjective conscious experience, private narration, and public narration) that need to be separated and then interrelated to have a coherent working map of what is too often simply referenced as the mind or mental processes. The Map of Mind[1,2,3] allows for a proper descriptive metaphysics for psychology that has been missing. The final substantive task of this book is to link the descriptive metaphysical map of mental domains with metatheoretical formulations of Behavioral Investment Theory, the Influence Matrix, and Justification Systems Theory that assimilate and integrate the major paradigms and allow for the clear elucidation of the dynamic informational interfaces that operate between the domains.

References

Chalmers, D. (2007). The hard problem of consciousness. In M. Max Velmans & S. Schneider (Eds.), *The Blackwell companion to consciousness* (pp. 225–235). Blackwell Publishing.

Chiappe, D., & Vervaeke, J. (2018). The experience of presence in the Mars Exploration Rover mission. *Presence: Teleoperators & Virtual Environments, 27*(4), 400–409. https://doi.org/10.1162/pres_a_00337

Friston, K. (2010). The free-energy principle: A unified brain theory? *Nature Reviews Neuroscience, 11*, 127–138. https://doi.org/10.1038/nrn2787

Gallagher, S. (2017). *Enactivist interventions: Rethinking the mind.* Oxford University Press.

Goldstein, E. B. (2020). *The mind: Consciousness, prediction, and the brain.* MIT Press.

Harris, S. (2020). *Making sense: Conversations on consciousness, morality, and the future of humanity.* Harper Collins.

Henriques, G. R. (2003). The tree of knowledge system and the theoretical unification of psychology. *Review of General Psychology, 7*(2), 150–182.

Henriques, G. R. (2011). *A new unified theory of psychology.* Springer.

Hull, C. L. (1943). *Principles of behavior: An introduction to behavior theory.* Appleton-Century.

Koch, S. (1961). Psychological science versus the science-humanism antinomy: Intimations of a significant science of man. *American Psychologist, 16*(10), 629.

Merriam-Webster. (n.d.). Consciousness. In *Merriam-Webster.com dictionary.* Retrieved August 8, 2022, from https://www.merriam-webster.com/dictionary/consciousness

Nagel, T. (1974). What is it like to be a bat? *The Philosophical Review, 83*(4), 435–450.

Newborn, M., & Newborn, M. (2003). *Deep Blue: An artificial intelligence milestone*. Springer.

Prete, F. R. (2004). *Complex worlds from simpler nervous systems*. MIT Press.

Sokol, J. (2017, May 23). The thoughts of a spiderweb. *Quanta Magazine*. Retrieved August 8, 2022, from https://www.quantamagazine.org/the-thoughts-of-a-spiderweb-20170523/

Varela, F. J., Thompson, E., & Rosch, E. (1991). *The embodied mind: Cognitive science and human experience*. MIT Press.

Weiskrantz, L. (1997). *Consciousness lost and found*. University of Oxford.

CHAPTER 12

A Metatheory of Mind[1]

In the first part of the twentieth century, biology was a fragmented science in much the same way that psychology is fragmented today. Fundamentally incompatible versions of evolution were being advanced, and there was a vitriolic split between the "naturalists" and the "experimental geneticists." The naturalists studied plants and animals in the wild and emphasized various models of evolution, whereas the geneticists worked in labs on inheritance and tended to think of major mutations as a driving force of change. Darwin's theory of natural selection was just one of many approaches, and it was not well integrated with genetics. However, over a 20-year period, the situation would change and a metatheoretical paradigm for biology would emerge. The "modern evolutionary synthesis" would become one of the great accomplishments of the twentieth century.

In a manner that directly parallels the way the modern evolutionary synthesis merged the selection science of Darwin with the information science of genetics, Behavioral Investment Theory effectively merges B. F. Skinner's concept of behavioral selection with the key insight from the cognitive revolution that the nervous system functions as a kind of information processing system (Henriques, 2003). According to BIT, the brain and nervous system of an animal is functionally organized to enable it to realize relevant paths of behavioral investment. Skinner's operant selection shows how there is a constant, iterative feedback loop between actions and consequences. Although this process is often taught in terms

G. Henriques, *A New Synthesis for Solving the Problem of Psychology*, Palgrave Studies in the Theory and History of Psychology,
https://doi.org/10.1007/978-3-031-18493-2_12

of rewards and punishers, the fact is that the behavioral selection process is much more nuanced.

For example, when you are writing with a pen, the ink on the page serves as a reinforcer for you to keep writing. That is, the ink is something you anticipate and expect, and its production is a salient variable you track that directly impacts your writing behavior. To see this, consider what happens when the pen stops producing ink. Do you just go on writing? Not at all. Your writing behavior quickly changes, as you might shake the pen or grab a scratch sheet and try to free the ink via scribbling. If that fails, then you proceed to get another pen. The point is that the ink was a salient variable that served as a stimulus that regulated your action. Putting this in the language of the Map of Mind^{1,2,3}, we can say that the selection processes that take place between the animal-environment in the domain of Mind^{1b} shape the structure of neurocognitive information in Mind^{1a}, which in turn structures the behavioral investments emitted in the future.

The larger point is that because it merges neuro-information processing with behavioral selection, BIT parallels the basic structure of the modern evolutionary synthesis. Moreover, framed by the larger UTOK metapsychology and its descriptive metaphysics and big picture natural science ontology, BIT is structured to directly link radical behaviorists and neurocognitive functionalists via the concept of mental behavior. That is, behavioral selection operates on neuro-informational processes in a manner that generates mental behavioral evolution. Via the vision logic of the natural world given by the ToK System, we can say that just as the evolutionary synthesis provides the Matter-to-Life joint point to provide a metatheory for living behavior, BIT provides the Life-to-Mind joint point for mental behavior.

The Core Principles That Frame Behavioral Investment Theory

The central insight of BIT is the idea that the nervous system functions as an "investment value" system that enables animals to make predictions about the effective expenditure of work effort toward outcomes that are evaluated on cost-benefit, risk-reward ratios that emerge as a function of both phylogeny and ontogeny. Several prior publications have shown how BIT provides the causal explanatory framework for the Mind plane of existence and does so in a way that effectively assimilates and integrates major domains of inquiry. Specifically, BIT can be considered a (1) Neurocognitive, (2) Behavioral, (3) Biological, (4) Physical, (5)

Developmental Systems Metatheory for understanding the complex adaptive behavior of animals with brains.

The capacity to effectively assimilate and integrate these perspectives makes BIT a cogent and comprehensive metatheory. As reviewed in *A New Unified Theory of Psychology*, BIT consists of six core principles, which are as follows: The *Principle of Energy Economics* frames mental behavior in terms of energy expenditure (or work effort) that emerges as a function of investment value dynamics framed by costs versus benefits or risks and rewards. It grounds the perspective in the physics of energy (work effort) and entropy (the constant drift toward disorder) in a way that orients to the principle of least effort. The *Principle of Evolution* states that modern-day animals are the product of eons of evolutionary forces, and they carry that history with them. Whereas the principle of energy economics incorporates the basic physical processes governed by the second law of thermodynamics that must be operating for animals to survive, the evolutionary principle is a statement of the processes by which these systems were built across the generational history of the animal via natural selection, along with some forms of epigenetic transmission.

The *Principle of Genetics* is the notion that genetic differences influence development in a way that reliably produces individual differences in characteristics and predisposed aptitudes and dispositions toward patterns of behavioral investment. The *Principle of Neuro-Computational Control* represents the idea that the nervous system functions as an information processing system that carves out paths of behavioral investment by attempting to predict and control the flow of resources. The *Principle of Learning* pertains to how animal investments are shaped by experiences, and the fact that animals learn to allocate their behavioral investments depending on the associations and contingencies to which they are directly exposed, as well as what they observe happening to others. Finally, the *Principle of Lifespan Development* states that animals develop along a trajectory and there are various genetically and hormonally regulated life history stages that require and result in different behavioral investment strategies for survival and reproductive success.

The six principles of behavioral investment can be thought of as being akin to the four questions posed by the Nobel Prize-winning ethologist Nikolaas Tinbergen, which are: (1) What are the mechanisms underlying animal behavior and how is the behavior elicited in relationship to recent learning? (2) How did the behavior develop during the animal's lifetime? (3) What is the function of behavior in terms of survival and reproduction?

and (4) How did the behavior evolve over the generations? The principles of BIT align well with Tinbergen's four questions. The principle of energy economics orients the scientist to view mental behavior as something that the animal is doing, and framing that in terms of expending work effort toward a potential outcome in a manner that is consistent with the laws of physics. The neuro-computational control principle orients us to the underlying mechanisms that are guiding the overt behavioral investments. The learning principle also aligns here, both in terms of the immediate feedback (affordances and punishments in the environment) and historical experiences. The developmental principle also emphasizes the animal's history (in terms of accumulation of occurrences) and current life history stage (i.e., juvenile versus adult). The evolution and behavioral genetic questions orient the researcher toward both the distal and proximal generational history of the animal.

Coordinated Movement Is at the Root of Behavioral Investment Theory

We can obtain a better sense of how these six principles function to generate a metatheoretical framework for animal-mental behavior by exploring how and why animal movement is at the root of BIT, and how to think about animals as behavioral investors, and by demonstrating how BIT and its six principles can frame the current approaches to empirical research in animal behavioral science. Almost all plants are autotrophs, meaning that they can obtain their energy directly from the environment. (Note that there are some exceptions in botany, such as in carnivorous plants, like the Venus fly trap.) Via photosynthesis, plants convert energy from the sun to do the necessary metabolic work to maintain their complex organization. In contrast, animals need to eat other organisms for their energy supply. To accomplish this, they need to detect where their food source is and move in that direction. This gives rise to a central claim from BIT, which is that *the animal's capacity for autonomous movement as a whole unit is the fundamental building block of Mind.*

This idea of the root of mindedness aligns with Aristotle's concept of the animal soul being characterized by a loop of sensation and movement. We must note that cells and plants both sense and move; however, if we add the emphasis on sensation and movement of the animal mediated by the neuro-muscular system, then we can proceed with appropriate specificity. We should note that this is not a novel claim. For example, in *The*

Evolution and Function of Cognition, Goodson (2003, p. 5) captured this insight: "Motile life [had to contain] certain elements. It had to be able to move away from harmful energy sources and toward beneficial ones …. We may consider differential responsiveness to pertinent energies as the foundation upon which all progressive steps have been based." More recently, Barrett (2020) emphasized the point that brains are not for thinking but for moving. After summarizing how the brain works to predict changes and effects to coordinate movement, she then moves to budgetary considerations (2020, p. 8):

> So imagine a tiny Cambrian creature drifting in the current. Up ahead, it senses an object that might be tasty to eat. What now? It can move, but should it? After all, moving takes energy from the budget. The movement should be *worth the effort*, economically speaking.

The life cycle of a sea squirt exemplifies the connection between the brain, energy expenditure, and coordinated movement (Llinás, 2001). After conception, the sea squirt grows into a tadpole-like creature, complete with a basic brain, spinal cord, eye, and tail for swimming. As such, it provides an example of a creature that, in terms of behavioral complexity, resides at the very base of the Mind dimension. Interestingly, unlike most animals, its movement is not for predation or finding mates, but only for location. It seeks a suitable place to live out the rest of its life. Once it finds a suitable home, it attaches itself to the surface of it, never to move again. Then something fascinating happens. Because it no longer moves about, it no longer needs a brain, as it eats via filtering food from the water that flows around it. What does it do with the brain it no longer needs? It digests it for the calories. Sea squirts literally eat their own brains.

Although free movement can obviously yield benefits (i.e., toward the good and away from the bad), it also takes much internal organization to be enabled. It requires a computational control center, which turns out to be costly in terms of metabolic energy. Consider that, although the human brain only takes up about 2% of a human's total body mass, it consumes upwards of 20% of our metabolic energy (Shulman et al., 2004). Not only is brain tissue expensive, but muscles also burn many calories, especially when active. Moreover, movement itself is risky because it greatly increases the novelty in the kinds of environments one will encounter. That is, one can get lost, fall, or stumble into an area with a predator. This means that we need to consider a guidance system that directs movement toward the good and away from the bad across many different dimensions of consideration.

BIT allows us to readily draw a connection between movement and the construct of motivation. To understand motivation in a straightforward way, we can ask: What are the kinds of things that animals and people invest in? There are almost an infinite number of things one could be doing. Yet, the actions of both animals and people are far from random or chaotic. They are functionally organized toward ends, considering various options and constraints. The word motivation stems from the Latin word "movere," which means to move. We can note that if movement is costly at some level, then it follows that it must be valued, which is why BIT characterizes the entire nervous system as an ***investment value system***. In short, investment and valued movement toward or away from certain goal states closely link us to the concept of motivation.

The evolutionary theorist and cognitive psychologist David Geary (2005) developed a formulation that emphasized the structure of nervous system processing in terms of the "motive to control" the flow of resources. Geary explicitly argued that his motive to control formulation directly aligned with BIT. He argued that "the brain and mind of all species has evolved to attend to and process the forms of information, such as the movement patterns of prey species, that covaried with survival and reproductive outcomes during the species' evolutionary history. These systems bias implicit decision-making processes and behavioral responses in ways that allow the animal to attempt to achieve access to and control of these outcomes, as in prey capture" (Geary, 2005, pp. 23–24).

The point is that we begin with coordinated movement of the animal as the fundamental principle of Mind and the functional organization of the nervous system. This sets the stage for animal action to be guided by motivational principles that can be framed in terms of the attempts to control the flow of resources toward affordances or desired outcomes and away from stressors or other problematic states. This sets up the argument that animals invest their actions in an internal value system that is assessing costs and benefits. And it means that we can consider animals as functional behavioral investors that are investing in the actions we can observe from the outside.

Animals as Functional Behavioral Investors

Mental behaviors can be thought of as describing what animals do at the level of Mind, both in terms of the directions of their overt actions and in terms of the function of how they process neuro-information. Grounded in BIT, we can frame these doings as investments that function to move animals toward certain outcomes and away from others. The description of a praying mantis setting himself up to ambush his prey is one example of a pattern of behavioral investment. A bee pollinating a flower, a bird singing a song to attract a mate, a seal dodging a great white shark, and a wolf mother defending her pups are additional examples.

Behavioral investment lines up well with how humans intuitively understand animal behaviors. Consider that for a period as I worked on this book, I would write in the morning from about 6:00 am to 10:00 am. Each day I would come into my office and turn on the light to my fish tank. When I did so, the fish would become activated and swim up to the top of the tank. Why do they behave this way? First, eating is a classic example of a mental behavioral investment pattern. Second, environmental cues have primed them to be prepared to eat when they see me approach the cage and turn the light on. Like Pavlov's dogs, they have been shaped to respond to these cues. Their mindedness was clear to see as I observed their movement.

Because you are an animal, the concept of investment should be readily applicable to understanding your own minded behaviors. Take a moment and frame your activities via the question: What are you working to do? What are you investing in? Reading this book, answering emails, cleaning the house, and practicing a musical instrument are all everyday examples of "doing." Also note how we ask others to *pay attention*, which suggests that attention is costly, and we spend it for desired outcomes. We can organize this claim by several key points, especially if we liken the investment value process to a vector. First, it suggests that your mental activities should have a direction. That is, they should be oriented toward some outcome that you are investing in. Second, these investments should vary in degree of intensity. Some you work hard and long for and will take many risks to achieve, and others are barely worth the effort. Third, your investments are nested in a hierarchy of motivational goals. Fourth, we should be aware that investment occurs across time scales, from the micro-movements of one's eyes toward relevant information in the environment that takes milliseconds, to planning to get a doctoral degree, which can take decades from start to finish. Finally, we should be able to see your investments as being balanced as a function of predicted benefits and costs/risks at some level.

It is also crucial to note that, although much of your conscious thinking can be framed by investments, many of your investment calculations are done with no conscious effort or awareness. Indeed, intuitively you will tend to spend you actions in accordance with what is known as the principle of least effort (Zipf, 1949). That is, all else being equal, animals and people tend to expend the least amount of time and effort to achieve the desired outcome. For example, when you go to the store and pick up some milk, your unconscious behavioral investment system intuitively calculates the shortest drive, the easiest way to get from the parking lot to the store, and, if you are familiar with where the milk is, the shortest route to it. The principle of least effort is well understood by marketers and product makers. One of the most powerful ways to sell something is to sell it in terms of convenience, that is, making resources available in a way that saves time, effort, money, or risk (Wu, 2018).

Nikolaas Tinbergen's questions make clear that animal investment can be understood either in terms of neurobiological mechanisms and learning during the animal's lifetime or in terms of more distal evolutionary causes. An analysis of deep-sea squid mating behavior serves as a useful example in understanding how current patterns of activity can often be understood as a function of evolution. Researchers discovered that male squid regularly engage in "same sex" behavior, and the findings were covered in the *New York Times* (Gorman, 2011). This made headlines in part because we humans are often debating about same sex behavior, and whether it is "natural" (i.e., occurs in nature). Indeed, as Gorman notes, same sex behavior is quite common in the animal kingdom. Our focus, here, however, is on why the squid might be engaging in such mental behaviors, and why animal researchers found it a perplexing question. The reason is that scientists expect animal behavioral investments to "pay off" in terms of either survival or reproduction, either directly or indirectly. Therefore, their finding was a bit of a surprise. At first glance, the same sex behavior seemed unlikely to benefit the squid in terms of either survival or reproduction.

How did the researchers make sense of this mental behavior? Via a "reproduction and cost-to-benefit ratio" of action investment. The researchers noted that male and female squid are almost identical in terms of structure and appearance, and thus distinguishing between the sexes in the dark depths of the ocean is extremely difficult. Because the cost of the sperm and the mating behavior was not high, they concluded that the overall tendency was understandable in terms of investment relative to costs and benefits. As noted in the article, it is "far better to waste a few million sperm than miss out on a chance to reproduce" (Gorman, 2011). The point here is that animal actions are presumed by animal behavioral scientists to reflect a form of cost-benefit valuation process that has evolutionary roots. These brief descriptions highlight that it is natural to describe the behavior of animals in terms of behavioral investments. However, what is more important is that BIT aligns well with how animal behavioral scientists formally study, research, analyze, and explain animal behavior.

Behavioral Investment Theory Frames Scientific Research on Animal Behavior

The examples offered suggest that BIT provides a general way to understand animal behavior. That is useful, but the more stringent test pertains to the way scientists conduct research. If BIT works as a genuine metatheoretical framework for the science of animal mindedness, then it should be able to frame how researchers explore and test animal behavior. To test this prediction, the first six articles of the most recent issue of the flagship journal on animal behavioral science were examined at the time of this writing (*Animal Behaviour*; May 2020). Five of the six articles were directly framed by BIT and are summarized below. By that I mean that the authors explored animal behaviors as investments that emerged as a function of: (a) expenditure of energy as a function of benefit, cost, or risk; (b) evolutionary history and reproductive fitness; (c) behavioral genetics; (d) neurocognitive functional analyses; (e) associative learning to enable adaptive prediction; or (f) developmental lifespan considerations. The one article that was not readily captured by BIT was on digestion and the presence of bacteria in an animal's gut depending on the food it consumed.

The first article was titled "Dynamic terminal investment in male burying beetles," and it explored the behavior of beetles investing in future offspring depending on its current state and place in the lifespan (Farchmin et al., 2020). The research was grounded in the "dynamic terminal investment threshold model," which posits that "the propensity of an individual to terminally invest in response to an immediate survival threat, such as an infection, depends on other factors that alter an individual's residual reproductive value" (p. 1). We can see that the theory that grounds this research is directly in line with the principles of Behavioral Investment Theory, and even its basic language. The researchers injected male beetles at two different ages with bacteria that would signal a threat to their immune system. According to the predictions, those males who were both older and immune challenged would show a longer period of parental care, contribute to producing greater larvae, and consume less food. In other words, the researchers found that as the current life circumstances became vulnerable, more energy would go into the next generation. This is an example of thinking about mental behavior specifically from the vantage point of energy economics, evolution, and developmental life history (i.e., BIT Principles 1, 2, and 6).

The second article was titled "Virile crayfish escalate aggression according to body size instead of weapon size." It studied signaling, fighting, and dominance patterns in crayfish (Graham et al., 2020). The study demonstrated that both size and claw strength, rather than claw size, were key in determining winners of fights. Moreover, they found that claw size was not a determinant in escalating aggression, but body size was. This finding connects with BIT, especially when viewed through the lens of Geary's motive to control lens, which emphasizes how patterns of variation frame information processing and choices in accordance with resource acquisition associated with survival and reproductive success. The article was thus concerned primarily with BIT Principles 2 (evolution), 4 (neurocognition), and 5 (learning and adaptation).

The third article was on the microbial gut biome of monkeys and was thus not directly applicable to BIT. The fourth article was titled "State-dependent learning influences foraging behavior in an acoustic predator" (Hemingway et al., 2020). It deepened the analysis of the fact that many different animals have been shown to exhibit preferences for food options associated with greater past need. The study explored the extent to which this tendency was present in frog-eating bats. The researchers trained the bats on two kinds of cues that would lead to identical food rewards. However, one cue was present when the bat was food-deprived, whereas the other was present when the bat was pre-fed. The researchers found, consistent with their predictions, that the bats strongly preferred the ringtones associated with greater prior food deprivation. Can we explain this finding from the vantage point of BIT? Yes. Recall that the nervous system is an investment value system that is shaped in part as a function of the contextually expected rate of return. The ringtone associated with the food-deprived state was thus tagged as having more value to channel the direction of the behavioral expenditure. This study thus aligns with BIT Principles 1 (energy economics) and 5 (learning).

The fifth article was titled "How does individual variation in sociality influence fitness in prairie voles?" It examined levels of social contact and exchange and corresponded them to fitness indicators (Sabol et al., 2020). The researchers found that "females and males with an intermediate number of social connections had higher mating success overall and, for the analysis with all connections, produced more offspring. Males with many or few social connections also had the lowest average body mass" (p. 39). Although we have not yet explored the Influence Matrix in this work, it is an extension of BIT applied to social motivation and emotion and the relational world of social animals in general and human primates in particular. It maps social motivation first in terms of social influence (i.e., the capacity to influence others in accordance with one's interests) and second in terms of the process dimensions of power/rank, affiliation, and autonomy-dependency. The last process dimension is theorized to be characterized as a dialectic and optimal social influence will be associated with both interdependency and a sense of autonomy, whereas the extremes of dependency and counter-dependency (i.e., excessive independence) are less adaptive. Although there are only tenuous parallels between human social motives and prairie voles, it is nonetheless clearly the case that the Unified Theory offers an initial frame for understanding these results that is consistent with BIT. In addition to potentially connecting to the autonomy-dependency line on the Influence Matrix, this article is an example of both BIT Principles 2 (evolution) and 5 (learning).

The sixth article examined was titled "From ridge tops to ravines: landscape drivers of chimpanzee ranging patterns" (Green et al., 2020). These researchers were extending the analyses from optimal foraging theory. Optimal foraging theory refers to the idea that animals tend to maximize the ratio of energy outputs to inputs when foraging. It is thus directly consistent with the logic of BIT. This study used recent developments in tracking technologies to explore how chimpanzees foraged and specifically examined the landscape in terms of rugged terrain and human-made trails. They found that the chimpanzees tended to use existing trails, and the authors interpreted the findings based on cognitive and energetic benefits. Again, this would be an example of BIT Principles 1, 2, and 5.

These articles were chosen simply as a function of the time when I was writing this section. The more general prediction I am asserting is as follows: *Modern animal behavioral research can be readily interpreted through the metatheoretical lens afforded by BIT.* That is, animal behavioral patterns are "mental behaviors" that can be interpreted based on the six BIT principles and their extensions. Others are encouraged to explore this prediction to assess the comprehensiveness of the framework.

BIT and the Evolution of Mental Behavior in Four Stages: From Reacting to Learning to Thinking to Talking

In the section that follows, we explore the evolution of mental behavior in four steps, from reacting to learning to thinking to talking. This four-step pattern in the evolution of animals has been seen by others. For example, in his excellent book *Darwin's Dangerous Idea*, Daniel Dennett (1995) identified four different kinds of organisms. He first identified "Darwinian creatures" that are shaped by natural selection, which include cells and plants and animals with simple nervous systems like sponges. He called the second class of creatures "Skinnerian organisms," which included creatures like crabs, fish, and frogs that have the full suite of classical and operant learning capacities. Dennett identified the third class of animals as "Popperian organisms," named after the philosopher Karl Popper because of his emphasis on prediction in science (which leads to tests that can falsify ideas). These are animals like whales and elephants and chimpanzees, who can think about future outcomes and make decisions accordingly. Finally, there are what Dennett called the "Gregorian organisms," named after the psychologist Richard Gregory. These are the humans and their capacity for symbol use, and reason-giving on the social stage, and are thus called *talking animals* and refer to human persons.

A similar although less colorful grouping was proposed by Aunger and Curtis (2008), as they classified the evolution of different kinds of animal-mental behavioral patterns (Fig. 12.1). They argued for a division and typology of behavioral control that moved across evolutionary time from reactive (e.g., sponges and worms) to motivated (e.g., vertebrates) to executive (e.g., primates), noting that the kind and complexity of responses increases over time. They did not include humans in their model, but placing humans as talking animals in the hierarchy would logically follow. Below, their model of behavioral production shows how they viewed the

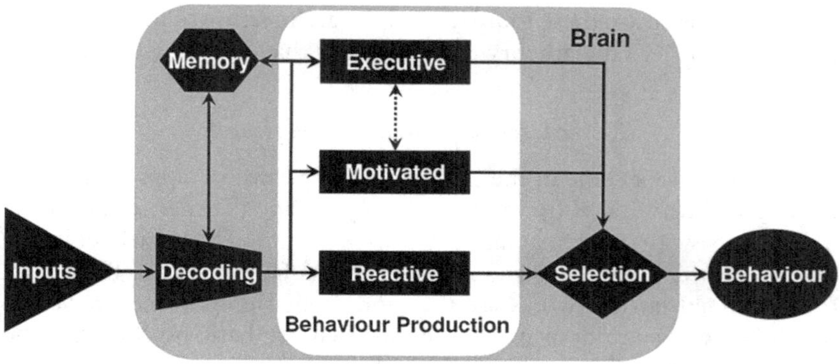

Fig. 12.1 Aunger and Curtis' (2008) behavioral production model (Reproduced with permission)

evolution of mental behavior and how it was regulated by neurocognitive processes.

As we will see, this model overlaps significantly with a model of the neurocognitive architecture of the human mind that stems from BIT. This diagram allows us to see how to approach mental behavior from the vantage point of neurocognitive functionalism. Information comes in from the world and is "decoded" and processed at various levels of behavioral production. The reactive level goes straight into an action selection and then results in behavioral output, such that the stimulus is tightly connected to the response. Motivated behaviors are processed at a higher level of organization and serve to regulate, feedback, and inform action selection. These behaviors are more flexible, and goal directed. Executive behaviors are still higher, as they serve to regulate motivations across time. Of course, all these neurocognitive processes are mediated by the brain. There are (mental) behavioral outputs that then shift the animal–environment relationship.

By combining Dennett's (1995) analysis with that of Aunger and Curtis (2008) an interesting picture of both mental evolution and the neurocognitive functionalist arrangement of the human mind emerges. Namely, we should be able to identify the evolution of the brain–behavior relationship in animals across four different stages. We can give them straightforward, functional labels: (1) reacting; (2) learning; (3) thinking; and, in humans, (4) talking. Because of the way evolution builds on what comes before, we should be able to see these four layers in the architecture of human neurocognition. BIT is a metatheory of mindedness that affords us this lens.

The Emergence of Reacting Animals

Sponges represent some of the earliest animals to evolve, appearing in the fossil record between 900 and 800 million years ago (Turner, 2021). They have very basic nervous systems, which are made up of neuronal networks, but are not centrally organized into a brain. As framed by the PTB, neurons arranged into networks are the fundamental units of neuroscience, which is the seventh floor of science, just beneath basic psychology. The most basic neuronal network consists of (a) an input or afferent neuron that translates an external stimulus into a neural signal, and (b) an intermediate, processing neuron that takes that input and provides feedback and output to (c) an efferent neuron that sends its signal to a muscle or a gland. In the landmark book *The Organization of Behavior*, Donald Hebb (1949) proposed a theory of cell assembly that would become a cornerstone in modern behavioral neuroscience. He postulated that "when an axon of cell A is near enough to excite a cell B and repeatedly or persistently takes part in firing it, some growth process or metabolic change takes place in one or both cells such that A's efficiency, as one of the cells firing B, is increased" (p. 62). This is commonly stated as "neurons that fire together wire together."

Eric Kandel (2009) empirically demonstrated the validity of Hebb's postulate and won a Nobel Prize for his work. He studied the siphon-withdrawal reflex of a sea slug, and showed that, indeed, neurons would shift their connections to foster learning. This became a model for the most basic forms of learning, which consists of the processes of habituation and sensitization. Habituation is a decrease in a reflex response resulting from repeated presentation of an initiating stimulus. Consider, for example, how upon first entering a hot shower, you notice the temperature. However, after a period your body and brain adjust, such that you are no longer aware of the heat. You have become habituated to it. Conversely, sensitization refers to the process by which an animal learns to increase its reflexive responses to noxious or novel stimuli. A powerful example of sensitization is found in PTSD when individuals can become triggered by noises or smells. Their nervous systems have become sensitized because the powerful and traumatic events have "seared" these stimuli into the nervous system. Such foundational learning processes form the base of the Mind dimension.

Although neuronal nets, reflex patterns, and habituation and sensitization are present at the Life-to-Mind joint point, we are not yet at a fully functioning Mind[1] system of animal mental behavior, which corresponds to the eighth floor of the PTB. Two things need to happen for the process to move from the distributed neuronal networks and reflexive behavior of creatures like jellyfish into the world of creatures like praying mantises, which are full-blown Mind[1] animals in the taxonomy given by UTOK. First, we need to develop the emergence of a centralized nervous system in the form of a brain. And then we need what Trestman (2013) calls "complex active bodies" that encase those brains and can be deployed in much more complicated action patterns.

In *Other Minds: The Octopus, the Sea, and the Deep Origins of Consciousness,* Godfrey-Smith (2016) tells a fascinating story of what may have been the foundational processes that sparked the evolution of Mind and mindedness (i.e., the complex adaptive mental behavioral landscape populated by animals with brains and complex active bodies that engage in sensory-motor loop behavioral patterns). First, he noted the evolution of bilaterally symmetrical animals, which have a front and a back, left and right, top and bottom. Consider, for example, the difference between a worm, like a planaria, which has a set of eyes and moves in a particular direction, and a jellyfish. The bilateral design of the planaria sets the stage for coordinated movement in a way that would necessitate a centralized control center.

Godfrey Smith highlighted the fact that most scholars think of the functional organization of the nervous system in "sensory motor" terms that are yoked together. That is, it both detects inputs and regulates outputs. This was the original starting point for BIT. However, there is also a second view, one that emphasizes the initial separation of the problem of coordinating action from the problem of sensation. Indeed, Smith argues that there is evidence, according to the biologist Detlev Ardent, that the nervous system has two origins, one of which is a sensory system that tracks light and the other of which is an action system that guides reflexive responses. As Godfrey-Smith (2016) notes, there is a possibility that (p. 39–40)

> At some stage, the two systems begin to move within the body, coming into new relations with each other. Arendt sees this as one of the crucial events

> that took bilaterarians forward in the Cambrian. A part of the body-controlling system moved up toward the top of the animal, where the light sensitive system sat. This light sensitive was…only guiding chemical changes and cycles, not behavior. But the joining of the two nervous systems gave them a new role. What an amazing image: in a long evolutionary process, a motion controlling brain marches up through your head to meet there some light-sensitive organs, which become eyes.

Godfrey Smith further documents that there are three major evolutionary lines of animals characterized by the fusion of sensory processes with motor processes into a centralized control center called a brain. These are the mollusks (including cephalopods), the arthropods (including insects like the praying mantis), and the vertebrates (from fish to birds to mammals). It is in these three phylogenetic lines that we see the full emergence and flowering of Mind[1].

Centralized brains and complex active body arrangements are also associated with a shift from reflexes and the associative learning patterns of habituation and sensitization into more complicated reactions. Examples of such reactions are found in what ethologists call "fixed action patterns" (FAPS). A fixed action pattern is a highly predictable behavioral response that is "reflex like" in nature. Such responses are specific to species and are "released" in the presence of a specific stimulus. Classic examples of FAPS include the aggressive displays from stickleback fishes in response to the red bellies of their rivals, and digger wasps repeatedly inspecting a nest over and over if its routine is disrupted. What is striking about FAPS is how automatic they are and how relatively fixed they seem to be. For example, a graylag goose was found to keep rolling balls regardless of the consequences (balls overflowing or rolling away again and again). More recently, ethologists have extended this concept into "modal action patterns" (MAPS). A MAP is one that is like a FAP but has a bit more flexibility and is somewhat more open to learning and feedback.

The idea of a reactive base of the mind that engages in relatively fixed or modal action patterns ingrained into the system and triggered in relatively nonconscious ways should sound familiar to you as a human being. The linkage with humans is found most directly in our habits and what are called automatic procedural memories. Unlike FAPS, which emerged fully formed at birth, habits are learned over time and via association and consequence. Virtually all of them are better considered as MAPs. However, as you are likely aware, they can operate in rigid, repetitive, and unthinking ways.

Consider, for example, my mental behavioral investment patterns following the installation of a new dishwasher in our kitchen. Because the new dishwasher had a handle on it, we could not fully open the drawer adjacent to it unless we first opened the dishwasher. That was where we kept our silverware, and the inconvenience was such that we had to move the silverware to a new drawer. As you might suspect, my habit system had developed the action pattern of heading to the original drawer whenever I needed silverware. In the months that followed, I would often find myself mindlessly going to the old silverware drawer, even though the self-conscious parts of myself were very aware that the silverware had moved. I would open the drawer, only to see a mishmash of rarely used items instead of the expected silverware. I would then become conscious of my procedural habits, curse, and move to the correct drawer. As this story highlights, we humans are clearly capable of building procedural action systems that have many similar features to FAPs/MAPs.

The Emergence of Learning Animals

In *Metazoa: Animal Minds and the Birth of Consciousness*, Godfrey-Smith (2020) formally lays out the case that we can trace the concepts of mind and sentience to the "Cambrian explosion," which happened from around 560 to 520 million years ago. It was a crucial period in the evolution of the animal kingdom, and it is when animals had complex sensory motor networks and bodies that could engage in much more sophisticated movement patterns. Territory defense, predation, and the avoidance of becoming a predator's next meal became central aspects of the adaptive fitness game animals were playing. Godfrey-Smith (2020) succinctly puts the interactive landscape as follows: "From this point on, the mind evolved in response to other minds" (p. 36). Responding to other animals is what likely gives rise to a new kind of mental behavior and what we can call the learning animals.

Learning animals exhibit the classic complex active responses that make animals so obviously different from plants. This is the world of what Dennett calls Skinnerian animals, and also corresponds to the level of motivated behavior in the Aunger and Curtis model. BIT allows us to see how these are connected. Consistent with 4E cognitive science, BIT gives rise to a cybernetic or control theory model of learning. There are three essential components of a control system: (1) an input sensor; (2) a reference goal; and (3) an output mechanism. Control theorists use the following formulation to define the relationship between the key variables: input–reference goal => output.

A thermostat connected to an air conditioner is a simple example of a control system. The temperature gauge is the input sensor, the temperature you set the room at is the reference goal, and the output mechanism is the addition of cool air. These simple ingredients allow one to maintain a comfortable room temperature. When you set the inside temperature at 74 degrees, you are setting the reference goal. Because the room is cooler than the outside, warm air enters, so the temperature in the room will heat up. The difference between what the input sensor is registering (say, 78 degrees) and where you set the reference goal triggers the air conditioner to turn on and cool air is added to the room. It will do this until the room cools down and the reference goal is reached, at which point it turns off. The room will then start to heat again, and this process thus maintains the room at the equilibrium of the set point.

We can apply this model to how animals behave in a goal-oriented manner. BIT corresponds an animal's perceptions with the inputs, its motivational states with the goals, and emotions with the outputs, which function as energizing action tendencies. This can be represented as the Perception–Motivation => Emotion (P – M => E) control theory formulation. It is translated to mean that a *perception of an actual state relative to a motivational state leads to an emotional state.* Figure 12.2 provides a diagram that can help clarify what is meant by each of these terms and show how they are interrelated by this formulation.

We can readily correspond this model with Aunger and Curtis' behavioral production model. The first level involves sensation. The dotted line represents how, for reactive animals, sensation can trigger fixed action patterns and motor responses. The P – M => E layer corresponds to the motivated layer of behavior in the Aunger and Curtis model. Perception is a higher-level mental process than sensation and occurs via the integration of sensory inputs that result in a meaningful interpretation of an object or event. Perception is a consequence of both bottom-up processing, which refers to the pattern of sensory inputs, and top-down processing, which refers to the individual's knowledge, memory, and expectations. The basic outline of how perception works is this: Through lived experience, the mind/brain builds perceptual categories of objects and events. These categories emerge from interaction with the object and events and

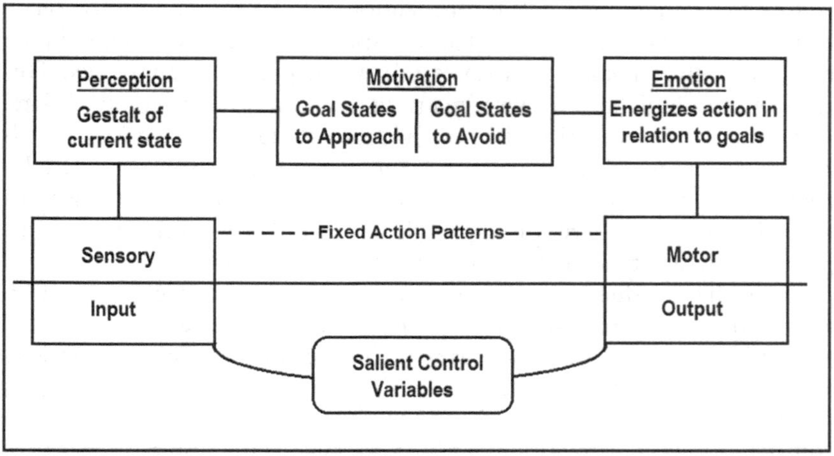

Fig. 12.2 The P – M => E learning control theory formulation

affordances that form templates or schema that enable the animal to determine the situation it is in and make predictions about what might happen next.

Motivation, the "M," refers to the goal states that the animal is working toward attaining or avoiding. These are the beneficial affordances and aversive stressors present in the environment, referenced against interior body states (e.g., needs for food, oxygen, safety). The reason that there are two broad classes of approach and avoidance goals is that the basic templates emerged due to evolutionary processes. Specifically, there were animal–environment relationships that either positively or negatively correlated with survival and reproductive success, and this separation gives rise to approach and avoidance tendencies that frame how an animal learns. Consider, for example, the basic human motives as represented by Maslow's hierarchy of needs. We can easily use an evolutionary frame to understand why humans are motivated to approach shelter, good food, status, sex, and being a well-liked member of a group. Likewise, we can readily understand why humans work to avoid injury, rancid food, extreme temperatures, and predators like tigers or sharks. The former states were positively associated with survival and reproductive success, whereas the latter were negatively associated with evolutionary fitness.

Emotions are framed in this formulation as "perceptual response sets," which means that they are activated to respond to perceived changes in the motivationally activated goal states relative to perceived appraisals. This view of emotions lines up with many theories and much empirical data. For example, Rolls (2013) defines emotions as states elicited by rewards and punishers which have particular functions, which primarily includes energizing action to obtain or avoid the rewards and punishers. Consistent with the fact that there are approach and avoidance motivational templates, there are two broad emotion systems, often labeled positive and negative. The reduction of a discrepancy between perceptions and an approach goal state activates a positive emotional state (e.g., satisfaction, joy). It is also the case that increasing the discrepancy between perceptions and an aversive state also activates positive affect, although of a slightly different tenor (e.g., relief, relaxation). In contrast, decreasing the discrepancy between perceptions and an aversive state results in negative emotions (e.g., fear, hurt), and so does increasing the discrepancy between perceptions and an approach state, although again of a slightly different tenor (e.g., frustration).

Bringing emotions into the equation gives rise to the question of Mind^2, and the experience of subjectivity. Does the presence of emotions mean that there is necessarily a subjective felt experience of being? No. Emotions do not *necessarily* require subjective feeling states. The word can be thought of as energized motion, and there are many good reasons to believe that action patterns emerge first, and subjective feelings are secondary. Nonetheless, we can foreshadow the analysis of the next chapter and say that learning animals are the likely place where Mind^2 has its roots. The experiences of pleasure and pain seem to function as nature's signals to approach and avoid. Toward that end, a quote from Richard Dawkins (1989) helps us link together the connections between foundational feelings (like pleasure and pain) and flexible behavioral patterns that are more responsive to environmental feedback.

> One way for genes to solve the problem of making predictions in rather unpredictable environments is to build in the capacity for learning. Here the program may take the form of the following instructions to the survival machine: 'Here is a list of things defined as rewarding: sweet taste in the mouth, orgasm, mild temperature, smiling child. And here is a list of nasty things: various sorts of pain, nausea, empty stomach, screaming child. If you should happen to do something that is followed by one of the nasty things,

> don't do it again, but on the other hand, repeat anything that is followed by the nice things'. The advantage of this sort of programming is that it greatly cuts down the number of detailed rules that have to be built into the original program; and it is also capable of coping with changes in the environment that could not have been predicted in detail. (p. 57)

If we break down Dawkins' description of the evolution of learning, we see that there are three connected elements. First there is the capacity of the animal to perceive where they are. Second, there is an inventory of the kind of things that are good (rewarding) or bad (nasty), which connect to the interior state of the animal (i.e., drinking water is a good thing when one's internal state is registering thirst). Third, there is experience of good/pleasure or bad/pain that energizes the animal to move toward or away from things. As Dawkins notes, this model gives rise to a much more flexible system of behavioral adaptation. However, we will hold off on exploring this claim further, as sentience and the domain of Mind[2] is the subject of the next chapter.

The Emergence of Thinking Animals

The learning level of mental evolution situates the animal in its environment and sets the stage for general seeking and approach or withdrawing and avoidance behavior patterns. It can be thought of as a kind of participatory relationship that results in a dynamic feedback loop between perception, motivation, action, and feedback in the agent-arena context. However, now imagine an animal that is engaged in foraging and comes to a fork in its path. It must either travel left or right, and the contingencies that will follow are very different. Because there are two different paths, it cannot engage the environment in an immediate and participatory way. Thus, feedback is disassociated from the current context. What is such an animal to do?

If the animal has any familiarity with the context, the best thing it could do is run a simulation. Ideally, it will have traveled both paths before. If so, it might have what Edwin Tolman called a "cognitive map" of the environment. In the 1930s, Tolman's work was at the cusp of the emerging neo-behavioral movement, when psychologists were beginning to argue strongly that thought processes mediated animal action and needed to be included in the equation. Tolman had demonstrated that if you allow rats to gain exposure to a maze for several trials and then place a reward in the maze, they learn to locate and return to the food much faster than rats with no prior experience. Tolman argued that this demonstrated that "latent learning" had taken place absent any specific reward and that the rat had developed a cognitive map of the maze.

In his excellent book *The Mind within the Brain: How We Make Decisions and How those Decisions Go Wrong*, A. David Redish (2013) recounts how Tolman's work was incorporated into the cognitive revolution and served to ground the concept of a "schema," which is the idea that animals develop a model of some aspect of the world and how they might interact with it. This sets the stage for us to become clear about what "thinking" is foundationally about. Specifically, motivated learning happens when the animal is dynamically interacting with the world, being actively shaped by reinforcing and punishing events. Thinking takes place when animals develop perceptual schema that models the animal–environment relationship such that they can run simulations on possible paths of investment and function to select paths that are predicted to yield the highest value.

In direct accordance with the model of mental evolution being laid down here, Redish (2013) builds toward deliberate decision making by starting with reflexes and then moving toward the emergence of emotion and the role that the emotions play in action selection. Redish then explains how modern cognitive behavioral neuroscience can map the processes by which animals deliberate and make choices. He explains how birds and mammals regularly engage in exploring their environment and building cognitive maps. These schemata are then searched in memory and used to map or simulate the current situation. Then evaluations are made regarding the predictions of best possible outcomes. This requires the animal to inhibit responses and plan a sequence of actions to get to an anticipated goal.

Redish reviewed how behavioral scientists have long noted that when rats arrive at a "T-point" in a maze and need to either go right or left, they appear as if they are mulling over their choices. Research has validated this commonsense interpretation. Redish and his colleagues tracked brain activity of rats at those points. He reported the results as follows:

> We found that during those paused 'vicarious trial and error' events, the hippocampal representation swept ahead of the animal, first down one potential future choice, and then down the other. Just as my friend was imagining what it would be like to take one job or another—so, too, the rat was imagining what would happen if it went running down the leftward

> path or down the rightward path. The most important observation we made was that these [neurologically mediated] representations were sequential, coherent, and serial…Our rats really were imaging their future. (p. 50–51)

BIT helps us to place this in the larger picture. Actions are expensive. The task of the nervous system is to process information and generate models of animal–environment relations that enable the most viable path of investment to be realized. As such, the evolution of the mind as a neurocomputational control system and mental behavior more generally can be traced in terms of greater sophistication in mapping possible outcomes. The deliberative or thinking or executive mind is all about extending the animal beyond its current situation and projecting possible situations and outcomes, simulating behavioral paths, and evaluating outcomes and then selecting actions, satisfying the time constraints.

The Evolution of Human Language and the Emergence of Talking

We have now established the basic neurocognitive architecture in animals that undergirds their mindedness. It moves from reflexes and fixed action patterns into a more open dynamic system regulated by perception, motivation, and emotion that enables instrumental goal achievement and is shaped by the consequences of the actions. Then "higher cognitive capacities" emerged that involve behavioral simulations, mental manipulations of schema, and evaluations of relevant information to generate predictions and orient toward future paths of behavioral investment. We have not yet discussed much in the way of sociality and the role that relationships, cooperation, care, and competition play in the evolution of mental behavior. Given that the social world played a major role in both the demands of the adaptive landscape and in benefits of more advanced thinking in our evolutionary line, this is an important gap that must be filled in. Consider, for example, that a mother needs to be able to empathize with the needs of her offspring and thus must start to simulate outcomes from the perspective of another. The role the relational world plays in the animal and human mind will be explored in Chap. 15.

More relevant to the current discussion, the social world of our primate ancestors was undoubtedly central to the evolution of language. The evolution of language is a notoriously controversial topic in the academic world, having been forbidden as a topic of discussion by the Linguistic Society in Paris in 1866 because speculation was rampant, but evidence was scant. Over the past 30 years, the topic has engaged many scholars, and works like Steven Pinker's (1994) *The Language Instinct* attracted much attention and sparked important debates.

The basic outline from higher cognitive processes into language seems to involve increasing capacities for abstract representation and mental time travel played out in a social world with much capacity for shared attention. It seems likely that the human capacities for hunting and tool use were important elements that drove the evolution toward increased encephalization, creating a feedback loop. Better brains led to more complicated cooperative patterns, and better hunters and gatherers were more effective at obtaining food and protection. The social dimension was surely central, and it is highly likely that our hominid ancestors started to develop much better capacities for sharing joint attention and intention. All of this sets the stage for memetic gesturing and the symbolic tagging of shared thoughts.

From the vantage point of the Unified Theory, as was discussed in the chapter detailing JUST, the key tipping point was the shift from symbols and gestures into the full, open, symbolic-syntactical language that generated propositional statements that could be questioned. This capacity for full question and answer dialogue seems to be present only in Homo sapiens, although there is much debate about the linguistic capacity of our close cousins, like the Neanderthals and Denisovans. Given that very little of what constitutes language becomes fossilized, the history remains largely a mystery. What the archeological, anatomical, genetic, and neurocomputational data do say very clearly is that by 75,000 years ago a major transition in behaving had happened in the Homo sapiens line, and we had emerged as a singularly unique animal that could talk and give reasons for our actions in a community linked by a shared verbal system of understanding. Chapter 16 explores Mind^3 and the Culture-Person plane in greater detail.

The Neurocognitive Functional Architecture of the Human Mind

With the outline of the evolution of Mind in place, we can now shift gears to generate a schematic of the neurocognitive architecture of the human mind. Figure 12.3 provides a schematic of the neurocognitive architecture of the human mind based on BIT. There are several aspects of this diagram

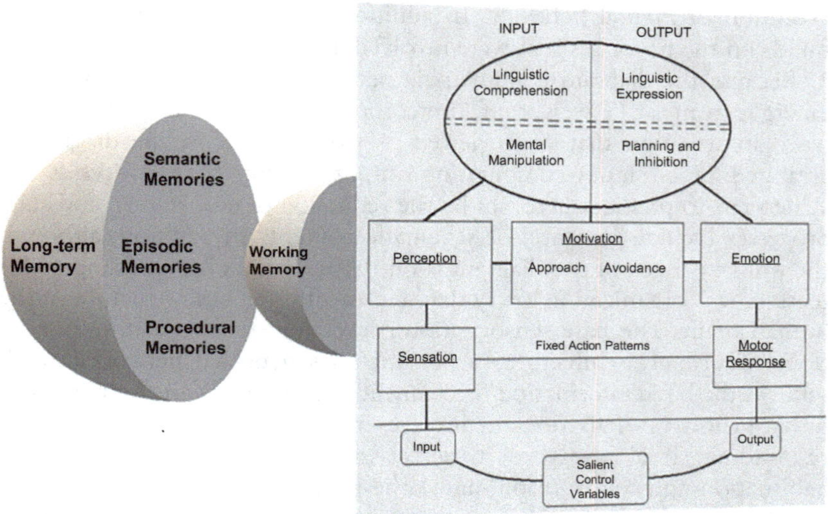

Fig. 12.3 The architecture of the human mind as mapped by BIT

that need to be unpacked. First, the figure is grounded in the six principles of BIT, such that it is a neurocognitive system that is coordinating mental behavior as value-based investments that can be understood as a function of energy economics, evolution, genetics, learning, and developmental history. Second, because it is a cognitive and functionalist map, it includes the domains of inputs, computational control, memory, and outputs, all of which operate in dynamic feedback relations with the environment. As reviewed in *A New Unified Theory of Psychology*, it is informed by many diverse approaches. Perhaps most obviously, it represents the evolutionary history of the neurocognitive system in four steps. That is, it divides the neurocognitive architecture into the four levels of reacting, learning, thinking, and talking.

Starting with the basic structure, we can see that the right side is arranged "vertically" in that it proposes four different levels of mentation. These are the: (1) sensory-motor and procedural level (i.e., reactive); (2) perceptual-motivational-emotional level (i.e., learning); (3) mental manipulation, simulation, and planning (i.e., thinking) level; and (4) linguistic comprehension and expression level (i.e., talking). As should be apparent, these levels of mentation correspond directly with the four stages in the evolution of mental behavior. In addition, each level incorporates and builds on the processes that were involved in the level beneath it.

Because the diagram is framed via neuro-information processing that emerges from a BIT viewpoint, it incorporates the three meanings of the word information that we reviewed previously. That is, the diagram is arranged as an input-computation-output system, such processes are structured to predict outcomes by the reduction of uncertainty, and such processes include semantic or schematic content. In addition, although the primary focus of the diagram is on the information processing architecture, it is organized in a way that aligns with the basic structure of the human brain. The base sensory-motor level corresponds to the foundational neuronal architecture, which includes neuronal networks which serve as the basic information mechanistic units that link the brain to cognitive processes. This domain involves reflexes and automatic ways of responding. It also involves basic, classical learning processes such as habituation and sensitization, and basic procedures that can be enacted automatically. The processes that are operating at this level are largely unconscious, that is, there is no necessary experience of subjectivity involved.

The second level involves perceptions, motives, and emotions that guide more flexible behavioral responses that are contingent and shaped based on feedback and experiences. As reviewed earlier, this is tied together via the P – M => E control theory formulation, which is framed in a way that is consistent with cognitive neuroscience, consciousness studies, and behavioral science. To see how this connection is made, we can turn to the "salient control variables" listed at the bottom of the right-hand side, which refer to the key pieces of information from the environment that are entering the system. The term "control variable" is offered as a way of linking two different conceptions of mental behavior, one of which is the traditional behavioral view, in which the environment is controlling behaviors and can be concretely manipulated by the behavior analyst to shift frequencies in response patterns. The other view comes from perceptual control theorists like William Power (1973), who argues that overt actions are better conceived of as a function of the animal attempting to control its perceptual inputs. That is, perceptual control theory takes the point of view of the animal rather than the environment and explains the relationship of operant behavioral processes from the inside.

Consider, for example, the task of making yourself an egg. According to perceptual control theory, through experience, you have an image of an ideal egg, and that becomes the motivated reference goal state that you are working toward producing. Once you crack the egg on to the pan, there are key variables, such as the whiteness of the egg white as it cooks and the texture of the yolk that you are tracking. You add heat to move the raw egg toward the perceptual representation of the ideal egg and remove the heat when it matches. The traditional behavioral science view argues that behavior changes as a function of various stimuli. For example, we could hypothesize that the key variables that were controlling your egg cooking behavior were the color the egg white and the texture of the yolk. A clear egg white and fluid yolk tends to reinforce cooking behavior, whereas a white egg white and a hardening yolk tends to elicit behavioral tendencies to remove the egg from the heat. The diagram allows for the view from either inside or outside the animal, and we can consider the salient control variables from the inside in terms of perceptual control theory and from the outside in terms of conditioning.

The Thinking and Talking Levels and the Correspondence with Intelligence

Our focus here is on humans, and the top two levels are represented as a circle, and correspond to the human cortex. The circle is divided in a way that represents the major divisions of the cortex. The back-to-front divide represents the back three lobes of the brain (i.e., occipital, parietal, and auditory) which deal primarily with perceptual and linguistic inputs and the integration of information across sensory modalities. In contrast, the frontal lobe coordinates output sequences, and engages in planning and inhibition (both verbal and nonverbal) across time. The left-to-right divide captures the fact that there is some specialization between the hemispheres, with the right hemisphere tending to be more holistic, contextual, and perceptual and the left being more focused on particulars, logical sequences, and linguistics.

A useful way to conceptualize the functional arrangement of thinking and talking in humans is to consider intelligence, as it is assessed by the best available assessments. Although intelligence is a controversial topic, much of this controversy is tangled in the dynamics of intelligence testing and the implications that intelligence testing has had for making claims about social groups and the way such tests have been employed by institutions. In terms of thinking about intelligence from a purely scientific perspective, it is well established that we can measure "mental abilities" with a degree of reliability and validity that is equal or superior to almost any other mental construct, including personality traits. Because the architecture of the human mind diagram is grounded in a neurocognitive functionalist model of mental behavior, we can make the assertion that it should correspond with the concept of intelligence as measured by the major assessment instruments.

The Weschler Adult Intelligence Scales (WAIS) are the most popular individually administered tests of intelligence, and many scholars consider the WAIS to be the gold standard of adult intellectual assessment. As such, we can use it to determine alignment with the neurocognitive functional architecture given by BIT. Directly relevant to our current considerations is the fact that decades of research on the WAIS have revealed four factors or domains of intelligence. These factors form four key indexes, which are thought to map specific kinds of information processing and problem-solving capacities. Two of the indexes are "major" and two are minor. One major domain is called the Perceptual Reasoning Index (PRI). This refers to the capacity to mentally manipulate objects accurately and perceive patterns and sequences in designs. A common subtest of this index is the block design task in which an examinee is given a picture of red and white blocks along with a group of actual red and white blocks and needs to recreate the image, all while being timed. The primary skill is the ability to mentally represent and manipulate the boxes in an accurate and efficient way. This corresponds to the (nonverbal) thinking mind and is, broadly speaking, more anchored to the way the right hemisphere processes information in humans.

The second major index on the Weschler is the Verbal Comprehension Index (VCI), which is a measure of an individual's vocabulary, fund of verbal information, verbal concept formation, and capacity for comprehension of verbal information. Tests of vocabulary and fund of information along with assessments that ask the examinee to identify the conceptual relationships between words (e.g., explaining how a chair and a table are similar) make up this index. We can identify it as the capacity and fluency of the talking mind, which is more anchored to left hemisphere processing in humans. In short, the two major domains of measured intelligence, Perceptual Reasoning and Verbal Comprehension, align directly with the "nonverbal thinking" and "talking" competencies depicted in the diagram.

A lesser domain of intelligence is called the Processing Speed Index, which refers to how quickly a person can perform a sensory-perceptual task that takes very little "thinking." For example, an examinee might be asked to scan a page and cross out shapes that match a form and be measured to see how many they can cross out in a minute. We can align this skill with the first reacting layer of the diagram. The second lesser domain is called working memory. Working memory refers to the capacity to hold and accurately manipulate information. Whereas the PRI focuses on the complexity of the task and kinds of manipulation, the Working Memory Index focuses more on capacity to hold information. For example, a common working memory task is called digit span, which relates to how many numbers someone can repeat back, either forwards (in the order they were given) or backwards (when the order is reversed).

The Domains of Memory

This brings us to how memory is depicted by the diagram. The diagram is arranged in a way that corresponds to Atkinson and Shiffrin's (1968) well-known and long utilized three-stage information processing model of memory. This includes: (1) a sensory-perceptual "iconic" phase that lasts from one to three seconds; (2) a working memory phase that lasts between three and 30 seconds; and (3) a long-term memory storage system that can store information across the lifespan. The right side of the diagram corresponds to the iconic memory phase and, as we will see, it lines up with the metaphor that we can think of consciousness as something which is placed under a spotlight or on a display screen of perceptual attention and subjective experience. Just behind that screen is working memory. Working memory refers to information that was recently recalled and can be held and manipulated and integrated with the sensory-perceptual screen. Working memory is quite limited in time and capacity, lasting only for about 20 seconds. The classic number of objects that it could store was "seven, plus or minus two," spelled out in a famous paper by George Miller (1956). Subsequent recent research has suggested that this estimate was a bit generous (e.g., LeCompte, 1999), but the work nevertheless remains a classic.

Baddeley's (2007) multicomponent working memory model offers an excellent map of this domain that is consistent with the model of the mind we are operating from. Baddeley's model proposes that four different functional elements go into the working memory system, which are (1) an attentional controller, which directs one's focus and three temporary storage systems, namely (2) a visuospatial sketchpad, (3) a phonological-verbal loop, and (4) a system that links short-term with long-term memory storage. Given our discussion above, we can already see that the phonological loop connects to the talking level of mentation and the visuospatial sketchpad to the perceptual reasoning or nonverbal thinking domain.

Baddeley's attentional controller refers to a central executive and lines up with the general goals and intentions of the person. To see the role of the attentional controller in perception and memory, consider an experiment that asked participants to imagine going through a house and later asked what they saw (Pichert & Anderson, 1977). Prior to the exercise, one group was asked to consider whether they would like to live there. A second group of participants was asked to imagine themselves as a burglar who was scoping out the place for a future robbery. Not surprisingly, when they were later asked to recall salient items, they reported seeing very different things in the house. Those who had focused on the possibility of living there emphasized the size of the rooms, the placement of the furniture, and the functional utility of the house. Those who looked at the house through the eyes of a burglar focused on the items that could be stolen and how they might enter or exit the house. The experiment provides us with an excellent example of the role of the attentional controller in what is perceived and remembered.

Finally, there is long-term memory, which exists over much larger spans of time and has a much larger storage capacity. As shown in the diagram, long-term memory in humans can be divided into semantic, episodic, and procedural memory systems. To obtain a clear understanding of these three memory systems, consider the following questions: (1) Do you know if wearing helmets while riding your bike is a law or not? (2) Do you recall when you first were learning how to ride a bike, or perhaps a time when you fell off and were hurt? and (3) Could you get on a bike now and ride it with no difficulty? If you knew the answer to the first question about helmet laws, you retrieved it from your semantic memory system, which stores factual knowledge about rules, norms, math or logic, and historical events. Answers to questions such as "Who was the third president of the United States?" "How many protons are in a carbon atom?" and "What is four cubed?" are stored in your semantic knowledge system.

Episodic memory is different. It holds perceptual and emotional experiences, normally as visually sequenced images from the point of view of the person. In terms of accessing them over time, such memories are generally stored based on their affective valence, such that the greater the emotional charge the greater likelihood of recall, although this is not always the case. There are some rare people who have an "eidetic" memory, such that they almost never forget anything. But most people quickly lose events that do not have emotional significance, especially as time passes. It would not be unusual to have a memory of the first time you rode a bike or when you had a serious injury because these carry emotional charges. However, it would be odd if you could remember the seventh time you rode a bike on a perfectly average trip around the neighborhood.

Both semantic and episodic memories are often referred to as "declarative" memory systems because they can be brought into awareness and reported on. That is, from an exterior behavioral point of view, people can make declarative statements about them. Procedural memories are different. People may have declarative knowledge of whether they can ride a bike, but they will not have any knowledge about how they do so. They either can get on and ride or they cannot. That is, they have either instantiated the key procedures in their neurological and muscular systems to coordinate the needed variables to perform the task or they have not or have lost the ability. Such action patterns are stored in procedural memory, which is often referred to as an "implicit" memory system because one cannot introspectively report much information about it.

The fact that procedural learning can take place completely independently of the more conscious, declarative memory systems was made clear by the famous case of HM in the 1950s. He was a patient who suffered from serious epileptic seizures. The doctors operated on him and ablated key areas in his hippocampus, a section of the brain that is crucial for transferring working memories into long-term storage. This meant that after about five minutes had passed, HM would have forgotten all the new events that had occurred. Thus, when the doctors came to meet him each day, every encounter was experienced as if it was happening for the first time. However, researchers found they could teach HM procedures, such as drawing in a mirror. Although he would deny he had any recollection of doing such activities, he was able to learn procedures almost as effectively as someone who had full conscious recall abilities. This demonstrated that procedural operations are separate from declarative memory systems.

The architecture of the human mind diagram enables us to generate a depiction of the neurocognitive functional processes that guide or mediate overt human mental behavior. It carries with it the evolutionary history of layered mentation across the domains of reacting, learning, thinking, and talking. It also shows how increasing cognitive capacities for simulating possible paths of behavioral investment resulted in working and long-term memory systems that allowed animals in general, and humans in particular, to extend those patterns across many different situations and time periods. The structural and functional organization of the diagram allows us to bridge from those phylogenetic considerations to empirical explorations of cognitive functioning, as revealed by intelligence testing and modern models of memory.

Conclusion

Our task in this final section of the book is to build off the descriptive metaphysical systems given by the ToK, PTB, and Map of Mind[1,2,3] that enable us to specify the ontological referents associated with mental behavioral processes and shift to provide metatheoretical structures that afford us a more causal explanatory framework for understanding the dynamic processes associated with the informational interfaces between the domains. This chapter summarized how BIT gives us a general theory of Mind, explicated as the property of mindedness and the domain of Mind[1]. First, directly paralleling the modern evolutionary synthesis, it is located as the joint point between Life and Mind on the ToK System and enables a bridging function between neuro-information processing models and Skinnerian behavioral selection. Second, BIT explicitly locates the evolutionary origin of mental behavior in the movement of the animal as a whole. Consistent with a 4E cognitive science view, the animal is an embodied agent in the arena, and mental behaviors are the complex adaptive processes represented by the Mind dimension of complexification.

We reviewed how BIT bridges between the key paradigms that make up the mind, brain, and behavior sciences, and how it delineates six core principles that work together to provide a metatheoretical architecture that captures how modern animal behavioral researchers frame the processes by which animals invest their actions to control the flow of resources. We also demonstrated how BIT effectively characterizes the major steps in mental evolution, from reacting to learning to thinking to talking. And then we shifted to show how it is structured to frame a working model of the neurocognitive architecture of the human mind that is consistent with much research and mid-level theory in cognitive psychology.

Consistent with work by Chalmers (2007) and others in the philosophy of mind and consciousness, we can frame all of the above in terms of the relatively easy problems of consciousness. That is, the theoretical formulations put forth are grounded in data available to an exterior epistemological framework. Specifically, what we are trying to explain is mental behavior defined in terms of the way animals exhibit functional awareness and responsivity that can be observed and measured from the outside of the system. Yet, there remains a key aspect of mental life that is not really being accounted for in the account of mental processes given in this chapter. We have not discussed what it is like to be an animal from the inside. The hard problem of subjectivity is, indeed, a hard problem. However, it is one that is made exponentially more difficult because of the massive metaphysical confusions and problems with language and referents that stem from the Enlightenment Gap, our convoluted philosophy of mind, and the problem of psychology. As such, UTOK is well positioned to considerably advance our understanding of subjective conscious experience. Like most naturalistic approaches, UTOK posits that Mind[2] emerges from—and is a special kind of—neurocognitive processes. As such, the implication is that the metatheoretical framing of Mind[1] should also provide a structural and functional grounding for understanding Mind[2]. We tackle this question and associated complexities in the next chapter.

References

Atkinson, R. C., & Shiffrin, R. M. (1968). Human memory: A proposed system and its control processes. In K. W. Spence & J. T. Spence (Eds.), *The psychology of learning and motivation* (Vol. 2, pp. 89–195). Academic Press.

Aunger, R., & Curtis, V. (2008). Kinds of behaviour. *Biology and Philosophy, 23*(3), 317–345.

Baddeley, A. (2007). *Working memory, thought, and action* (Vol. 45). Oxford University Press.

Barrett, L. F. (2020). *Seven and a half lessons about the brain.* Houghton Mifflin Harcourt.

Chalmers, D. (2007). The hard problem of consciousness. In M. Max Velmans & S. Schneider (Eds.), *The Blackwell companion to consciousness* (pp. 225–235). Blackwell Publishing.

Dawkins, R. (1989). *The selfish gene.* Oxford University Press.

Dennett, D. (1995). *Darwin's dangerous idea.* Simon and Schuster.

Farchmin, P., Eggert, A., Duffield, K. R., & Sakaluk, S. (2020). Dynamic terminal investment in male burying beetles. *Animal Behaviour, 163*, 1–7.

Geary, D. C. (2005). Motive to control and the origin of mind: Exploring the life-mind joint point in the tree of knowledge system. *Journal of Clinical Psychology, 61*, 21–46.

Godfrey-Smith, P. (2016). *Other minds: The octopus, the sea, and the deep origins of consciousness*. Farrar, Straus and Giroux.

Godfrey-Smith, P. (2020). *Metazoa: Animal life and the birth of the mind*. Farrar, Straus and Giroux.

Goodson, F. E. (2003). *The evolution and function of cognition*. Lawrence Erlbaum Associates Publishers.

Gorman, J. (2011, September 20). Amorous squid seeks partner: Any sex will do. *New York Times*. Retrieved August 9, 2022, from https://www.nytimes.com/2011/09/21/science/21squid.html

Graham, Z. A., Padilla-Perez, D. J., & Angilletta, M. J. (2020). Virile crayfish escalate aggression according to body size instead of weapon size. *Animal Behaviour, 163*, 9–15.

Green, S. J., Boruff, B. J., & Grueter, C. C. (2020). From ridge tops to ravines: Landscape drivers of chimpanzee ranging patterns. *Animal Behavior, 163*, 51–60.

Hebb, D. (1949). *The organization of behavior: A neuropsychological theory*. Psychology Press.

Hemingway, C. T., Ryan, M. J., & Page, R. A. (2020). State-dependent learning influences foraging behaviour in an acoustic predator. *Animal Behaviour, 163*, 33–38.

Henriques, G. R. (2003). The tree of knowledge system and the theoretical unification of psychology. *Review of General Psychology, 7*(2), 150–182.

Kandel, E. R. (2009). The biology of memory: A forty-year perspective. *Journal of Neuroscience, 29*(41), 12748–12756. https://doi.org/10.1523/JNEUROSCI.3958-09.2009

LeCompte, D. C. (1999). Seven, plus or minus two, is too much to bear: Three (or fewer) is the real magic number. In *Proceedings of the human factors and ergonomics society annual meeting* (Vol. 43, pp. 289–292). Sage.

Llinás, R. R. (2001). *I of the vortex: From neurons to self*. MIT Press.

Miller, G. A. (1956). The magical number seven, plus or minus two: Some limits on our capacity for processing information. *Psychological Review, 63*(2), 81–97.

Power, W. (1973). *Behavior: The control of perception*. Aldine Transaction.

Pichert, J. W., & Anderson, R. C. (1977). Taking different perspectives on a story. *Journal of Educational Psychology, 69*, 309–315.

Pinker, S. (1994). *The language instinct: How the mind creates language*. Harper.

Redish, A. D. (2013). *The mind within the brain: How we make decisions and how those decisions go wrong*. Oxford University Press.

Rolls, E. T. (2013). *Emotion and decision-making explained*. Oxford University Press.

Sabol, A. C., Lambert, C. T., Keane, B., Solomon, N. G., & Dantzer, B. (2020). How does individual variation in sociality influence fitness in prairie voles? *Animal Behaviour, 163*, 39–49.

Shulman, R. G., Rothman, D. L., Behar, K. L., & Hyder, F. (2004). Energetic basis of brain activity: Implications for neuroimaging. *Trends in Neurosciences, 27*(8), 489–495.

Trestman, M. (2013). The Cambrian explosion and the origins of embodied cognition. *Biological Theory, 8*(1), 80–92.

Turner, E. C. (2021). Possible poriferan body fossils in early Neoproterozoic microbial reefs. *Nature, 596*, 87–91. https://doi.org/10.1038/s41586-021-03773-z

Wu, T. (2018, February 16). The tyranny of convenience. *The New York Times*. Retrieved August 9, 2022, from https://www.nytimes.com/2018/02/16/opinion/sunday/tyranny-convenience.html

Zipf, G. K. (1949). *Human behavior and the principle of least effort: An introduction to human ecology*. Addison-Wesley Press.

CHAPTER 13

Mind^2: Subjective Conscious Experience in Animals and Humans

One day I came into my home office only to find that I had left the top of my fish tank open, and one of the fish had jumped out and landed on the floor. It was now dead, and its body lay dry and stiff, its lifeless eye staring up at me. As I disposed of it, I felt guilty and sad and—perhaps as a way of generating an intellectualizing defense against those feelings—I found myself wondering what the fish's last moments were like. I have asthma, and I know what it is like to work hard for breath. I have a distinct memory of being about ten and wrestling with my older brother in the lake in front of my grandparents' house, whereupon he proceeded to dunk me, such that I reached a panic point and had a surge of adrenaline and thrust myself out of the water, violently swinging my fists. Thankfully, my attempted blows did not land on their target. "Take a chill," my brother said. "Screw off," I replied. Getting out of the lake and heading up to the house, I simultaneously felt the relief of the air in my lungs and the humiliation of being dominated.

Bringing this to the domain of Mind^2, my point is that I know—intimately—what it *feels* like to be me. I know this in a way that is qualitatively different than knowing how it would be to be a fish or even another person. This is true in at least two important ways. First, I know my qualitative experiences *directly*. With no effort, I am simply thrown into my qualitative world, and have direct access to my experiences in a way that can be described as "hyper-present." This is the essential nature of

G. Henriques, *A New Synthesis for Solving the Problem of Psychology*, Palgrave Studies in the Theory and History of Psychology, https://doi.org/10.1007/978-3-031-18493-2_13

first-person perspectival knowing. Second, I can only know the subjective conscious experiences of others indirectly, via a third-person behavioral view. This is true in many ways, but it becomes especially apparent from some angles. For example, I have a very hard time considering the experiences of a fish. Indeed, as we will see, the kind of conscious experiences fish have is a subject of significant scientific debate. Not surprisingly, this disagreement is in no small part a function of the fact that such experiences cannot be observed from the outside.

But the subjective divide can be starkly apparent even with other people whom I know well. Indeed, although I am a clinical psychologist whose job it is to be an expert in empathy, I can still sometimes struggle with fully grasping even the simple subjective, qualitative experiences of others. Consider, for example, that when we were growing up my older brother loved it when my mother would make calves' liver and bacon. In contrast, I have always loathed the taste, texture, and smell of liver. Although I intellectually understand the fact that my brother enjoyed that meal, to this day I have a hard time conceiving exactly what his first-person experience is like. Many questions arise as I try to genuinely empathize with his point of view. Are his sensations and perceptions of the smell, texture, and liver fundamentally different than mine? Or is it the case that his sensations are largely the same, but his emotional evaluation of them is different? Did he learn to like the meal? Could I?

These reflections highlight the complicated nature of subjective conscious experience. Thankfully, we have already done much work in addressing the concept of consciousness that will be of assistance to us. In this chapter, we are focused on clarifying the evolution and emergence of Mind[2] in the animal kingdom. We will see that there is an important distinction between early and late appearing models of consciousness in animals. Scholars who argue for the early appearance of subjective experience posit that it might date back over 500 million years ago to the Cambrian explosion. They argue that creatures like praying mantises likely have a conscious experience of being in some basic senses of the word. Other scientists argue for a later appearance. They claim that subjective conscious experience requires higher brain structures which are not present in praying mantises, and argue that subjective conscious experience is likely to be present only in some birds and mammals, and thus evolved only a few hundred million years ago, or even more recently.

Our zoomed-out view provides us with a frame for a plausible interpretation that potentially reconciles these views. Specifically, we will follow the levels of mental behavioral evolution afforded by BIT and proceed to distinguish between “the base of sentience” that likely emerges at the learning level and a fuller “subjective conscious experiencing self” that likely emerges at the third level of thinking and deliberation. The proposal is that the base of sentience goes far “down” the phylogeny of animal kingdom and is likely associated with flexible learning patterns shaped by flashes of pleasure or pain. However, this is just the beginning of sentient experiences and does not constitute the full suite of subjectivity.

The full component of what we called the “experiential self” in the Updated Tripartite Model of human consciousness likely requires a more complex “global workspace” that is probably present only in some birds and mammals. The argument will be that the distinction between the base of sentience and the experiencing self represents a helpful differentiation in developing a richer descriptive metaphysics and ontology of Mind^2. In the next chapter, we will extend this conception of Mind^2 into the relational world of attachment, competition, and cooperation using UTOK's Influence Matrix.

The Nature and Evolution of Subjective Conscious Experience in Animals

The nature and scope of consciousness in animals has long been debated, and a wide variety of positions have been taken. During my education as a psychologist, I was repeatedly informed that René Descartes and George Romanes took diametrically opposed views on the nature of animal consciousness. It was often claimed—with some justification—that Descartes thought that only humans were conscious and that all other animals were machine-like automata. Drawing on Descartes' analysis, in 1689 Malebranche wrote:

> In dogs, cats, and other animals, there is neither intelligence nor a spiritual soul in the usual sense. They eat without pleasure; they cry without pain; they believe without knowing it; they desire nothing; they know nothing; and if they act in what seems to be an intelligent and purposive manner, it is only because God has made them fit to survive, and has constructed their bodies in such a way that they can organically avoid—without knowing that they do so—everything that might destroy them and that they seem to fear. (Translated from Huxley, 1896, pp. 218–219)

In contrast, the nineteenth-century psychologist George Romanes argued that many animals had rich mental lives. A close friend of Darwin's, he was one of the founders of comparative psychology and wrote several important books on animal intelligence. He opened his 1883 book *Mental Evolution in Animals* by breaking up mental capacities into 50 different levels of intellectual, emotional, and conscious development. Adult humans, with their capacity for language, culture, and explicit self-reflection, occupied the top rung. The highest animals listed were dogs and apes. Romanes characterized them as reaching rung 28, and as having an "indefinite morality" and the capacity to experience remorse, shame, deceit, and the ludicrous. Birds were a few levels down, appearing at 25. He posited that they could understand words and pictures and experience terror. Just beneath them were bees and ants, which could communicate ideas and feel sympathy. Several steps down, at level 18, were worms and insect larvae, which Romanes argued experienced primary instincts and the emotions of surprise and fear.

At first blush, it appears that Romanes and Descartes did indeed advocate for radically different views on the nature of animal consciousness. And it is the case that there are real differences between them. However, the lack of shared understanding regarding what is meant by "mind" or "consciousness" or "experience" or "awareness" or "self" hides the fact that there is much greater overlap in their views than appears on the surface. Thankfully, with UTOK's Map of Mind[1,2,3], we can make sense of the terrain and see the previously hidden correspondence.

Romanes devoted a significant portion of his book on mental evolution to delineating what he meant by both mind and consciousness. Consistent with our analysis of the epistemological gap-and-portal, he began with a discussion on the difference between the first-person subjective and third-person objective points of view. Directly aligned with the ToK System's framing of the Mind dimension, he considered "mind" as being accessible from both the subjective interior and behavioral exterior vantage points. Not surprisingly, he struggled somewhat to precisely define the mind from the interior. He defined consciousness largely in terms of (a) subjective experience and (b) the appearance of an animal making a choice or deliberate action. He conceived of the subjective conscious mind emerging out of reflexes in a way that overlaps with BIT's evolution of mental levels. Romanes mapped out 50 levels of mental evolution and he believed that conscious/sensory experiences start to appear at level 14. He listed pleasure and pain as the first kind of inner experiences (he labeled them as "sensations"), which, as we will see, is also consistent with the model developed here.

Romanes also argued that there were significant and qualitative distinctions between animals and humans. Almost the entire upper half of Romanes' 50 levels are occupied by humans only. Self-conscious reflection starts at level 34, which is six levels higher than the highest nonhuman animals. As we highlight this fact, a striking correspondence arises. Namely, once we place their ideas into the metaphysical taxonomy given by the Map of Mind^{1,2,3}, we see that Romanes' frame overlaps significantly with Descartes'. If we start with Descartes' famous dictum, "cognito ergo sum," it is clear that he is referring primarily to the domain of Mind^3. Indeed, Descartes thought about consciousness in terms of how perceptions and feelings *become ready for reason,* meaning that perceptions and feelings are already present in the analysis. This means that Descartes' conception of consciousness is found in the relationship between Mind^2 and Mind^3 and is most essentially found in the jump from Mind^2 to Mind^3. Understanding that this jump is Descartes' primary referent, we can see why he would claim that there is no consciousness for animals. But when we examine Romanes' levels of mental evolution, we can see that this is exactly his conclusion as well. Clarity and alignment between the positions is achieved when it is understood that, for Romanes, consciousness refers to the jump from Mind^1 to Mind^2, in contrast to Descartes using consciousness to refer to the jump from Mind^2 to Mind^3.

This analysis can be deepened when we more deeply consider the claim, as expressed in the quote by Malebranche, that Descartes' model suggests that animals experience no pleasure or pain. A careful reading of the quotation shows that Malebranche adds the qualifier "in the usual sense." Malebranche is emphasizing that, although they seem to have purpose and intelligence and experience fear, the core claim is that they are this way but "without knowing that they do so" and they lack a "spiritual soul." These qualifiers point to the reflective knowing or secondary access and awareness about perceptual experience as being the primary referent (i.e., the jump from Mind^2 to Mind^3, such that the individual has access to and can reflect on experience). The conclusion is bolstered by being reminded that Descartes was, of course, aware of sensations and perceptions; however, he generally considered them bodily experiences that were understandable from a mechanical perspective. Thus, it is not clear that Descartes would have said that animals do not experience the sensations of pleasure or pain at all. Rather, it is that they could not reflect on the meaning of that pain and suffering in the same way that humans might. *The point here is that it is very likely that Descartes considered many animals as having a Mind*[2].

This analysis of Romanes and Descartes through the descriptive metaphysics of the Map of Mind[1,2,3] suggests that, rather than being diametrically opposed, Romanes and Descartes are, in fact, aligned in the basic schematic of mental processes mapped by UTOK. The basic form for each is that Mind[1] first evolves as a function of basic sensory and motor reflexes, and it is highly likely that animals with simple nervous systems and no brains like jellyfish do not have any Mind[2] experiences whatsoever. As we will see, animals like praying mantises that have complex active body plans and brains may well experience pleasure and pain, although the question remains unanswered. However, the evidence builds, such that as we get to vertebrates (and some mollusks like octopuses) we can be confident that there are Mind[2] processes. By the time we are at mammals such as dogs, the evidence is overwhelming that they both experience pleasure and pain and deliberate on inner experiences by running mental simulations about various paths of possible future investments.

This formulation is also directly consistent with modern work on animal consciousness. There is an emerging picture filled with converging lines of evidence that allows us to make confident assertions along these lines. Indeed, in 2012 a group of scholars generated *The Cambridge Declaration of Animal Consciousness*, which asserted:

> The absence of a neocortex does not appear to preclude an organism from experiencing affective states. Convergent evidence indicates that non-human animals have the neuroanatomical, neurochemical, and neurophysiological substrates of conscious states along with the capacity to exhibit intentional behaviors. Consequently, the weight of evidence indicates that humans are not unique in possessing the neurological substrates that generate consciousness. Nonhuman animals, including all mammals and birds, and many other creatures, including octopuses, also possess these neurological substrates.

The Declaration shows that there is widespread agreement among scholars that many animals have Mind[2] and rich mental lives. Yet it is also the case that virtually everyone agrees that humans have a different kind of mind. The point here is that perspectives ranging from Descartes, to Romanes, to UTOK and modern animal behavior science all converge. This analysis adds yet another example of how the failure of the

Enlightenment to generate a clear map of the various meanings of mental processes resulted in endless confusion and debate that can be cleared up by the current work.

Animal Consciousness in Two Steps

In *The Ancient Origins of Consciousness: How the Brain Created Experience*, Feinberg and Mallatt (2016) offer a helpful framework for considering how SCE potentially evolved. These authors refer to SCE as "sensory consciousness," which they note is also sometimes referred to as phenomenological consciousness, perceptual consciousness, or sentience. They highlight that the concept is often described as a singular notion in terms of SCE, by which they mean the Nagel definition of what it is like to be something. However, they explain that there are potentially at least three different kinds or domains of qualitative experience and identify three mental processing pathways that might be associated with these different domains.

First, there is "exteroception," which refers to inner experiences or sensory qualia that are experienced as mental images of the outside world. For example, when you see a coffee cup, with its color and shape, or hold it in your hands and feel the warmth. These are examples of exteroception, experienced perceptions of incoming sensory data that represent and model the outside world. Next there is interoception. This refers to the sensory inputs from the body that include things such as the position of the body and the states of the body that are closely associated with drives or needs. Examples here are thirst, hunger, and pain. These provide information about the state of the body and are closely tied to homeostatic processes, which are the processes that keep the complex machinery of the body in the necessary parameters (i.e., not too hot, not too cold, balance of sleep and wakefulness, etc.). The authors note that interoception can be either localized, as in the case of experiencing "sharp pain," or more body-wide or global, such as fatigue, nausea, or oxygen deprivation that gives rise to "air hunger." As my narrative about being dunked makes clear, intense air hunger signals a crucial homeostatic imbalance and quickly energizes a motivational state to find a behavioral investment pathway to fill that need and restore equilibrium.

The third kind of sensory consciousness that Feinberg and Mallatt (2016) identify is called "affective." They describe it as follows (p. 138):

> Affective consciousness is global and involves the entire 'self'. It assigns the basic affective states (of good and bad feelings) and it is also responsible for complex emotions in humans of sadness, joy, shame, despair, fear, and so on. The affective-limbic aspect relates more directly to internal motivations, drives and behavioral responses than do the other two aspects of sensory consciousness. Positive affects (liking pleasure) motivate us to approach a rewarding stimulus and negative affects (dislike, displeasure, discomfort) motivate avoidance or escape from a noxious or threatening stimulus.

In other words, affective consciousness involves positive or negative feeling states that monitor the relationship between the exterior and interior and orient the animal to approach or avoid and provide feedback on how things are going.

Feinberg and Mallatt's three domains may have a somewhat familiar ring to them. If you recall the learning stage of mental evolution from the previous chapter, the reason will become clear. There are strong parallels between these three domains of conscious experience and the "perception relative to motivation activates emotion" control theory learning formulation (i.e., P – M => E). Perception corresponds to exteroception and the mapping of the outside world, which allows the animal to develop a representation or model of what is "out there." In the learning formulation, this state is symbolized as "P" and it is referenced against the interoceptive information about the homeostatic needs and bodily state and position of the animal.

Interoceptive information and its relation to perception corresponds to the motivational state, or M, in the formula. The motivational state can be framed in terms of homeostatic processes and focus on affordances to be approached and or stressors that are to be avoided, given one's state (e.g., thirst that orients the animal's attention toward finding water). The emotions correspond to processes that "energize motion" toward affordances and away from stressors. Indeed, given Feinberg and Mallatt's analysis, we could rewrite the P – M => E formulation as the Exteroception–Interoception => Affective Valence formulation.

To see this dynamic learning system in action, we can picture a thirsty cat that stumbles across a bird bath in a neighbor's backyard. The cat's body has many homeostatic mechanisms that can activate the motivational template of thirst. This sets the stage for the cat to be scanning its perceptual environment for sources that would quench its desire. Upon seeing the bird bath and recognizing it as a potential source of water, it would feel a jolt of positive affect, which would energize it to move toward. Upon finding the water and drinking from it, positive emotions create associative connections between the bird bath and the satiation. The next time the cat is thirsty and nearby, it will be intuitively guided toward the bird bath as a low-investment high-return option for quenching its thirst. If the cat regularly drinks from the bird bath time after time, eventually a procedural habit will form.

If one day a new family with a loud and aggressive dog moves in, and it is outside while the cat is attempting to poach water, a whole different behavioral investment system will be activated. The perception of the dog will likely activate an intensely strong avoidance motivation, which will activate a fear response, and the cat will run away. The cat's behavioral investment system now has competing approach-avoidance frames for this water source, and the next time it experiences thirst, the cat's investment value system will be in conflict. Should it find a new source of water altogether? Is it worth the risk to try again? The felt fear that the cat re-experiences when approaching the site of the surprise dog attack will be weighed against previous successful satiation attempts, and the cat may avoid the yard altogether (or perhaps still approach, but this time more cautiously). Either way, the emotions the cat experiences serve as markers that guide it toward a best decision given the new information about the environment the cat has internalized and assimilated.

As noted by the *Declaration of Animal Consciousness*, there is virtual consensus in the scientific community that cats have SCEs along the lines of what is described here. That is, they have mental images of the outside world and experiences of the body, along with urges to act, and they simulate possible outcomes in what we might refer to as the "mind's eye." A recent study with crows demonstrates that they also have the basic ingredients of being able to recognize their own subjective experiences (Nieder et al., 2020). That is, the researchers demonstrated that the neurological correlates of SCE were found in crows and that their behavioral choices for reward were dependent upon those markers, providing strong evidence that crows have SCEs and make decisions based on what they can access in their subjective experience.

What is much less clear is when Mind^2 appears in mental evolution as we move "down" the phylogenetic tree, into reptiles, fish, and invertebrates, such as cuttlefish or insects. Feinberg and Mallatt (2016) analyze the likely presence of SCE in terms of two sets of criteria, one set that explores affective consciousness (which may imply the interior states or felt experiences of pleasure and pain) and the other that explores sensory-exteroceptive consciousness. For affective consciousness, they consider (a) operantly learned responses to punishment and reward; (b) behavioral trade-off; (c) frustration behaviors; (d) self-delivery of drugs to reduce pain (analgesics); and (e) approach behaviors toward reinforcing drugs. They document that the fruit fly has demonstrated positive responses on each of these five indicators. Other invertebrates, such as jumping spiders, crabs, and crayfish, have also shown these kinds of behavior patterns.

For sensory-exteroceptive consciousness (i.e., mental images of the outside world), the authors identified eight different criteria as follows: (a) complexity of nervous system; (b) levels of neuronal organization; (c) presence of multiple sensory hierarchies; (d) isomorphic organization (i.e., body plan represented in brain organization, as the somatosensory cortex); (e) reciprocal interactions and pathways between domains; (f) multi-sensory convergence and possible sites of conscious unity; (g) memory regions; and (h) selective attention mechanisms. The authors conclude that many insects meet six of these criteria, only lacking in the size and complexity of the nervous system. They argue that squid and octopuses meet all of these criteria, as do birds and mammals.

Feinberg and Mallatt (2016) note that many scholars disagree with their conclusion that invertebrates like lampreys (early appearing jawless fish) have either sensory or affective conscious experiences. The primary reason is that these "lower animals" (e.g., insects, crustaceans, squid) lack the key brain domains that have been found to be necessary for consciousness in humans. Feinberg and Mallatt respond to this criticism by hypothesizing that the brain systems that support Mind^2 have migrated over the course of evolution. Specifically, they argue for a two-step model of consciousness across our evolutionary lineage. The first step emerges approximately 520 million years ago, in the context of and following the Cambrian explosion. These authors focus mostly on exteroceptive sensory consciousness (especially vision) and the possible brain areas that allow for multisensory convergence and possible sites of conscious unity. They claim that the portion of the brain called the optic tectum was present in the ancestors of vertebrates, and it served as the place that formed visual images, which these authors believe may have been the first qualia.

These authors argue that complex adaptive movement and learning was necessary and was associated with the evolution of exteroception. They put the argument in terms of predictive processing and argue that these mental images were broad and accurate enough so that they allowed for the animal to make better predictions about what was going to happen and coordinate their responses accordingly. Then, a significant upgrade happened at "step 2," which involved the way consciousness emerged in birds and mammals. They argue that there is a shift in the primary brain site from the optic tectum to the dorsal pallium and then into the cortical structures. There are also many behavioral differences associated with birds and mammals, in terms of creativity, planning, and social relations.

For example, the nature of problem solving on land, such as the need to get water and the capacity to recall where water was located, may well have driven higher cognitive capacities. The UTOK provides a zoomed-out meta perspective and vocabulary that allows some conceptual framing that might set the stage to resolve the differences. We can readily note that the two steps identified by Feinberg and Mallatt align with the "learning" and "thinking" levels of mental evolution identified by BIT. However, prior to explicating that alignment, it is useful to briefly review another recent account of the emergence of Mind² in animals.

The Emergence and Evolution of the Sensitive Soul

A somewhat similar account of the early appearance of animal consciousness is found in Ginsburg and Jablonka's (2019) excellent work *The Evolution of the Sensitive Soul: Learning and the Origins of Consciousness.* These authors grounded their analysis in Aristotle's concept of the soul as he laid it out in *De Anima*. They noted that although Aristotle lacked a modern evolutionary view—and thus had a relatively unsophisticated categorical taxonomy compared with our current understanding—it nevertheless remains the case that his analysis of the functional forms of the three levels of the soul in terms of the (1) nutritive/vegetative, (2) sensitive/animal, and (3) rational/human levels is a powerful lens through which to see behavioral patterns in the living world. These categories become especially powerful when they are updated and modernized via the ToK System's frame of Life-Organism, Mind-Animal, and Culture-Person.

In Ginsburg and Jablonka's analysis, we clearly see that modern science lost sight of the animal soul. In alignment with the UTOK's analysis of the Enlightenment Gap, they highlight the difficulties associated with scientifically tackling SCE via what they call "Kant's epistemological gap." Their analysis of the Kantian epistemological gap represented a blend of the ontological and epistemological aspects of the hard problem of consciousness. Given how often epistemological and ontological aspects of consciousness are confounded, it is not surprising that they are a bit tangled. The primary difficulty that Ginsburg and Jablonka emphasize is the problem of mechanism, and they quote Kant, who noted the impossibility of understanding a dynamic, complex adaptive system like animals and humans via the mechanistic matter-in-motion paradigm of natural science. Thus, Ginsburg and Jablonka are referencing our lack of good epistemological grounds for understanding the ontological mechanisms that give rise to conscious experience.

Ginsburg and Jablonka (2019) tackle the issue of mechanism by making explicit that the first step needed in understanding SCE is the move from the material dimension into the living dimension, with all the goal-directed, autopoietic, complex dynamic structures entailed therein. That is, the conceptual jump to get to Mind^2 must first be placed in the Life plane and its complex adaptive dynamics, rather than the complicated, mechanical Matter plane. They then differentiate the domain of the mental from the domain of life (although they do not used the word mental directly). They do so by explicating why consciousness is not synonymous with life, but is part of a different, subsequent emergent plane and requires activity of a nervous system. In the language of UTOK, they first place Mind^2 in the Life dimension and then they place it in the Mind dimension.

As part of their argument, they rightfully note that the concept of consciousness is only meaningful if it can be lost. They explain that although a cell is autopoietic and dynamic and processes functional forms, it is unclear and perhaps conceptually meaningless to ascribe SCE to a cell. To make this point, they ask: *What would it mean to say that a cell loses consciousness but still goes on living and behaving in the same way?* In contrast, creatures like mammals lose consciousness every time they fall into a deep sleep. This is a crucial argument, and it allows us to see clearly that there is an important difference between functional awareness and responsivity that is seen from the outside and the subjective conscious experience of being as felt from the inside.

The authors proceed to offer a summary of what they call the "emergentist consensus" that has gained strength in the last two decades among evolutionists, neurobiologists, behavioral biologists, comparative psychologists, and neurocognitive consciousness researchers. They list the key characteristics of SCE as consisting of seven different features as follows: (1) global activity and broadcast accessibility; (2) the bound and unified nature of experience; (3) the role of selection, learning, and attention; (4)

the fact that SCEs have intentionality or are about things; (5) the temporal binding of SCE and its relationship both to the present and to past memories and future anticipations or imaginations; (6) the fact that SCEs have an affective valence and orient the body to approach or avoid; and (7) the fact that SCEs are associated with a sense of agency and the notion of a "self." The authors proceed to review major approaches to understanding consciousness put forth by luminaries like Francis Crick, Rodolfo Llinas, and Gerald Edelman. They state that, although there are important differences among these perspectives, there is also an emerging picture of SCE that is coming online. Given our metatheoretical level of analysis, it is affirming to note that the emergentist consensus is highly consistent with the view of Mind² offered by the UTOK metapsychology.

The book's central thesis is that "unlimited associative learning" (UAL) is the key neurocognitive capacity that is associated with SCE. They define UAL as the capacity of an animal to ascribe motivational value to a novel, compound, non-reflex-inducing stimulus, or action, and use it for the basis of future learning. This is highly consistent with BIT's notion that the nervous system is an investment value system, and that the learning layer of processing emerges out of the reflex layer. Much as Feinberg and Mallatt argue, they see the Cambrian explosion as the key event where the capacity for animals with complex active bodies and the necessary central nervous system development intersected along with the capacity for UAL, resulting in a massive shift in the animal kingdom during that period. Although the authors agree with the timeframe, they disagree with the way Feinberg and Mallatt break up sensory consciousness into the three domains of the exteroceptive, interoceptive, and affective. According to their view, "all conscious, subjectively experienced states are sensory, all involve motor-sensory-motor loops, and all involve memory for compound patterns, and all are valued/stabilized" (p. 380). As suggested by this quote, the authors argue that UAL is the basic requirement for a minimally conscious animal precisely because it requires binding between exterior, interior, and affective valence.

The UTOK's Outline of Mind²: From the Base of Sentience to a Perceiving, Experiencing Self

We are now able to take stock of a broad view of mental behavioral evolution and the emergence of SCE, and its components. From BIT and our analysis of Mind¹, we know that the nervous system is an information processing system that generates schematic maps and models of the animal–environment relationship and attempts to anticipate and control events in accordance with the principles of behavioral investment. In tracing mental evolution, the first layer takes the form of reflexes, automized control systems, habituation, and sensitization, along with fixed or modal action patterns. In humans we can see this as being the domain of habits, automatic actions, and procedural memories. Then a shift emerges from reactive reflexes to more flexible learning patterns. This shift coincides with the Cambrian explosion when we get animals with complex active bodies that have fully developed brains and operate in dynamic participation with the environment and with each other, especially in the form of mating and prey/predation relations. This kind of flexible behavioral investment requires a more centralized command center that orients toward salient control variables in the environment. This sets the stage for consolidating and integrating mental processes, which is one of the core functions of conscious experience.

The arguments reviewed suggest that these "learning animals" likely represent the first step in the emergence of Mind^2. The UTOK proposal is that it is here that we find the key ingredients for the "base of sentience." We can see this from our zoomed-out metapsychology view by connecting the dots between the P – M => E control theory approach to learning; Feinberg and Mallatt's analysis of exteroception, interoception, and affective valence; and Ginsburg and Jablonka's work on the UAL. A key aspect of consciousness is that it integrates information and broadcasts it to allow for dynamic coordination in a world that is not easy to predict. In emphasizing these notions, we can suggest that what is emerging at the core of consciousness is a system that yokes together the external and internal worlds to energize movement toward anticipated affordances and away from anticipated stressors.

If we return to Romanes' view of mental evolution, we see obvious candidates for this concept in pleasure and pain. Similarly, if we recall from the history of psychology Thorndike's original formulation of the law of effect, he connected the learning principles of reinforcement and punishment with pleasure and pain. We can note that pleasure and pain have some unique features that make them prominent candidates for the emergence of sentience. First, they combine or bridge sensation and action. Indeed, the link is so strong that they almost seem inseparable. However, this is not exactly true. As the psychologist Nicholas Humphrey (2006) has documented in his writings, there are people who can have sensations of pain but have no affective reaction to it. That is, they feel the sensations but do not have the intense emotional dislike and urge to remove them. In addition, they also clearly tie to survival and reproductive success. Indeed, in Henriques (2011), I argued that BIT gave rise to the pleasure pain parallel fitness principle, which emphasized how pleasure could be framed as a signal that oriented toward that which was positively correlated with survival and reproductive success, whereas pain emerged from the negative side of the correlation.

The outline in place suggested by the P – M => E formulation and strengthened by Ginsburg and Jablonka's analysis is that there are centralized broadcast centers that integrate across networks and give rise to what might be "flashes" of felt experience that are organized by affective valence. Of course, questions remain as to exactly how and when such flashes occur and appeared. For example, it is currently largely a matter of opinion as to whether the praying mantis would experience such flashes, as opposed to a strong argument that can generate consensus. The argument is stronger that creatures like fish do. The specifics of this question await future research, as the experts dive further into this area. I use the word flashes here because it will connect to the idea that conscious experience will expand in time via memory. Specifically, the first step will be a sensory-image flash that will last a few seconds and then extend in some animals with more advanced nervous systems into a much longer working memory that lasts 20 seconds or so and affords the animal the capacity to simulate the environment and develop long-term memories.

The relevant point is that UTOK boxes in Mind^2 as follows: SCE begins as an alignment and integration between the senses that tracks the exterior world, maps the interior world, and broadcasts affective evaluations of the "goodness" or "badness" of the situation, which serves as a basic guidance system to motorically approach affordances or avoid stressors. This base of sentience is the first evolutionary step and is associated with flexible learning. However, this is not a fully integrated, conscious, or "experiential self," but rather a minimally sentient entity (see Godfrey-Smith, 2016). If we come at SCE "from the top" starting with human conscious experience and move from Mind^3 self-conscious justification processes into the domain of Mind^2, we see that inner life is much richer than just brief flashes of pleasure and pain that orient approach and avoidance behavior. Rather, there is a complex network of perceptual categories, bodily sensations, mental imagery and plans, emotional reactions, and longer, more general mood states. That is, there is an "inner mind's eye," such that there is a screen of experience that is being perceived. To understand this fuller experiential self, we need to consider what Feinberg and Mallatt identify as "step 2" in the emergence of animal consciousness.

As we saw in the previous chapter, BIT argues that there is an important shift in the levels of mental evolution from the basic learning patterns into more advanced thinking, planning, inhibiting, and deliberating processes. Aligning this with the two-step argument, the picture that emerges is that mental evolution takes a significant step in birds and mammals (and perhaps, in parallel, creatures like octopuses) that can be considered "thinking." This level requires a much more elaborate working memory system that gives rise to a clear distinction between the perceiver and the mental image being perceived, all taking place on a stage of active working memory. This means that the animal can both extend itself into the past via accessible images brought into attention by memory retrieval and project itself into the future via simulation. If we return to the cat and the bird bath example, we can consider how different a cat's inner life likely is from a praying mantis'. That is, it seems possible that the praying mantis would experience jolts of pleasure and pain or flashes of mental images. However, these would be brief and function to coordinate the animal in the immediate moment. In contrast, the cat has a potentially deliberative inner process that is much more unified, extended, and elaborate. The contrast between the cat and the praying mantis raises another issue, which is that we need to consider that there are likely various domains of conscious experience that are elaborated in different ways in different animals.

The Various Dimensions of Animal Consciousness

As exemplified by Romanes' 50-step ladder, a single dimension of animal consciousness consisting of stacked levels from reflexes to self-reflective reasoning has long been a working frame for many scholars. The stacking of mental complexification does have some validity, and it is shared by the ToK System depiction and the four levels of neurocognition present in BIT. However, it is only a general schematic that paints with broad brush strokes, and there is much rich detail that is potentially lost in such a depiction. The sheer diversity of the kinds of animals that likely have consciousness, from octopuses to pigeons to killer whales, should serve as an indicator that any simple ladder-like conception of animal consciousness is unlikely to do justice to Mind^2 in the animal world. This becomes more apparent when we consider the wide variety of brain structures across the animal kingdom implicated in potentially supporting consciousness, and when we consider that consciousness is likely tightly associated with sensory-perceptual systems, learning, and memory, and thus tied to the kind of adaptive problems that animal lineage has faced.

Consistent with this analysis, Birch et al. (2020) recently offered an important critique of the ladder model of animal consciousness. They consider the extant research on the mental lives of elephants, crows, and octopuses and conclude that it is very likely that each kind of animal has rich mental experiences, but that they are likely different across several domains. If we return to Feinberg and Mallatt's differentiation of sensory consciousness into the exteroceptive, interoceptive, and affective domains, we can see why there are likely many different domains of mental experience. Indeed, from this vantage point, we can expect that different patterns of interaction and different adaptive contexts (e.g., social or nonsocial) will drive different levels of conscious capacity in different domains.

Birch et al. (2020) articulate why we should expect different domains of conscious experience and identify six possibilities (noting that there could be many others) in the form of: (1) perception (vision); (2) perception (touch); (3) emotion; (4) unity and integration; (5) memory across time; and (6) selfhood. They argue that crows, elephants, and octopuses likely have very different levels of conscious experience across these domains. For example, they hypothesize that an elephant would likely have high levels of emotionality, unity, temporality, and selfhood. In contrast, crows seem very high on vision and somewhat high on emotion and temporality. Octopuses, on the other hand, may be very high on perceptual touch, but may have very little in the way of a sense of self. This is an important point and congruent with the model of Mind² afforded by the UTOK.

Summarizing the Evolution of SCE

Studies of animal consciousness have been fraught with problems in terminology, as well as in difficulties grappling with both the epistemological and ontological problems associated with scientifically studying Mind². Thankfully, the zoomed-out view afforded by UTOK allows us to develop a clear outline of how SCE likely emerged and evolved. Specifically, there is good reason to frame the evolution of Mind² as a two-step process that corresponds to the learning and thinking levels of mental evolution. SCE is likely not present in the reacting, reflexive, fixed action level of animal mentation, such as seen in a sea slug. The argument here is that the beginnings of SCE emerge with the learning level, such that pleasure and pain can be thought of as nature's first qualia that integrate and coordinate flexible learning patterns based on association and consequence. These "mental flashes" model the animal–environment relationship and orient approach or avoidance behaviors.

Then, as the cortical systems grow, working memory comes online that gives rise to the distinction between perceiver and perceived and allows for deliberation, inhibition, and planning. This second step involves the "inner mind's eye" of the experiential self. Here the SCE shifts from brief temporal flashes into more extended, expanded, and unified capacity to not only experience the present but actively recall the past and project the animal into the future. This mental simulation affords the animal capacity to develop predictions beyond the immediate moment. For example, a rat can now simulate both arms of the T-maze as it arrives at the choice point, or a cat can decide about risking a drink at the bird bath. We can consider this the presence of an "experiential self" that becomes more differentiated into a perceiver that perceives the inner world. That is, with this second step we have the emergence of something akin to an inner theater of experience that has material on stage that is being seen or heard by an audience. We will return to this metaphor shortly.

This summary aligns well with Antonio Damasio's work. Over the course of several books, he has delineated a model of mind, self, and consciousness that is embodied and grounded in feeling states. In his first book, *Descartes' Error: Emotion, Reason, and the Human Brain*, Damasio (1994) argued that thought is grounded in emotion, which in turn emerges from somatic markers, embodied feeling states that orient and motivate the animal or person. In subsequent works, he extended this analysis to how consciousness and the self evolves. In *The Feeling of What Happens*, Damasio (1999) proposed the idea that the self is a system that models the relationship of the animal to the environment. First, there is a "proto-self," which represents a nonconscious map of the state of the animal as it moves in the environment. Emotions, according to Damasio, are energized action patterns that move the animal toward and away from goal states. Then a "core self" emerges as a function of modeling the proto-self. This is where the base of feeling states emerges. As it grows and expands based on the complexity of the animal's nervous system, an "extended self" appears. Damasio argued that there is some evidence for an extended self in other animals, but it is most developed in humans. Aligning this with UTOK, we can consider the proto-self as being at the level of fixed action patterns, just before Mind². The core self is grounded in the base of sentience and then grows into a perceiver-perceiving experiential self. This gets extended in higher mammals and takes off with human language, ultimately becoming the narrating ego.

More recently, in *The Strange Order of Things: Life, Feeling, and the Making of Cultures*, Damasio (2018) focuses on the concept of feelings as the fundamental units of subjective experience in a way that is directly aligned with the UTOK metapsychology. Specifically, he links subjective feeling states to homeostatic processes, such that they serve as general broadcast signals regarding the state of the body and its motivational orientation. For Damasio, feelings are "evaluative qualia" that signal whether things are going well or not. And they serve to orient the animal to maintain constructive homeostatic processes. In other words, feelings are valence qualia that emerge in relationship to the interior needs and exterior situation. Over time, this evolves into a core experiential self that models the world in increasingly complex ways.

The Structure, Function, and Phenomenology of Consciousness

With the outline of the evolutionary emergence of SCE in place, we now can turn more directly to models of the structure and function of consciousness and how to align it with the domain of Mind² in humans. Specifically, we need to: (1) consider the essential architecture that constitutes consciousness; (2) delineate the function of consciousness more clearly; and (3) consider how theories of structure and function correspond with everyday human experiences. In this section our language will shift somewhat from SCE to consciousness and phenomenology. The reason is that these are theories of consciousness, and they involve various aspects of cognitive processes, functional awareness, and the subjective experience of being in both animals and humans.

Integrated Information Theory: The Basic Structure of Consciousness

What exactly is the structure of SCE? That is, if we start with phenomenology, we can ask: *How can qualitative experiences be realized in the material world?* The fundamental architecture that constitutes SCE is the core question that drives Integrated Information Theory (IIT). It approaches the issue by highlighting the fundamental properties of experience and by framing how such properties might emerge in the material world. It also affords us a powerful mathematical formalization that yields a measure of integrated information, known as phi. In a general statement on the theory, Oizumi et al. (2014) describe IIT 3.0 as follows:

> Integrated information theory (IIT) approaches the relationship between consciousness and its physical substrate by first identifying the fundamental properties of experience itself: existence, composition, information, integration, and exclusion. IIT then postulates that the physical substrate of consciousness must satisfy these very properties. We develop a detailed mathematical framework in which composition, information, integration, and exclusion are defined precisely and made operational. This allows us to establish to what extent simple systems of mechanisms, such as logic gates or neuron-like elements, can form complexes that can account for the fundamental properties of consciousness. Based on this principled approach, we show that IIT can explain many known facts about consciousness and the brain, leads to specific predictions, and allows us to infer, at least in principle, both the quantity and quality of consciousness for systems whose causal structure is known. For example, we show that some simple systems can be minimally conscious, some complicated systems can be unconscious, and two different systems can be functionally equivalent, yet one is conscious and the other one is not. (p. 2)

This useful summary highlights the core of IIT. I agree with the authors that there are strong arguments for considering consciousness to be a kind of integrated information, and the properties of integrated information are well specified by IIT. Indeed, the basic structure of the ToK System, with its depiction of the evolution of complexification, aligns with this formulation, and IIT helps provide more clarity about that structure. SCE involves both information integration and exclusion, and this can be effectively quantified by phi via the mathematical formalisms provided by Tonini and colleagues. This is a powerful addition to the scientific arsenal and informs us about the kind of consolidating and integrating and excluding that ground the basic structure of Mind^2.

I say "basic structure" because although IIT provides a necessary formulation for SCE, it is not one that is sufficient. That is, it fails as a comprehensive frame for SCE because it does not effectively account for the properties that make SCE unique. Rather, it simply states that SCE is a kind of integrated information and then it proceeds to swap out consciousness for integrated information. That IIT fails to provide the necessary descriptive metaphysics to frame SCE is seen in how over-inclusive it can be in considering entities as conscious. According to IIT, a very simple mechanism, such as a "photodiode" (to use their example) can be considered "minimally conscious." The photodiode consists of a detector and a predictor that has a memory storage system. It is a system that will turn on if it receives at least two inputs from internal or external sources and will vary based on past experiences. The authors classify it as conscious. They write: "Simple as it is, the photodiode system satisfies the postulates of IIT: both of its elements specify selective causes and effects within the system (each element about the other one), their cause-effect repertoires are maximally irreducible, and the conceptual structure specified by the two elements is also maximally irreducible" (Oizumi et al., 2014, p. 19).

The problem here is that IIT confuses the necessary structural conditions of consciousness for SCE itself. Since the experience of being in the world is what we are trying to explain, then an entity like a photodiode is not "minimally conscious." Put differently, IIT ends up confusing structural systems that have integrated information and exhibit functional awareness and responsiveness with SCE. Functional awareness and responsivity are what we see in Mind[1]. Indeed, much of Life can be characterized by functional awareness and responsivity and exhibits biological intelligence. So too can much in the world of artificial intelligence. However, if the referent is SCE, then functional awareness and responsivity is a necessary but not sufficient frame. To use Ginsburg and Jablonka's (2019) analysis, what would it mean to say that the photodiode lost its conscious experience of being? This point highlights why it is crucial to have a broad descriptive metaphysical framework so that we can effectively box in SCE. It seems clear that based on this analysis, integrated information is necessary but not sufficient for Mind[2]. Rather, we need to combine IIT with work in the evolution of consciousness like that of Ginsburg and Jablonka to effectively box it in.

Global Workspace Theory: The Broadcast Function

Global Workspace Theory (GWT) is a cognitive theory of consciousness originally proposed by Bernard Baars. In *The Theater of Consciousness: The Workspace of the Mind*, Baars (1997) laid out the basic idea, which is that the function of consciousness is to broadcast salient aspects of the landscape in a way that allows for brain-wide sharing and coordination. Using the metaphor of a theater for SCE is appealing. First, it accords with Ginsburg and Jablonka's reminder that consciousness can be lost, with the analogy being when the theater is dark. A dream would be akin to the lights on the stage or screen flickering or flashing fuzzy images. And when the person wakes up, the spotlight or screen appears in full and what is on it represents what the animal has conscious access to in that moment. Of course, the theater metaphor is potentially problematic to the extent that it is taken to mean that there is a specific audience member that is watching what is on the screen, which would give rise to the homunculus problem. We cannot, of course, explain consciousness by positing a little person inside one's head that is watching what is happening. GWT avoids the homunculus problem because the audience in the model is the rest of the brain and the information on the screen or in the spotlight is broadcasting that message out to simpler, nonconscious portions of the neurocognitive system.

Figure 13.1 depicts how Baars mapped the *Theater of Consciousness*. The "spotlight" refers to the focal point of conscious attention. The material that is under the spotlight represents the focal elements that the rest of the brain has access to and thus can be coordinated around. We can think about the physical structure of the building as being the body and the brain. That which is under the spotlight is on the screen of awareness and everything else in the theater (i.e., exterior sensory input, the concept of self, long-term memories, rules of grammar, automatic behaviors, etc.) is considered "unconscious." As noted, working memory represents the "stage" upon which conscious awareness operates, although it too is unconscious. The backstage is considered the "contexts," which can be considered the perceptual, motivational, or expectational schema that frame the material given the voluntary behavioral control processes that are guiding the animal. The behavioral and attentional control processes can be thought of as the processes that are directing where the light is shown. The audience in the model represents the neurocognitive domains that are absorbing the information that is being broadcast.

One way to conceive of how SCE is modeled is to imagine that the theater is hosting a talent show that allows one act to perform under the spotlight at a time. This metaphor allows us to consider that there are several potential "stars" that represent different acts that exist in competitive tension to be brought into the spotlight one act at a time. This connects to the fact that consciousness is a serial processor that focuses on one concept, event, or entity at a time. The coordinated competition references the fact

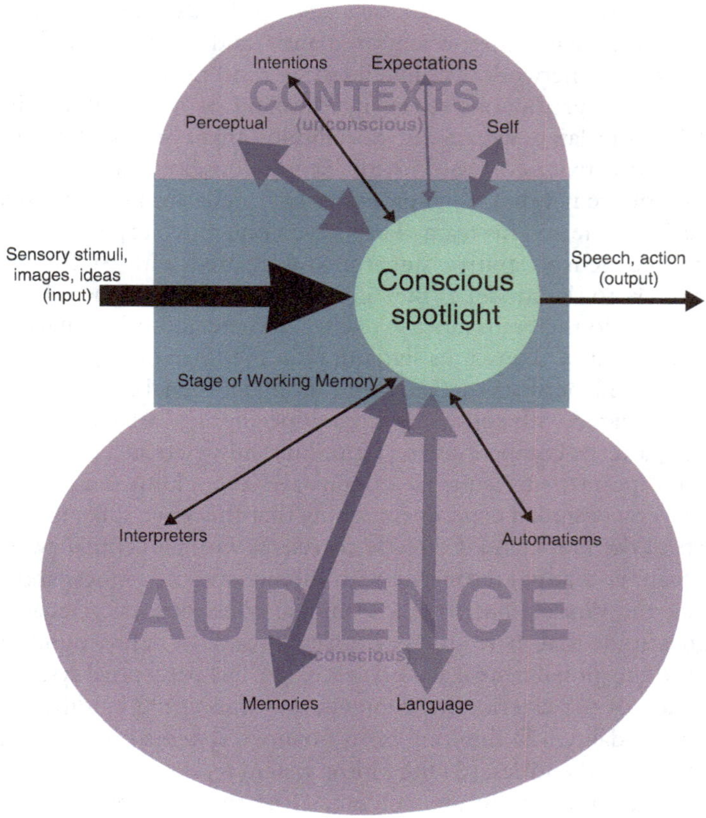

Fig. 13.1 Baar's schematic for Global Workspace Theory (Reprinted with permission)

that there are attentional selection pressures both "from below" in terms of sensory inputs and "from above" in terms of motives and expectations that impact what is on stage. Although the architecture of the human mind diagram shared in the previous chapter is one that is grounded in a Mind[1] neurocognitive functional view, we can nonetheless make clear connections with it and Baars' conception. Specifically, the right-hand side of the diagram can be thought of as the sensory-perceptual affective screen of awareness, which some call sensory memory. It lasts for between three and 30 seconds, and represents the center of the spotlight, whereas the working memory corresponds to the stage that holds the screen, and then the long-term memory represents semantic and schematic information stored across the network over much longer time spans.

GWT makes predictions regarding the kind of neuronal activity that should be correlated with SCE. Specifically, there should be interconnected brain activity correlated with having conscious access to subjectively experienced events. Such processes have been studied extensively by Stanislas Dehaene and his team. They have done much experimental work showing the neurocognitive signatures associated with SCE. Dehaene (2014) refers to the model as the "global neuronal workspace theory" to make the link to the brain more explicit. Dehaene and others have identified what he calls a "conscious signature" in the form of a spike in neuronal activity and networked communication between brain areas that is reliably associated with conscious access. Specifically, Dehaene argues that neuronal patterns compete for recognition, and when successful, there is an ignition point that results in a brainwave spike, known as a P3 wave. Many cleverly designed experiments show that this wave spike links different parts of the brain, and it directly correlates with perceptual awareness.

In a manner that has resonance with Bircher and colleagues' arguments regarding the various dimensions of animal consciousness, Dehaene and colleagues argue that there are five primary streams of neurocognition that connect to the global neuronal workspace: (1) The perceptual systems provide models of the exterior environment and thus could be differentiated based on modality, (2) the long-term memory systems provide access to stored events and entities, (3) the motor system represents action selection patterns that anticipate future outcomes, (4) the emotion system tracks values and the current homeostatic states, and finally (5) the focus of attention is usually framing the inputs, but also can, in humans, become under the spotlight as an individual wonders what they should be paying attention to. What becomes conscious has a strong enough bottom-up activation coupled to attentional focus, whereas there are subliminal aspects that are not strong enough, preconscious (strong enough, but no attention), or disconnected (important, but not able to become the focus of attention).

GWT is a popular theory of SCE for a reason. It provides a compelling answer to the question of what conscious experience is for; namely, consciousness provides an "access function" that allows for brain-wide information sharing. Specifically, it allows for the process of amplifying aspects of relevant information about specific events or objects to be actively shared with the rest of the brain. In so doing, GWT provides a reasonable account of nature and relation between serial computational processes and parallel connectionist processes. Most of the brain's communication systems are operating in distributed, parallel connectionist networks, but there is a specific access point that allows for material to be serially processed and broadcast to allow for focused attention and coordination. If we return to the talent stage model, much activity might be going on in the audience or backstage in parallel, yet the spotlight affords one act to take the stage and share its content with the whole theater.

GWT also aligns quite well with many lines of research. If we return to Baars' schematic, we can see that perceptual schema serve as the "contexts" for sensory input. It can be helpful to think of the relationship in terms of sensation being a "bottom-up key" and the perceptual templates as being a "top-down lock." Modern cognitive science suggests that the brain works as a predictive processor that is trying to determine what it is perceiving and anticipate what is going to happen next. As such, it is trying to match incoming sensory signals with perceptual templates that allow it to confirm a hypothesis of what the object is. This highlights two aspects of consciousness. One is how much what we see depends on what we focus on, and the other how what we experience will flip back and forth depending on that matching process.

To see the role of attentional focus, consider that if I invite you to pay attention to your big toe, you can shift the focus and now bring those sensations onto the screen of SCE. Just how important attentional focus is comes into view with the work by Simons and colleagues and their famous "gorilla experiment" on inattentional blindness (e.g., Simons, 2000). In the setup, participants are asked to keep track of a ball being passed between participants for about a minute. The task is quite easy, and most count the right number. However, at the end of the experiment, the participants are then asked if they "saw the gorilla." Many say no. Yet, if you re-examine the footage, one plainly sees a man in a gorilla suit walk calmly into the center, pound his chest, and then walk on. Many who see it a second time will have trouble believing that it so obvious the first time. This shows the power of top-down attentional focus. It frames what you are looking for and screens out other events.

We can also see how this "template matching" process frames SCE via the well-known "duck-rabbit" illusion, which has the distinction of now being a brand name of a beer. Depending on how you look at the drawing, it is perceived either as a duck or a rabbit. According to GWT and many other approaches, the reason it is experienced as one or the other is that the key-into-lock relationship is igniting the brain and bringing one hypothesis or the other onto the stage and broadcasting it to the rest of the brain. This example shows that GWT can help us understand Mind[2] processes as flashes of unified experience that occur because of a hypothetical matching process between a "backstage" template and sensory input that then functions to broadcast the focal to the rest of the neurocognitive system. In the next section, we advance this analysis by showing how we can enhance GWT by aligning it with the Updated Tripartite Model (UTM) of human consciousness introduced in Chap. 5.

Mapping Human Phenomenology Via the Merger of GWT and the Updated Tripartite Model

The UTM divides human consciousness into the experiential self, the ego (or private narrator), and persona (or public self). Aligning this model with GWT, we can consider the material in the conscious spotlight as being the focal point of the experiential self. In the UTM, the attentional filter divides the experiential self from the sub- or nonconscious neurocognitive domains of Mind[1]. We can consider GWT affording us a detailed description of the attentional filtration processes at work, both in terms of the bottom-up sensory inputs and top-down expectations and in terms of how the material on the screen of awareness is a function of where an individual points their attention.

As noted in Chap. 5, the experiential self can be divided into two different streams or core aspects, which are found in its name. First, there is the raw perceptual experience that presents itself to the individual in the form of qualia, perceptions, images, and feelings. The other refers to the self and how the individual is modeling what is relevant for the self over time. If we look at the basic structure of attention, we see these two streams represented in the distinction between two major neurocognitive systems (Horn et al., 2013). One is the task network system, which is active when an individual is paying attention to a particular object or set of events, or is engaged in a task that requires behavioral control. The other is the default mode network, which largely functions offline to process material in parallel. The material that is being processed by the default mode network is generally associated with the self. That is, we can think of the default mode network as modeling relevant events and aspects of the agent–arena relationship that are important for the self over time. Specifically, this involves explicit thoughts about the self, thoughts about others relating to the self, and future and past events that are deemed relevant to the self. The UTM of the experiential self thus is a model that suggests that experiential awareness is toggled back and forth between what is happening in the agent–arena relationship in the here and now via the task network and is processing what is relevant for the individual's interests over time via the default mode network.

GWT affords us a way to generate a basic metaphorical picture of this process, and show, via Dehaene's work, how it aligns with the brain's architecture and the neurocognitive correlates of SCE, especially of the task-focused sort. However, while GWT enables us to generate a basic frame regarding the perceptual awareness that is tied to the experiential self, it is also the case that GWT is underdeveloped when it comes to Mind³. UTOK tells us that if we are going to fully understand human consciousness, we need to add self-conscious narration to the picture. In UTOK, this is the puzzle piece provided by Justification Systems Theory, which was summarized in Chap. 5.

To highlight the way self-conscious awareness and narrative interpretations can feedback in human phenomenology, consider this account by John Horgan (2018), which opens his book *Mind-Body Problems: Science, Subjectivity, & Who We Really Are*:

> I am walking near a river on a hot summer day. My left hand grips a fishing rod, my right a can of worms. One friend walks in front of me, another behind. We're headed to a spot on the river where we can catch perch, bullheads and large-mouth bass. Weeds bordering the path block my view of the river, but I can smell its dank breath and feel its chill on my skin. The seething of cicadas builds to a crescendo.
>
> I stop short. I'm me, I say. My friends don't react, so I say, louder, I'm me. The friend before me glances over his shoulder and keeps walking, the friend behind pushes me. I resume walking, still thinking, I'm me, I'm me. I feel lonely, scared, exhilarated, bewildered. (published online)

Horgan was profoundly impacted by this event, and it stayed with him throughout his life and related directly to his quest to explore the mystery of the mind–body relationship.

A way to characterize Horgan's story is that he had a flash of narrative insight regarding the logical structure between his inner experience, his egoic narrator, and the condition of his life. Prior to his insight, he, of course, had some self-reflective awareness capacities. He could have answered straightforward self-referential questions, such as "What are you doing and why?" He knew he was going fishing with his friends because that is what he liked to do. What changed for Horgan was a fundamental shift in his self-conscious justification narrative and how he intuitively gripped the world. That is, Horgan's insight transformed the working model he had of himself as an agent in the world. Specifically, he realized there was a complicated feedback loop between his conscious experience and self-reflective awareness and that changed how he thought of himself and the world around him. Another way of saying this is that he became self-aware of what Douglas Hofstadter (2007) calls a "strange loop." Like the Escher hands that draw themselves, the strange loop of self-awareness refers to the fact that one's justification narrative feeds back on itself, such that cause and effect are massively entangled rather than clearly separated.

This raises the question about whether we could merge the GWT model with the UTM to generate a more complete rendering of the theatrical stage model of human consciousness. Such a model was developed by myself and my former graduate student (Henriques & Quay, 2016). It adds the fact that there is an internal narrator commenting on the perceptual events taking place on the stage of awareness. It also adds the idea that, in addition to a "brain audience" there is the audience in the world, the actual public stage that one's actions take place on. Thus, there is the persona that is managing aspects of what is seen by others on the outside. The combination of the ego and persona helps us see how language-based thought interact with Mind^2 perceptual experiences.

We can see how important the narrator is when we contrast Horgan's experience with a very similar experience reported by Carl Jung, who reported a similar self-conscious awakening:

> I was taking the long road to school … when suddenly for a single moment I had the overwhelming impression of having just emerged from a dense cloud. I knew all at once: I am myself! … Previously I had existed, too, but everything merely happened to me …. Previously I had been willed to do this and that: now I willed. This experience seemed to me tremendously important and new: there was "authority" in me. (cited in Ryckman, 2004, p. 75)

Both Horgan and Jung had similar moments of insight. They both "awoke" to the fact that they were themselves in a "strange loop" manner, and that changed their grip on the "self–world" model they were operating from. However, they interpreted the implications of this insight in different ways. Horgan found this to mean that he was alone and felt lonely, scared, and bewildered. In contrast, Jung interpreted this to mean that he had new authority and felt individuated and empowered. This example highlights how the perspective frames the propositional interpretations that emerge, which in turn feedback on the individual's experience.

This brief synopsis shows that the UTM can be merged with GWT in a way that is consistent with human conscious experience in the Culture-Person plane of existence. This is important because UTM bridges the work into the real world and allows for a much richer analysis of the domains of SCE, self-conscious reflection and narration, and the dynamic interfaces between the domains. We can readily hypothesize that Horgan felt self-conscious about his "discovery" and, since his friends did not provide him affirmation or empathy but instead brushed him off, that he would have felt confused and perhaps been inhibited in sharing his experience in the future. As such, this would create secondary anxieties and conflict. The point here is that we need theories of consciousness that stretch from the earliest beginnings of sentience into the inner mind's eye mapped by the global neuronal workspace into the relational world of social animals and finally into the self-conscious world of cultured persons.

Conclusion

Over the past three decades, scientists have developed a renewed interest in the concept of consciousness in both animals and humans. Global Workspace Theory, Integrated Information Theory, and others have captured the attention of many as affording new insights and leading to more advanced research paradigms. However, although much interesting work is being done in consciousness studies, there are still questions about how much genuine progress is being made. For example, Horgan's book argued that a "crisis" is emerging because there are many different "mind-body problems" that are all tangled together into a knotty mess, with no big picture viewpoint to sort out the issues and consolidate our understanding. Horgan believes the problems facing the field of consciousness studies are so significant that they will never be solved. From a UTOK perspective, Horgan's skepticism is understandable. It seems likely that, unless a course correction is made, consciousness studies will inevitably become entangled with the long reach of the Enlightenment Gap. Indeed, the shadow of the Enlightenment Gap swallowed psychology and the philosophy of mind, and there is every reason to believe it will swallow consciousness studies unless there are advances in our metaphysical understanding of ontology that can bridge matter and mind and clarify scientific relative to subjective and social knowing.

The UTOK metapsychology argues that a new path is needed, and we have laid out several key pieces already. The ontology of Mind, framed as the set of minded behaviors mapped by the ToK System, is now coming into view. The Cambrian explosion resulted in a fundamental shift in the way animals are organized. Brains and complex active body plans set the stage for a qualitative shift in sensory-motor behavior. Animals evolved from reacting to learning to thinking and became active behavioral investors that developed patterns of functional awareness and responsivity that allowed them to navigate the environment with increasing levels of sophistication.

The Map of Mind^{1,2,3} provides us with a clear descriptive metaphysical vocabulary that affords us a way to properly label and differentiate mind as behavior (Mind^{1b}), mind as cognition (Mind^{1a}), mind as subjective conscious experience (Mind^{2}), and mind as self-conscious reasoner (Mind^{3}). This chapter has provided a clear outline of Mind^{2} as emerging via a two-step process that overlaps significantly with the learning and thinking levels of mental evolution. The nature of animal movement was such that it required a fast, dynamic feedback system that oriented the body toward affordances and away from stressors. Pleasure and pain appear as valence signals that broadcast aspects of the relations the body has, both internally and relative to the external environment. Such signals allow learning animals to engage in complex active adaptive behavioral repertoires as they track their interoceptive homeostatic states. In Damasio's terminology, such events are the feeling of what is happening. This broadcast function starts an integrated information network that becomes differentiated into focal, serial recursive relevance realization processes that are well specified by GWT.

With land animals and larger brains, a greater capacity for SCE emerged. Specifically, the stage of working memory enabled images to be extended in time and a self–world modeling system was generated that can be characterized as an experiential self. This consists of two aspects of conscious experience, one that can be framed as SCE itself in the moment (i.e., pure awareness) and the other as the aspects of that experience that are deemed relevant to the self over time (i.e., motives, emotions, and imaginations of what the future might hold). The GWT enabled us to connect the model of Mind² to human consciousness. The UTM adds to it by clarifying the place and intersection with self-conscious justification on the social stage. In the next chapter, we deepen and build from this formulation of Mind² and extend it into the relational world of parent–offspring dynamics, friends, competitors, lovers, and the groups to which animals belong in a way that affords us a much clearer understanding of the intersubjective world.

References

Baars, B. J. (1997). *In the theater of consciousness: The workspace of the mind*. Oxford University Press.

Birch, J., Schnell, A. K., & Clayton, N. S. (2020). Dimensions of animal consciousness. *Trends in Cognitive Sciences, 24*(10), 789–801.

Damasio, A. R. (1994). *Descartes' error: Emotion, reason, and the human brain*. Penguin Books.

Damasio, A. R. (1999). *The feeling of what happens: Body and emotion in the making of consciousness*. Houghton Mifflin Harcourt.

Damasio, A. R. (2018). *The strange order of things: Life, feeling, and the making of cultures*. Pantheon.

Dehaene, S. (2014). *Consciousness and the brain: Deciphering how the brain codes our thoughts*. Penguin.

Feinberg, T. E., & Mallatt, J. M. (2016). *The ancient origins of consciousness: How the brain created experience*. MIT Press.

Godfrey-Smith, P. (2016). *Other minds: The octopus, the sea, and the deep origins of consciousness*. Farrar, Straus and Giroux.

Ginsburg, S., & Jablonka, E. (2019). *The evolution of the sensitive soul: Learning and the origins of consciousness*. MIT Press.

Henriques, G. R. (2011). *A new unified theory of psychology*. Springer.

Henriques, G. R., & Quay, M. (2016, March). The theater of consciousness: A new map. [Blog] *Psychology Today*, theory of knowledge. https://www.psychologytoday.com/us/blog/theory-knowledge/201603/the-theater-consciousness-new-map

Hofstadter, D. (2007). *I am a strange loop*. Basic Books.

Horgan, J. (2018). *Mind-body problems: Science, subjectivity, and who we really are*. Knot Press.

Horn, A., Ostwald, D., Reisert, M., & Blankenburg, F. (2013). The structural-functional connectome and the default mode network of the human brain. *NeuroImage, 102*, 142–151.

Humphrey, N. (2006). *Seeing red*. Harvard University Press.

Huxley, T. H. (1896). *Method and results. Collected essays* (Vol. 1). Appleton Press.

Nieder, A., Wagener, L., & Rinnert, P. (2020). A neural correlate of sensory consciousness in a corvid bird. *Science, 369*(6511), 1626–1629.

Oizumi, M., Albantakis, L., & Tononi, G. (2014). From the phenomenology to the mechanisms of consciousness: Integrated information theory 3.0. *PLoS Computational Biology, 10*(5), e1003588. https://doi.org/10.1371/journal.pcbi.1003588

Romanes, G. (1883). *Mental evolution in animals, with a posthumous essay on instinct by Charles Darwin*. Kegan Paul, Trench & Co.

Ryckman, R. (2004). *Theories of personality* (8th ed.). Wadsworth.

Simons, D. J. (2000). Attentional capture and inattentional blindness. *Trends in Cognitive Sciences, 4*(4), 147–155.

CHAPTER 14

Mind To Be and the Relational World

The Unified Theory's fourth key idea is the Influence Matrix, and it maps the relational process structures that guide humans in the relational world. To get a flavor for how we can track and analyze these kinds of mental behaviors, consider that one of the things I noticed as a parent was that my children were exquisitely sensitive to how much attention they received relative to their siblings. For example, when my son was three and my daughter five, if one of them received a slightly bigger piece of cake than the other, they were primed to call me out on that and draw attention to the fact that they had received the short end of the stick. According to the Influence Matrix, my children's attention to the size of their piece of cake was a function of a deep relational architecture that is rooted in a system that is shared with many social mammals.

We can see this relational pattern on display in the now famous "cucumber-grape" experiment developed by Frans de Waal and his colleagues (van Wolkenten et al., 2007). In the experiment a capuchin monkey is trained on a simple task, such that when he picks up a rock and gives it to the experimenter, he gets a cucumber in return. Put two monkeys in cages next to each other and have them both get cucumbers for the rocks and they will do it all day long, so long as they are hungry. However, something changes dramatically if you give one of the monkeys a grape instead of a cucumber. Monkeys desire grapes more than cucumbers, and when one monkey sees that the other receives a grape when it gets only a

G. Henriques, *A New Synthesis for Solving the Problem of Psychology*, Palgrave Studies in the Theory and History of Psychology, https://doi.org/10.1007/978-3-031-18493-2_14

cucumber, it triggers an obvious emotional reaction. The experiment has been shared on YouTube, and it is hard not to see the monkey's feelings of anger, frustration, and a sense of injustice as he rejects the cucumber, throws it at the experimenter, and grabs hold of the bars of the cage and shakes them in frustration.

How can we understand these actions from a Behavioral Investment Theory perspective? The first part of the experiment is simple. The hungry monkey values the cucumbers and has learned that grabbing the rock and handing it over generates that reward. It is easy to do and has few costs, and we can say with confidence that the action pattern was selected on the learning principle of return on investment. However, something dramatically changes the value of the action when the monkey sees the other monkey get a grape for the same effort. What might that be? We can start our interpretation by being reminded that the investment elements that guide behavioral outputs are not absolute values but are set by expectations and past actions. Consider if I hired a babysitter and he would have been happy to get $20 for the job but I give him $30 instead; we can predict that he would experience a jolt of positive affect. However, if the next week he returns and I say, "I have thought about it, and $20 seems fair," we can predict that he would likely feel somewhat disappointed and perhaps even reject the offer. This example shows how the anticipated return on our investments is evaluated relative to expected or past frames.

This idea of relative frame of reference is necessary to understand the monkey's rejection of the cucumber, but something needs to be added to it. After all, the monkey was given the same return that he had received previously. What changed? As we will see, the Influence Matrix offers a clear answer. It posits that social primates like capuchin monkeys and my children are attuned to their place in the social matrix of influence and relational exchange. It directly leads to the implication that humans track their investments and level of attention relative to what relevant others are receiving. Moreover, we can posit that the reason they do this is that it informs them of their value and status in the social field. On this account, my kids tracked the size of their piece of cake relative to their siblings because they were intuitively oriented to track signals of their relative importance to me and how that was realized in things like my allocation of attention and resources. In the babysitting example, we can engage in an additional thought experiment and recognize that the pattern holds there as well. For example, although he would have been happy with the $30 offer, the value of the offer would have significantly dropped if he had heard that I offered someone else $40 to do the same job.

What my kids, the monkeys, and the imaginary babysitter all have in common is that social comparison is a key feature of our world. As social animals, we compare what we are doing and getting to others and note who is "doing better" (i.e., upward social comparison) and "doing worse" (i.e., downward social comparison) and have feelings and engage in actions accordingly. As you might suppose, sometimes upward social comparisons result in individuals striving to work harder to achieve what those above have, whereas other times it is discouraging and causes people to shrink in disappointment. Likewise, downward social comparison sometimes results in us feeling better, as suggested by the phrase "life could be worse," and other times makes us want to separate ourselves from those "lower" on the status spectrum. Social comparison and a sense of fairness in tracking relational exchanges are central to interpreting the cucumber-grape experiment. This is evident in how Frans de Waal humorously summarizes the experiment as representing the fundamental motives that drove the "Occupy Wall Street" movement.

This chapter explores the basic architecture of the processes that guide us in the relational world. Its title, mind to be, is a play on words to give a double meaning. First, it refers to the centrality of the social world in shaping the human mind, both evolutionarily and developmentally. The human mind comes into being very much as a function of the intersubjective world in which it emerges. Second, the phraseology connects us to the Map of Mind^{1,2,3}, suggesting that there is a shared, implicit subjective field of understanding that emerges within and between humans as they interact with one another. As we will see, this implicit intersubjective space of shared attention and intention has been called a "we" space by Michael Tomasello. It can be tentatively framed as the domain of Mind^{2b}. This chapter shows how the Influence Matrix maps key aspects of that relational space, and how these processes manifest in human mental behavioral patterns intra-psychically and interpersonally. However, prior to mapping the human relationship system, we need to make some jumps from where we were in the last chapter. Specifically, we need to move from the evolution of neurocognition and consciousness into the experiential self and its connection to the relational world and human intersubjectivity. Apropos of the focus on relationality, to help make that bridge we turn to the work of John Vervaeke and explore how his metatheory of cognition intersects with UTOK.

Aligning John Vervaeke's Vision and UTOK

Dr John Vervaeke is a professor of philosophy and cognitive science who has long sought a consilient view of the human mind that can assimilate and integrate the cognitive sciences and many Eastern philosophical traditions (Buddhism and Taoism, especially). Consistent with the overarching theme of this chapter, we have worked to cultivate an intersubjective space of understanding. This alignment has been demonstrated in four educational series produced by *The Cognitive Science Show*. The first series, *Untangling the World Knot of Consciousness: Grappling with the Hard Problems of Matter and Mind*, built from Vervaeke's metatheory of cognition as recursive relevance realization and showed how it aligns with UTOK to give a new comprehensive way for thinking about the nature and evolution of consciousness (Vervaeke & Henriques, 2020).

The second series, *The Elusive "I": The Nature and Function of the Self*, included Christopher Mastropietro, and extended the synergistic analysis to construct a model of the experiential self that was consistent with cognitive science, clinical work, and existential considerations (Vervaeke, Henriques, & Mastropietro, 2021). In the third series, *Toward a Metapsychology True to Human Transformation*, we were joined by the educational philosopher and developmental psychologist Dr Zak Stein, and examined what transformation was and what kind of metapsychology was needed to frame it effectively (Vervaeke, Henriques, & Stein, 2021). In the fourth series, *Psyche-pathology and Well Being*, we were joined by Garri Hovhannisyan, and explored the nature of psychopathology and how to cultivate more optimal functioning and fulfillment (Vervaeke et al., 2022). Together, these educational series show convincingly that Vervaeke's integrative metatheory of cognition syncs up richly with UTOK, and they strongly point to the potential of the two systems' capacity to work together to provide a new metapsychological vision for the twenty-first century.

Vervaeke's 4P/3R Metatheory of Cognition

The previous two chapters demonstrated how BIT functions as a metatheory that can assimilate and integrate many different perspectives and generate a holistic account that enables us to achieve a broad and coherent neurocognitive functionalist account of animal mental behavior that also incorporates a frame for understanding Mind^2. Vervaeke's metatheory of cognition overlaps with this perspective, but his primary focus has been on the nature of cognition anchored in 4E cognitive science. Despite their independent origins, the two systems work seamlessly together to generate frames for understanding mental behavior, consciousness, and the experiential self. As such, we review Vervaeke's 4P/3R metatheory of cognition and show how it aligns with UTOK to generate a working model of the self that then evolves with the emergence of the relational world in mammals.

The 4 Ps of Cognition as Knowing. In everyday usage, cognition refers to specific kinds of "higher" mental processes, such as thoughtful deliberation, logical analysis, and problem-solving. However, in the context of cognitive science the concept of cognition carries a much broader meaning. For example, in an influential text by Neisser (1967), cognition is framed by information processing, defined as follows: "the sensory input is transformed, reduced, elaborated, stored, recovered, and used. [C]ognition is involved in everything a human being might possibly do; that every psychological phenomenon is a cognitive phenomenon" (p. 4). As someone trained in both philosophy and cognitive science, Vervaeke is well aware of the knotty conceptual issues at play, and his 4P/3R metatheory explicitly allows us to understand cognition as knowing and cognition as functional information processing.

It is useful to start with Vervaeke's taxonomy of four different kinds of knowing, which are as follows: (1) procedural; (2) participatory; (3) perspectival; and (4) propositional. Procedural knowing refers to knowing how to do something, exemplified by a recipe of action to control outcomes. A key feature of procedural knowing is that it is structured and can be performed like an algorithm, such that there are certain steps, rules, and sequences that can be followed to produce the result. Normally it is something that involves action, although one can follow procedures covertly, such as doing long division in one's mind. The key result of procedural knowing is instrumental power, in that the recipes for action allow one to control outcomes if done effectively.

The next kind of knowing is called participatory knowing. This refers to knowing how to dynamically act in the agent-arena environment. It is simultaneously one of the most basic and most profound kinds of knowing. One way to think about participatory knowledge is to consider the difference between being in a state of confusion and being in a state of flow. Flow is when you are in a groove and feel a natural "dance" between your actions and the environment. It is an example of high participatory knowledge. The opposite is when you feel awkward, self-conscious, clumsy, and out-of-step, which is when you are lacking in participatory knowledge. Participatory knowing can be framed in terms of knowing via effective conformity of one's grip on the agent–arena relationship and how to enact a role or task that is based on an identity rather than a script or recipe, which would be forms of procedural knowing. We can align procedural and participatory knowing with Mind^1 and capacities for functional awareness and responsivity, either in a specific context to achieve an end result (procedural) or referring to the general agent–arena relation (participatory).

Perspectival knowing refers to knowing via embodied perception. It consists of seeing the world and one's place in it via a specific point of view. Perspectival knowing refers to whether someone can see the salient aspects of the situation and make sense of the gestalt. When that happens, one achieves a sense of presence and realness. In contrast, when one is lacking in perspectival knowing, one has difficulty organizing the pieces of information into a coherent sense of the landscape, for example when someone tries to make sense of a scene, but finds the entire arrangement confusing. When aligning this domain with UTOK's Map of Mind, it relates primarily to the domain of Mind^2.

The first three p's are kinds of knowing that we share with other primates. In contrast, the last kind, propositional knowing, is uniquely human. At its broadest, it refers to explicitly knowing *that* something is true. Of course, other animals do have a form of semantic memory, but this is represented via procedural, participatory, and perspectival knowing processes. Propositional knowing in humans is done via symbolic language and justification. It connects directly to the domain of Mind^3, and the Culture-Person plane of existence, which we will be exploring in the next chapter. As such, it is the kind of knowing that has evolved most dramatically in the past 100,000 years.

The 3Rs of Functional Information Processing. Vervaeke's 3Rs of cognition refer to "recursive relevance realization." A key feature and function of neurocognition is determining what aspects of information are necessary such that one can model the situation and anticipate what is going to happen. This is the process of determining relevance. We can see why it is crucial when we consider that the environment is filled with essentially an infinite amount of information that could serve as inputs. The task that agents face is to discern the pieces of information that are relevant for making sense of the situation and both predicting what is currently the case and anticipating what will be the case going forward.

Vervaeke makes a compelling argument that there can be no general theory of relevance because what is relevant is always bound up in the context (Vervaeke et al., 2009). Vervaeke's insight is that a workable theory of how relevance is framed can be developed if it is tied to the concept of realization. Realization for Vervaeke carries twin meanings. It means both to realize as in to see, grasp, or understand, and to realize as in to make, create, or construct. As such, realization functions to serve as a bridging concept between perception and action. And it sets the stage for framing classic work in cognitive psychology on how agents engage in goal-oriented action.

Vervaeke grounds his relevance realization approach in the foundational work of Newell and Simon and their General Problem Solving (GPS) framework. In this framework, a problem is represented by four elements: (1) an initial state; (2) a representation of a goal state; (3) a representation of operations that might move the agent; and (4) path constraints that prevent problematic solutions. Hopefully, this frame sounds somewhat familiar. It overlaps substantially with the P – M => E learning and control theory formulation described in earlier chapters. Recall that "P" is the perception of the initial state referenced against "M"otivated goal states, which then energize motion to move toward solutions, given path constraints (i.e., "E"motion). As we will see, this frames the basic cognitive processing structure of the Influence Matrix.

Relevance realization emerges in part because there is the recursive property that allows the system to integrate information, sync up across layers of processing, and cohere. Recursivity is the third "R" in Vervaeke's formulation, and it refers to the nature of modeling and feedback loops both within the neurocognitive system and between the agent and environment. Much work in cognition and neuroscience shows that there are layers of neurocognitive processing, and modeling takes place across layers arranged in a hierarchy. There are also different domains of processing. For example, there are domains of vision, hearing, touch, emotional valence, and motor movements, and modeling takes place between them. As we will see, the way an animal models itself is crucial for how it models the world. Moreover, as an animal extends itself into the world over time, it builds a model of itself that is separate from the many possible environments it might encounter. This sets the neurocognitive ground for the experiential self, which in turn plays a major role in how social animals navigate the relational world. First, though, we need to understand how Vervaeke's metatheory deepens our understanding of Mind².

Focal Recursive Relevance Realization and the Three Kinds of Qualia

The primary topic of the first educational video series was subjective conscious experience. I initially became intrigued by Vervaeke's framing of Mind² when I heard him differentiate between two kinds of qualia, adverbial and adjectival. Qualia are normally framed as being a single kind of thing. However, based on much cognitive science and knowledge of meditation, Vervaeke questioned this assumption, and developed a cogent argument for separating them into adverbial and adjectival domains. Based on the convergence of our models in the series on *Untangling the World Knot*, we ended up adding a third kind, called valence qualia.

A key piece of evidence Vervaeke uses to justify the separation of adjectival and adverbial consciousness is the "pure consciousness event" (PCE) as experienced by advanced meditators. As described by Forman (1990), PCEs refer to states of pure experiential awareness that are reliably achieved by advanced meditators. The process of meditation that results in PCE is achieved by disciplined observation of the observing process itself. These states do not include any of the usual "gross" features of SCE. That is, when in these states, meditators are not aware of pain or colors or the window in their room. They are not even conscious of consciousness itself, at least not as the object of awareness. Rather, they are just conscious, full stop. Hence the term pure conscious events.

Vervaeke uses the PCE as evidence that the witness function of consciousness is different from the content (Vervaeke & Henriques, 2020). He developed the labels "adjectival qualia" for the content of conscious experience and "adverbial qualia" for the witness function that frames the experiences. Adjectival qualia refer to sensory properties of objects. Examples would include seeing French fries and tasting their saltiness. However, in addition to these sensory-perceptual adjectival properties, there is also the framing aspect of SCE. This refers to how conscious attention is directed. For example, it is the part that directs a person's attention to the French fries, as opposed to the fork next to their plate. Vervaeke argues that it is this adverbial indexing that gives rise to the "hereness, nowness, and togetherness" of conscious experience.

To help make this case, Vervaeke links adverbial qualia to salience tagging, which is the process by which our conscious attention indexes objects and events. Cognitive science has shown that indexing is different from the qualities or properties those objects have. For experimental evidence that indexing is different than perceiving adjectival properties, Vervaeke reviewed the work of Pylyshyn (2004) on Multiple Object Tracking. Subjects are asked to try to track multiple objects as they move around the screen. This is quite easy for three or four objects. However, once the number reaches seven or eight, the task becomes extremely difficult. Subjects begin to lose awareness of the specifics of the object, such as its shape or its color. That is, they lose the adjectival reference properties. However, they maintain awareness of where the object is and can successfully point to its final location. This analysis of adverbial and adjectival qualia aligns with GWT, and its frame that there is an attentional workspace that brings conscious content onto the stage or screen of awareness, such that it can be broadcast to other cognitive domains. The adjectival qualia represent the content on the screen, whereas the adverbial qualia represent the process by which the spotlight of attention is being guided to index and categorize the objects and properties attached to them.

Drawing this connection allows us to delineate additional ways to differentiate conscious cognition from nonconscious cognition. Recursive relevance realization is a general theory of cognitive process. However, what makes consciousness a different kind of cognition is that it focally "aspectualizes" a particular object, event, or idea, such that there is an integration of bottom-up and top-down processes in a way that generates a unified intentional field of integrated information. Aspectualize is Vervaeke's term for how adverbial qualia focus and index on specific content. The model that emerges is that SCE happens as a function of focused recursive relevance realization, whereby the adverbial qualia represent the aspectualizing elements of conscious attention, and the adjectival qualia are the properties experienced by that focal framing.

Where did this capacity for focal recursive relevance realization come from? How did it evolve? The discussion that emerged in *Untangling the World Knot* was how BIT provided a key link to recursive relevance realization. That is, it showed how the neurocognitive system could intersect at the point of learning and provide a basic metric of "caring" that would be an index of whether the animal was attending to and realizing affordances or encountering stressors. Pleasure and pain can be thought of as integrated amalgamations that include both sensory input and affective output elements. They also work to guide the animal in dynamic relation with the environment, and thus they track the agent–arena relationship.

When placed in the context of the four levels of mental evolution, the outline of a relatively coherent picture emerges. Specifically, pleasure and pain are the early broadcast functions of caring (i.e., investment valuing) and the energized emotions that guide approach or avoidance behavior at the learning level. This can be considered the fount of participatory knowing, as it reflects the dynamic dance of the agent–arena relation. It emerges from reflexes and relatively instinctual/genetically programmed procedural modal action patterns. It does so in conjunction with having to navigate the increasing mindedness of other animals, and solve problems related to predation and prey avoidance. In the model of Mind^2 that emerged in the first series, the valence qualia served as the foundational base of sentience, which then emerges into a more extended "inner mind's eye" in birds and mammals as they extend themselves across time. This gives rise to the adjectival and adverbial aspects of conscious experience.

The Witness Function, Thinking as Modeling, and the Emergence of the Experiential Self

The second video series, *The Elusive "I,"* shifted the focus to the concept of the self. Like the first video series, it began with a review of how problematic the concept of the self has been for psychologists and cognitive scientists to effectively define. The model of the self that emerged was framed via five basic steps. First, the model was grounded in a complex adaptive system model of neurocognition framed by recursive relevance realization that carved out paths of investment. This provided the basic frame for understanding animal behavior in terms of agent–arena dynamic relations, and grounds neurocognitive processes and the emergence of valence qualia.

The second step bridges into a more extended self as a necessary consequence of the capacity for animals to model scenarios across time and various situations (i.e., the domain of thinking specified in Chap. 12). Consider that the basic process of recursive relevance realization is shaped by the agent–arena relationship. That is, the cognitive processes are modeling, simultaneously, both the agent and the arena to decipher what is relevant and realize the necessary path of investment to move toward outcomes and affordances that the animal cares about. This involves working memory and the adjectival-adverbial inner mind's eye. Now consider what emerges when an animal can model events across time. Specifically, it can place itself in various simulated situations. What this means is that the self becomes separate and decoupled from the agent–arena relationship. This decoupling means that the animal will start to model itself as a separate entity from the environment. For example, consider the rat at the T-point in a maze. The path to the right is one image; the path to the left is another. But the rat will itself be relatively constant, and it will start modeling what its future self will experience in those different pathways. Thus, the process of extending across time down potential avenues of investment sets the stage for a separate model of the self.

The third step bridges the experiential self to the relational world. Specifically, it involves how animals are now required to model themselves in relationship to others with whom they engage in participatory relationships. This is because connecting with others requires complicated ways of representing the interests and mental states of others, but also requires ways of differentiating and separating one's own interests. This likely gets its foothold in evolution with mothers taking care of their offspring and is seen in things like attachment processes in mammals. Complicated social relations create the need for a self–other matrix and very likely deepen the ways the experiencing self functions to map the world in relation to others. As Carl Safina's (2015) work suggests, such mental capacities seem well developed in creatures like wolves, whales, and elephants.

The fourth step shifts into the world of the human, and our unique capacities for developing an implicit intersubjectivity, potentially framed via the Map of Mind^{1,2,3} as the domain of Mind^{2b}. To the extent that it is a valid descriptive metaphysical concept, it refers to the intersubjective space that emerges as two or more people get to know one another and can engage in increasingly sophisticated modeling and intuitive participatory relating with each other. As we will see, much research suggests that humans have remarkably advanced capacities for shared attention and shared intention. That is, humans readily generate a theory of mind in others and can effectively coordinate our actions in novel ways. The fifth step is propositional language and explicit, self-conscious recursive reflection that comes with reason-giving, arguing, and solving the problem of justification. While implicit intersubjectivity set the combustible stage, propositional language was the spark that set off the human mind's big bang and launched the emergence of the Culture-Person plane of existence. This final step differentiates the experiential self from the human ego and persona, as mapped by the UTM.

Figure 14.1 provides a map of the insights generated by the *Elusive "I"* video series. It depicts how layers of cognitive modeling emerge that function to generate models of both the world and the "Generalized Me" that models the self across time. It also places that in relationship to human consciousness via the inner mind's eye that functions as the adverbial qualia framing of the adjectivally experienced properties. On top of that primate self in humans is the justifying ego that manages the "legitimacy of the self" on the Culture-Person plane of existence.

We can think of this review as a kind of recursive relevance realization in relation to the argument we have been making about mental behavior and the nature of Mind^1 and Mind^2. Specifically, the fact that there were significant synergies between two different metatheories, one in psychology and one in cognitive science, on the nature of cognition, consciousness, and the self, is strongly suggestive that there is significant value to be realized in this formulation. The development of this synergy emerged via a back-and-forth dialogue, where we were recursively modeling both each other and our models.

This fact bridges us into one of the most positive synergies that emerged in the series, which was the connection between Vervaeke's recursive relevance realization and the fourth key idea in UTOK, the Influence Matrix. We can use Vervaeke's frame on cognition to make the prediction that

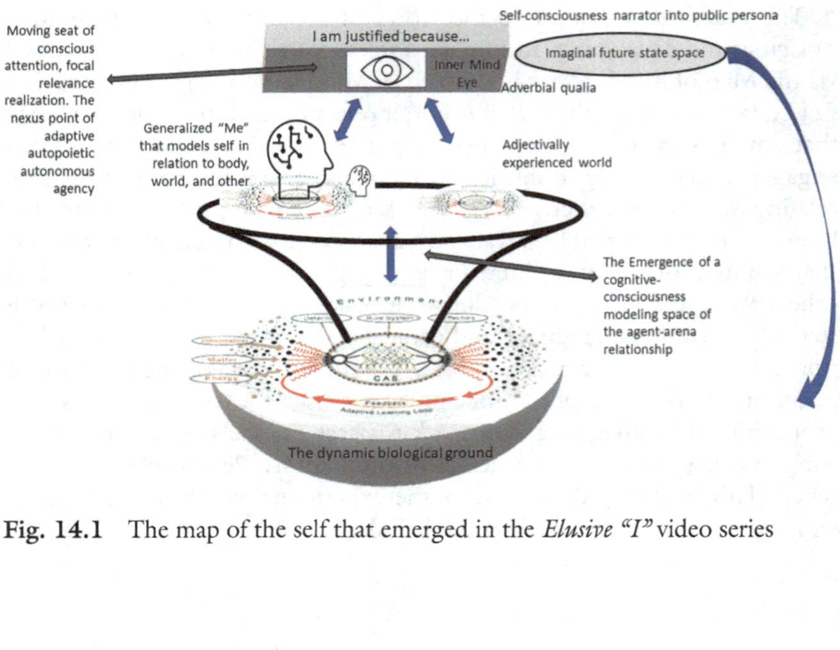

Fig. 14.1 The map of the self that emerged in the *Elusive "I"* video series

social primates like humans will be set up to both track relevant shifts in the relational world and model "self–other" relationships to foster shared perspectives and participatory knowing. Moreover, we can predict that certain kinds of shifts will be associated with specific kinds of emotional reactions. The cucumber-grape monkey experiment and my children's complaints about the size of their piece of cake are examples of the kinds of information that might be salient, and the kinds of reaction shifts that relational processes will engender. In short, we can suggest that, for social animals like humans, we can add a fourth "r" to Vervaeke's model and frame social exchanges as a process of recursive relevance realization in relation to others. Moreover, we can see that the key relational processes of self–other change that individuals track are mapped out by the Influence Matrix.

The Influence Matrix: Putting Self–Other Relations in a Recursive Relevance Realization Context

The Influence Matrix (Fig. 14.2) is a map of key aspects of the "human relationship system," which enables humans to map, model, and navigate the self–other dynamics in the agent–arena relationship. Like the architecture of the human mind diagram, it can be framed as a map of cognitive processing, broadly defined. Indeed, it can be placed in relationship to the architecture of the human mind in that it maps the "M"otivational and "E"motional self–other structures that are likely to be activated when changes in degrees of influence, relational value, or self–other process dynamics emerge.

If we return to Fig. 14.1, we see the "Generalized Me" that involves models of the self and others. These are what psychodynamic clinicians and attachment theorists refer to as "internal working models" of important others that people use to make sense of the relational world. In addition to these schema, there is, of course, also the actual, current interpersonal field and context that is presenting itself to the individual. The Matrix provides a way to map these intra-psychic and interpersonal processes. If we look at the Matrix through the lens of Vervaeke's 4P/3R theory of cognition, we can add richness by saying that the system is structured to identify and make salient relevant information regarding

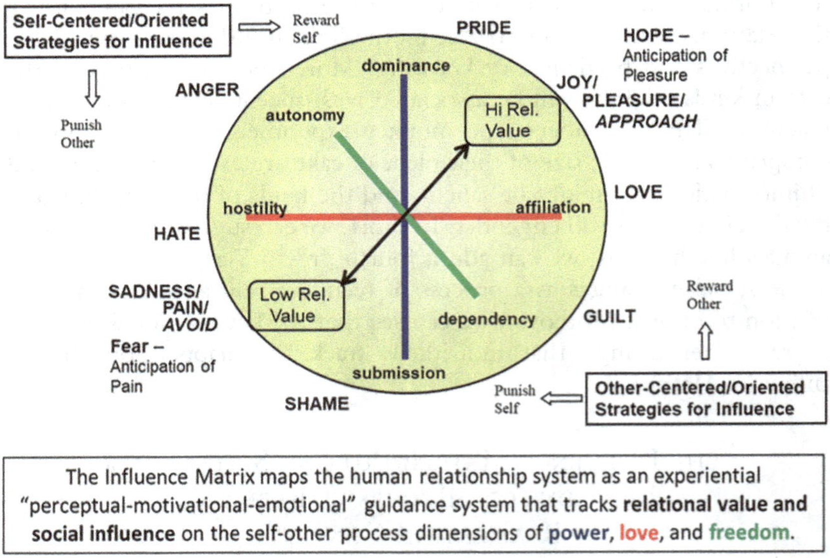

Fig. 14.2 The Influence Matrix

relational processes that sets the stage to realize paths of investment, given constraints.

Although I did not have this language when a constructed it, it is nevertheless how the Matrix is structured. The Matrix was built as an extension of BIT and the P – M => E learning formulation, and thus it has the structure of a general problem-solving system built into it. As such, it grows out of the idea that the nervous system is an investment value system that is seeking to approach and avoid particular states. It is also highly congruent with the idea that Mind[2] is anchored to positive and negative affect systems that orient toward objects, events, and changes both in the environment and in the body that are "good" and thus to be approached, versus "bad" and thus to be avoided. In addition, it is based on the idea that there are recursive self–other relational processes that are taking place either in the real interpersonal context or in the simulated internal working models that people use to make sense of the relational world.

When considering the model, it is important to note that, as is the case with social psychology more generally, social influence is a concept that refers to both a process (i.e., the way people influence each other) and a resource (i.e., the amount of influence one has with others). The central black line represents the fact that humans model, track, and are motivated in relationship to their perceived levels of social influence and relational value. In this context, social influence refers to the degree to which one can influence others in accordance with one's interests. Relational value refers to the extent to which one feels seen, known, and valued by important others. The overlap and difference between these domains and how they are related is described in more detail below.

In addition, there are three relational process dimensions, framed as power (i.e., the dominance versus submission blue line), love (i.e., the affiliation versus hostility red line), and freedom (i.e., the autonomy versus dependency green line) on the x, y, and z axes. The UTOK metapsychology posits that humans are tracking shifts in these process dimensions as self–other exchanges unfold. On the outside of the circle are emotions. As has been discussed, emotions can be considered perceptual response sets that are activated in response to the real or imagined changes in the relational landscape and energize action to shift the path of investment toward more desirable outcomes, via either negative emotions that generally orient toward threat avoidance or positive emotions that orient toward approaching affordances.

It is also fruitful to divide the Matrix into four quadrants. The upper right quadrant is characterized by relational positivity (i.e., high social influence and relational value and positive affects). In contrast, the diagonally opposite lower left quadrant is characterized by the inverse social position of relational negativity (i.e., low influence, low relational value, and negative social emotions). The lower right quadrant is characterized by "other over self" response sets, and is marked by the triad of the affiliative, dependent, and submissive poles. It is also characterized by the emotions of love, guilt, and shame. In contrast, the upper left quadrant is characterized by "self over other" response sets, and is marked by the triad of dominant, autonomous, and hostile poles. It is also characterized by the emotions of pride, anger, and hate (or contempt).

As was reviewed in *A New Unified Theory of Psychology*, the Matrix assimilates and integrates an enormous amount of literature from social psychology and various interpersonal and psychodynamic traditions. I will not repeat that review here, but Fig. 14.3 depicts the major domains of research that were covered.

The primary goal for the remainder of this chapter is to show how the Matrix maps the human relational world, such that we can appreciate how humans form internal working models and engage in recursive relevance

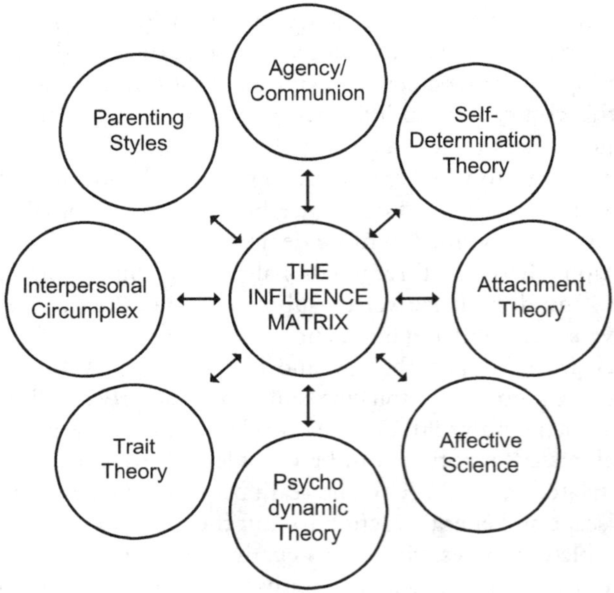

Fig. 14.3 The ideas the Influence Matrix assimilates and integrates

realization in relation to the real or imagined thoughts, feelings, and actions of others and the implications these have for the person's relational world. In addition, we review evidence that demonstrates that humans have evolved unique capacities for shared attention and intention and can create a nonverbal intersubjective "we space" that warrants consideration on the Map of Mind[1,2,3] as the domain of Mind[2b]. We start by reviewing the key elements of the Matrix, and then turn to how it frames the dance of attachment, shared attentional space, and the dynamic participatory processes that guide the relational world.

The Central Black Line: The Relational Value and Social Influence Barometer

The Influence Matrix starts with the idea that humans have a fundamental need for relational value and social influence. The Matrix posits that humans have archetypal templates for being seen, known, and valued in a positive light, as well as the converse (i.e., being dejected, devalued, neglected, or rejected), and that they track real or potential shifts in relational value and social influence. This capacity is represented by the central black line that runs down the lower left to upper right quadrants. It is called the RV-SI line, which stands for *Relational Value and Social Influence*. As noted, relational value refers to the extent to which one feels known and valued by important others, whereas social influence refers to the extent to which one can influence others' thoughts, feelings, and actions according to one's interests.

The two boxes signify "attractor templates," which represent organizing schema that serve as frames for the core of social motivation. The upper right box represents the positive forms of RV-SI that are approached or imagined and associated with the positive emotions (i.e., the images and feelings when one is loved, respected, honored, attended to, and valued by important others), and the lower left box represents the negative forms of relational exchange that are avoided and associated with negative emotions (i.e., the images and feelings when one is being rejected, disrespected, abandoned, ignored, ostracized, or devalued in some way). The line refers to the felt barometer of relational value and social influence, and the sense the individual has of changes in this crucial resource in one direction or the other.

For a concrete example of how to think about these templates, consider that when my daughter was about 12, she became inspired by the *Twilight* novel series and proceeded to write her own novel. It was excellent for a 12-year-old, and when we shared that we were impressed, she dreamed about getting it published and being famous. According to the Matrix, the potential for others both loving her work and loving her for her work is a basic or primary attractor or reinforcer. Put differently, it asserts that being seen, known, and valued by important others for both one's essential nature and for one's accomplishments is a central feature of our relational structure. My daughter did not need to have those signals and signifiers paired with food or any other reinforcer to learn that this is how she should feel. This becomes absurd if we play it out with a thought experiment. Imagine the case where my daughter gives the manuscript to her mother saying, "I hope that others will reject and criticize this as simple and naïve." And then her mother corrects her, saying, "No dear, in our culture we hope other people admire us for our work. You want them to love and appreciate your efforts," and proceeds to hand her a cookie.

Characterizing the central black line in terms of both social influence and relational value represents the one significant structural change to the

Influence Matrix that has been made since it was described in *A New Unified Theory of Psychology*. In that work, the templates in the upper right and lower left quadrants are simply labeled as high and low social influence. Relational value is only implied, as I had not fully teased apart the complicated interrelations between having social influence, being valued by others, and being seen and known. As such, I would now say that the social influence description is correct only in so far as there is a tight relationship between having social influence and being known and valued by others. I was aware that this was not always the case, but I had not developed the conceptual language to make the precise distinction.

I was introduced to the concept of "relational value" via Mark Leary's work on the sociometer approach to understanding the function of self-esteem (see, e.g., Leary & Baumeister, 2000). Leary analyzed the self-esteem literature and explored why it was such an important element of human psychology. He challenged the standard argument that self-esteem was simply good in and of itself and instead argued that we should think of self-esteem as functioning as a "barometer" that signals the extent to which one is valued by others. Thus, he considered self-esteem the barometer for relational value. Encountering Leary's work corresponded with an increasing realization on my part that it was crucial to distinguish the instrumental capacity to influence others from whether one was valued by others.

The distinction between social influence and relational value is obvious to anyone who has had a boss or dominant other that they feared. I was, of course, not blind to this distinction and usually emphasized the fact that a stable high social influence is marked by the balance of power and love. However, it was my work in the clinic room that made salient the fundamental distinctions between: (a) having high relative to low social influence and (b) being valued and (c) being seen and known and (d) valuing the other as important. It is the interrelationships between these four aspects of social relating that are key. That is, when they are aligned, we feel nourished and fulfilled, whereas when they are absent or misaligned (e.g., we have social influence but not relational value), we feel relationally malnourished and experience insecurity or longing. The upper right box represents the attractor goal state of having high social influence and being seen and known and valued by important others. When these are all positive, the approach goal state is realized with concomitant positive emotions. When they are all negative, the situation is a relational catastrophe.

Although social influence and the experience of being seen, known, and valued by important others overlap and are correlated, they are different variables, and different configurations create tensions. Consider, for example, the common clinical presentation known as the "imposter syndrome." This is the circumstance when a person has social influence and is valued by important others, but they do not feel known. It is an experience frequently reported by the students in our doctoral program. They have worked hard to get into the program and when they arrive, they are often exposed to many new ideas in a way that can be overwhelming. In addition, they interact with intelligent professors and other competent students, while often feeling confused or unsure. However, they want to be here and want to receive good evaluations. Even if they receive positive evaluations, often their sense is that if the professors only knew how confused they were much of the time, they would be "revealed" as imposters.

If we apply the Updated Tripartite Model of human consciousness to this presentation, we can say that such an individual is experiencing a public versus private split. Although they are getting signals that they are valued and have high social influence, they nonetheless feel insecure, and it is easy to see why. The felt perception is that if their true selves were revealed, they would be rejected. Thus, these individuals have social influence, but lack the felt sense of being truly seen and known. Consistent with this formulation, things usually improve for our doctoral students if they can develop a secure attachment with their advisor and share their insecurities. Since it is a common presentation that senior faculty have experience with, this revelation can be held in the context of what attachment theorists call "a safe haven." Being known in this way (i.e., revealing the felt sense of being an imposter), followed by the realization that this is quite common and not surprising to the faculty, shifts felt experience of the relational context. They can now be seen and known more fully and are still valued to the same extent.

This analysis of the imposter syndrome highlights why I both agree and disagree with Leary's argument that self-esteem functions as a barometer for relational value. He is right that self-esteem largely functions as a reflection of feeling valued for young children. That is, prior to the emergence of a stable, self-reflective identity that emerges in adolescence (i.e., a fully functioning, separate ego on the UTM), young children will judge themselves largely the way they perceive others to judge them. However, with the emergence of a reflective self and more stable, guiding self-concept, it is easy to see via the imposter syndrome how a disconnect can emerge between the inner and outer worlds.

Clinicians know that people can create massively rich "inner worlds," and much of self-esteem in adults is founded on whether the self-reflecting ego finds the "rest" of the individual valuable, likable, and competent. Indeed, in developing a comprehensive picture of the individual's sphere of influence and relational value regarding important others in the clinic room, my students and I assess for levels of RV-SI across developmental history in: (a) family of origin (i.e., parents, siblings, and extended family); (b) peers and friends; (c) romantic and sexual partners; (d) groups that one identifies with; and (e) the reflective self-concept that emerges in adolescence and can be framed as the egoic evaluation of the experiential self.

Feeling known and valued and having relatively high social influence across these five domains is a good general indicator of relational health. And much research supports the claim that good relationships are a central factor in mental health and well-being. For example, a large-scale, international study asked almost 3000 people about happiness, specifically what it was and what contributed most to it (Delle Fave et al., 2016). The results revealed a strong cross-cultural tendency to consider happiness in terms of both positive emotions and a sense of inner peace and harmony. Indeed, the authors noted that the notion of inner harmony was a much stronger theme among the participants than tended to be emphasized in the psychological science literature.

The researchers also looked at contextual factors associated with happiness and found two very strong ones. The first and most important was social/relational world. Good relations with friends, romantic partners, and family were seen to be central to happiness. In other words, people who felt known and valued by important others had higher levels of well-being and fulfillment. The second was biophysical health. The Unified Theory offers us a scientific framework to understand these important findings. The Influence Matrix identifies the core psychosocial need via the RV-SI black line. The Updated Tripartite Model helps frame and explain what is meant by inner peace and harmony. Specifically, it pertains to how the narrating ego relates to the feelings in the body and experiential self.

The Blue Vertical Line: Power, Rank, and Achievement Status

The Matrix further posits that there are three self–other process dimensions that are crucial to one's place in the social matrix and the way in which people participate in the dance of relational exchange. A process dimension represents the way social exchange occurs and the nature of the self–other positions and how they might change. The first two process lines were mapped by Timothy Leary's (1957) Interpersonal Circumplex Model over 50 years ago. The vertical line (or blue line) refers to when there is a hierarchy in the relationship, such that one individual is "above" the other in relevant ways. It is called the "power" dimension because a key feature is that there is an imbalance between self and other in terms of rank, control, or real or imagined resources. It is important to note both the overlap and potential difference between power and social influence as a resource. Someone who has substantial social influence is someone who can influence the actions of others. In the context of the Matrix, power refers to when social influence is realized via control or rank, as opposed to other avenues, such as care (e.g., young children have much influence with their parents, but do not have power in the blue line).

Because humans have incredibly rich and complicated relationships there are many shades and varieties of vertical relating. One of the most important distinctions to be made is whether the vertical differential is "direct" or "indirect." Direct power is when one has explicit authority or control over the other. For example, a boss–employee, master–slave, or parent–child are all framed by direct power relations, in that there is an established and expected hierarchy and place of authority in those roles. Indirect power relations emerge when there is competition regarding some limited resource that results in social comparisons of lesser or greater levels of attention, status, or other markers of success. Attractiveness, markers of wealth, degrees from Ivy League institutions, and even subtle displays of status in a conversation can be forms of indirect power. As suggested by these examples, indirect control then blends into social influence more generally.

Another crucial consideration when examining the nature of the vertical process dimension is the degree and nature of cooperation that frames the ranking relation. Consider, for example, a football game. Like all win-lose competitive games, football is vertically framed in that there is an explicit winner or loser. However, there is, of course, much cooperation that is built into the game. Both teams must voluntarily agree on the basic rules and framing and what constitutes victory. Similarly, it is crucial when considering dominant–submissive relations whether the relationship is forced or voluntary. The degree to which the vertical relation is voluntary usually relates to the broader RV–SI dynamics. People will voluntarily submit to leaders and rulers they see as good and worthy and have respect for or can benefit from. In contrast, when people feel forced into a submissive role, the powerful individual may have direct social influence, but he will not be valued. For example, slaves generally hate their masters. It is also worth noting that people will differ on the extent to which these hierarchies are emphasized. In terms of individual differences, those with authoritarian personalities seek to structure the world on this kind of hierarchy and be clear about who is above and who is below whom. In contrast, many hunter-gatherer societies work hard to eliminate virtually all forms of dominance and power differentials among the group.

Achievement and status are generally indirect forms of vertical relating. Recall my daughter's hope that her novel might be picked up and published, such that she would become admired. The amount of attention in the social environment is limited and it is competitive. One's book being chosen for publication and being admired by fans represent forms of indirect competition. It is also important to highlight that humans have remarkably complicated and fluid relationship dynamics. Most other primates have stable dominance hierarchies; however, there are many human relationships that fluctuate depending on context. The Matrix is framed such that it can account for both dispositional trait-like tendencies in self–other frames of being and more plastic, participatory, and dynamic exchanges.

The Red Horizontal Line: Affiliation and Hostility

The right-hand side of the red line is about the bonds we form in loving and liking relations, and the ways we cooperate. Prosocial elements come through the process of affiliation, which means to identify with and blend another's interests and desires with one's own. In *The Altruistic Brain: How We Are Naturally Good*, Donald Pfaff (2015) describes research in neurocognitive processes that show how the brain represents oneself and others and how, during processes of affiliation the brain will literally blend and generate static between the representation of self and other. Perhaps not surprisingly, some of the strongest contexts where I have felt this sense of blending my feelings with another have been with my children when they are ill. As is likely true of most parents, it literally hurts me to see them suffer.

As was the case with power, there are many facets associated with affiliation. The emotion associated with affiliation is love. Given the many different potential meanings of the word, it is important to understand how it is used here. To do so, we can return to some of the vocabulary used by the Greeks to define various kinds of love. Eros is romantic love and comes with sexual energy and a strong desire for physical contact. Storge refers to love between family members, such as my love for my children. Philia refers to the loving fellowship between good friends. Finally, agape refers to the loving feeling one has toward humanity writ large. These four domains of love line up well with the relational domains I assess when considering a person's experience of relational value across their lifespan. The first is the family of origin, which starts with their parents and then siblings and then moves into their extended family. This corresponds to storge. The second refers to the relationship with friends, which is philia. The third is with romantic partners (i.e., eros). Finally, there is the felt sense of belonging with groups or even humanity writ large (agape). So, love on the Matrix can be thought of via the various meanings associated with different kinds of affiliation and intimacy relations and motivational states.

Whereas affiliation is fundamentally about bonding and sharing interests, hostility is the inverse. It is about separating from the other and diverging in interest. The primary emotions associated with hostility are anger and its variations ranging from minor irritability to long-standing resentments to intense rage. We can identify two social emotions that are most diametrically opposed to love, which are hatred and contempt. Contempt refers to the sense of deep disapproval and disgust, and it involves viewing the other as diseased or contaminated or disfigured or ugly. It is one of the most fundamental forms of devaluation. Unsurprisingly, it is the one emotion more than any other that is corrosive to intimate relationships. Hatred is more active and orients the individual to engage and attempt to destroy the other. It is more associated with seeing the other as a fundamental obstacle or threat to one's safety, place, or value. Thus, hatred manifests in actions or fantasies that involve the destruction or obliteration of the other, whereas contempt often orients the individual to first blame and shame the other and then move away.

The Green Line: From Dependency to Interdependency to Autonomy to Counter-dependency

The green line on the Matrix represents the level and nature of social involvement and relational exchange. It is labeled freedom, which refers to the degree of freedom from social influence, obligation, and control from others. The adaptive advantage of freedom from control seems obvious in that if one is under the control of another, their interests are much less likely to be realized. However, it is also the case that humans are incredibly social creatures and depend on social engagement and exchange for many physical and emotional reasons. As any parent can tell you, humans are born into the world in a complete state of dependency. A newborn cannot even fully lift their head, let alone move about in an autonomous way to obtain food or meet other survival needs. As such, the key variable is the extent and nature of parental care, much of which depends on the desires and abilities of the primary caretakers.

Nonetheless, as any parent who has heard an infant cry in distress or coo in satisfaction can also attest, human infants have powerful capacities to influence caretakers and elicit investment. Care-eliciting behaviors can be conceptualized as expressions of the infant's dependency needs, which then plug into the parent's caregiving, affiliative tendencies. Broadly following Maslow, we can say there are two kinds of dependency needs, physiological and relational. Physiological needs refer to basic survival needs and include protection from harm, food, temperature regulation, and so forth. Relational needs include cuddling, eye contact, and the expression of positive emotion by the caretakers in the service of fostering a sense of emotional security. Relational needs then extend into social learning and socialization, to help the infant develop into an individual who can effectively operate in the social field.

As suggested by this summary, it is useful to frame the green line in terms of both development and processes of attachment. The first stage in Erik Erikson's model of psychosocial development is trust versus mistrust and relates to whether the infant can reliably anticipate being held, protected, loved, and responded to in a way that meets their needs. The second is autonomy versus shame and doubt, and it emerges as infants become increasingly mobile and thus are able to explore their environment with greater levels of independency. This model is enriched by considering John Bowlby's attachment theory and the way it was extended by

Mary Ainsworth to get additional insight into how differences in relational systems might develop (Ainsworth & Bowlby, 1991).

Attachment theory posits that many mammals have an inborn system that guides both parents and offspring in the dynamics of parental care and development. The basic idea is that many young mammals are highly dependent on their parents for care and survival and that the caretaker functions as a "secure base" from which to operate and explore the world. Building from Bowlby's work, Ainsworth proceeded to identify three different attachment styles that emerged in development: (1) secure; (2) insecure ambivalent; and (3) insecure avoidant. These attachment styles, examined from the vantage point of the Influence Matrix, can be understood as representing different socio-emotional strategies for influence. Securely attached children have their basic needs for relational value and social influence met and, consequently, feel more positive, safe, and comforted by the presence of the caretaker. In contrast, insecure children are theorized to not have their dependency needs met at least in some ways, which results in a general registering of low relational value, and this is associated with negative emotions, most notably general distress, fear, and sadness.

We can interpret the two insecure attachment styles as representing two different influence strategies on the autonomy–dependency axis. Ambivalent children adopt a hyper-dependent strategy, characterized by strong emotional displays of need and fear of abandonment from not receiving the necessary parental investment. In contrast, avoidant children adopt a hyper-autonomous strategy that can be understood as minimizing dependency needs and care-eliciting displays. Although dependency is an inevitable starting condition, it is nevertheless, by definition, a rather vulnerable state. If the interests or capacities of the individual on whom one is dependent change away from the individual, difficulties inevitably follow. In addition, achieving social influence via competition and altruism is an endeavor that takes time and energy that could potentially be spent doing other things. These opportunity costs occur in the best of cases. In worst case scenarios, social exchanges can result in individuals either being dominated and controlled or sacrificing without receiving any beneficial return. Consequently, individuals are theorized to be motivated toward self-reliance and the avoidance of excessive dependency on others.

Autonomy, which is defined as the capacity to function independently and be free from the undue influence of others, has been emphasized as a key psychological motive or need by several clinical theorists and researchers. For example, Carl Jung emphasized the importance of individuation, and the separation–individuation dynamic remains central to many psychodynamic theories. Similarly, autonomy versus shame and doubt is the second developmental task in Erikson's model of ego development. Carl Rogers argued that the fully functioning person had an internal locus of evaluation, and Marie Jahoda argued that self-direction and the freedom from the control of others were central to mental health. More recent psychological researchers, like Carol Ryff (1989), have argued strongly that a sense of autonomy is crucial to psychological well-being.

And yet with too much independence the opportunities for one's social needs to be met are greatly diminished. Indeed, extreme independence is likely to be a function of counter-dependence, meaning that the individual separates from others out of fear of failure, betrayal, rejection, or other costly social encounters. There is an important difference between healthy individuation and counter-dependence. The former is an "approach mindset" where an individual takes pride in discovering themselves and their capacities as a unique individual, whereas the latter is an "avoidance mindset," where the focus is on dangers to be protected against. According to the Influence Matrix, a balance between independence and dependency, what might be called a state of healthy autonomous-interdependence, is expected to be associated with optimal relational functioning.

Taken together, the four social motivational lines allow us to see generalizable patterns in human interaction that range from how people tend to make upward and downward social comparisons to patterns of dependency or counter-dependency to the dynamics associated with achievement motivation to the dynamics of friendship and resentment. Although such frames are useful and point to the power of the Influence Matrix, it is crucial to be clear that the Matrix is fundamentally a guide for understanding the dynamic dance of self–other interaction in the world. That is, it models how the human relationship system serves as a behavioral guidance system that tracks shifts in relevant information, which in turn orients to possible paths of investment, and is also engaged in recursive relational modeling, both in terms of the current interpersonal field and the internal working models that function to anticipate future states.

The Evolution of the Human Relationship System

Perhaps the easiest way to conceptualize the Matrix is to see it as a combination of attachment theory and the Interpersonal Circumplex, placed in an evolutionary developmental systems framework. In his important work *The Developing Mind: How Relationships and the Brain Interact to Shape Who We Are*, 3rd Edition, Daniel Siegel (2020) devotes a chapter to the attachment system and how it lays the groundwork for the emergence of the experiential self. After reviewing Bowlby's original work and its extensions via Main and Ainsworth, he describes the different patterns of attachment styles and how they emerge first with caregivers and are often sustained into adulthood. He then explores in rich depth the complex dynamics of the relational world and how human communication and states of mind emerge in a kind of "biosynchrony." After summarizing how research suggests that one of the core functions of attachment is that it helps the young to organize and regulate their feeling states, he summarizes the processes in this way (pp. 171–172):

> Patterns in the flow of energy and information within and between people comprise the fundamental components of a state of mind. In this way, attuned communication involves the resonance of energy and information between two people. For the nonverbal infant, this intimate, collaborative communication is mediated without words. This need for nonverbal attunement persists throughout life. Within adult relationships of all sorts, words can come to dominate the form of information being shared and this can lead to a different form or representational resonance. Such a verbal exchange can feel quite empty if it is devoid of the more primary aspects of each person's internal states. Infant attachment studies remind us of the crucial importance of attuned nonverbal communication in all forms of human relationships. It is a way of responding to more than merely external behaviors.

Siegel is making a central point that is implied in the title of this chapter, which is that for humans, the relational world plays a huge role in the development of mental behavior. That is, the processes by which the human mind will come into being is in large part a function of the relational world. Moreover, human mental behaviors take place in a relational world marked by something more than just overt actions or vicarious learning. Instead, in intimate exchanges, human minds sync up to form an implicit intersubjective field that can be considered a self–other participatory dance in the language of knowing afforded by Vervaeke's taxonomy. This carries substantial descriptive metaphysical implications regarding the nature of subjectivity and is the referent for the second meaning of the chapter title. Mind^{2b} is arguably a genuinely valid category of mental processes. You can see it in the participatory dances that you engage in with others you know intimately.

From an evolutionary perspective, the argument can be made that caring for offspring and the parent–child relationship was the first systematic, intimate relationship in our phylogenetic lineage. The attachment system is nature's solution to fostering the capacity for young to elicit care and get their dependency needs met and for parents to invest in their young with discernment. In humans, the attachment system motivates children to form deep and long-standing bonds with one or a few primary caretakers. It also motivates parents to care deeply for the well-being of their children and engage in the longest, most intense investment patterns seen in the animal kingdom.

Much reliable work has been done on deciphering the patterns of attachment. As suggested by the discussions of relational value and the freedom–dependency process line, researchers have identified two dimensions that frame how children who are insecurely attached develop different strategies; one is called "avoidance" and the other "anxiety." Children who are low in both feel secure (i.e., known and valued in the language of the Matrix). High avoidance refers to children who suppress dependency needs and are muted in their relational engagement (i.e., are counter-dependent on the green line). High anxiety refers to children who respond to vulnerability with exaggerated emotional displays (i.e., are hyperdependent on the green line). Finally, there are children whose attachment patterns are disorganized, and they exhibit a fearful and erratic pattern that combines and vacillates between high anxiety and high avoidance (they are in the lower left, low relational value quadrant). Figure 14.4 depicts the relationship between the two frameworks.

Although parenting and attachment first emerged in the mammal world, later social groups appeared that would create opportunities for long-term relationships between animals. This is especially true in social mammals like wolves, whales, elephants, and primates (Safina, 2015). In these animal groups, we see both competitive hierarchies that involve struggles for scarce resources, such as food, territories, and mates, and cooperative or affiliative alliances between individuals who engage in reciprocity and care. These patterns gave rise to the two dimensions of

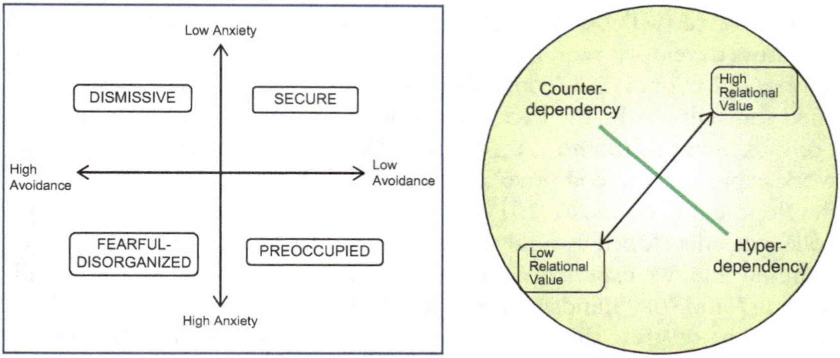

Fig. 14.4 The RV–SI and freedom dimensions align with attachment theory

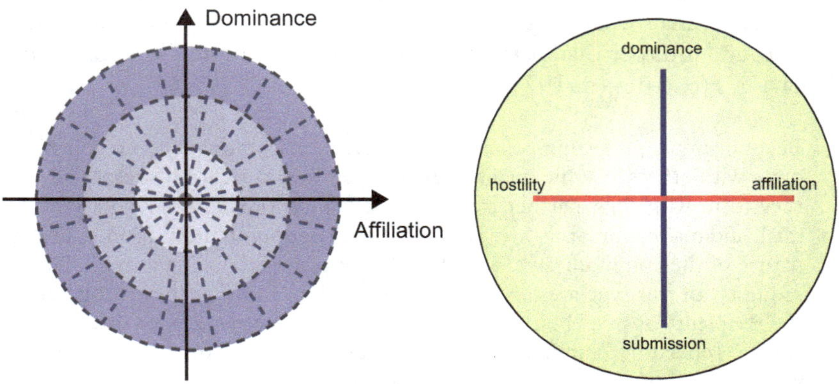

Fig. 14.5 The power and love dimensions align with the Circumplex model

dominance and affiliation mapped by Leary's Circumplex, which are the power and love dimensions on the Matrix. Figure 14.5 depicts the relationship between these frameworks.

A Significant Jump in Intersubjective Capacity

The Matrix suggests that different relationship systems, specifically the attachment system and the systems for competition and rank and reciprocity, become blended into a more dynamic whole in humans. Why?

According to UTOK, the complex relational fields that our hominid ancestors developed required a complex blending of the attachment/parenting and cooperative/competitive systems. This argument is strengthened substantially by the excellent work done by Michael Tomasello. His book *Becoming Human: A Theory of Ontogeny* summarizes decades of work exploring the cognitive and social aspects of human and great ape development (Tomasello, 2019). The program of research that he and his colleagues developed strongly suggests that at some point in our unique hominid line, we experienced a massive upgrade in our relational "mind-reading" and/or "mindsharing" capacities. Carefully designed empirical studies demonstrate that young children (i.e., ages 2–4) have remarkably advanced capacities for shared attention and shared intention. Although there are many potential examples to share, one of the clearest and most familiar is the way children are readily able to interpret pointing as indicating a state of mind and intention held by another. Tomasello (2019) emphasizes that what is going on in human development is an intrapsychic and interpersonal patterning that gives rise to a new kind of mental behavior. He writes (p. 197):

> Beyond simply preferring collaborative interactions, young children collaborate with others in some qualitatively unique ways as well. Most especially, they form with their partner a joint agent 'we' in order to pursue a joint goal, and maintaining this 'we' is part of their continuing motivation… The nature of the communicative acts that children [use are telling as well. The requests] of this type are fundamentally cooperative in that they recognize the recipient has free choice in the matter; they simply suggest or offer one choice. Indeed, attempting to force re-engagement would be inconsistent with the goal of reconstituting their mutually cooperative 'we.'

Tomasello identifies eight pathways that differentiate humans from the other great apes as follows: social cognition, communication, cultural learning, cooperative thinking, collaboration, prosociality, social norms, and moral identity. He lays out the case that, although there are elements of the capacities in other primates, they are tied together in human development to unique effect. Research demonstrates that by nine months of age, human infants show capacities for joint attention or "meshing with minds" with adults that are different and more advanced than other species. Then, between three and four, there are massive transformations in how human children model others and sync up and resonate with them, much as Siegel describes above. This is the ability to form a collective set of shared intentions that happens both with adults and, increasingly, with peers. Then, at about six years old, there is a transition to normative development that is characteristic of the Culture-Person plane of existence, which is the subject of our next chapter. This is apparent in how Tomasello ends his book (p. 343):

> The outcome of these ontogenetic transformations leading to uniquely human psychology—as far as we have followed them here—is the child of six or seven of age, who operates in her culture as a nascent person based on reason and responsibility. Reason and responsibility are normative notions: they involve standards one 'ought' to meet. In our view, the origin of normative force lies in the individual agent's sense of instrumental pressure—the sense that I ought to do x in order to obtain y—as a self-regulatory process. Then, in first entering into a joint agency, the young child transforms this individual self-regulation into social self-regulation, in which 'we' self-regulate 'me' and 'you' interchangeably. So now the question is what 'I' and 'you' as a part of 'we' ought to do. Then by six or seven years of age, the child starts to identify in addition with a cultural 'we,' which, upon internalization, executively self-regulates her and her compatriots' beliefs and actions normatively in the direction of collectively accepted group standards of rationality (reason) and morality (responsibility). From this point on, reason and responsibility represent the regulative ideals governing virtually all of the children's behavioral decision making, as they gradually become fully fledged persons in a culture.

The combination of Siegel's interpersonal neurobiological analysis, Tomasello's anthropological ontogenetic frame, Vervaeke's recursive relevance realization in relation, and the Influence Matrix's modeling of self–other process dimensions points to the notion that human interaction in intimate relational exchanges is marked by a shared implicit intersubjective dance that is central to the human experience.

Applying the Matrix to the Origin of Gender Roles

One of the most obvious contributions that the Influence Matrix makes in mapping out the human relationship system is that it affords us a way to understand hominid relational tendencies that are masculine and feminine in nature. As evidenced by the infamous Google Memo debacle a few years ago involving James Demore, there is much confusion in our society regarding the best way to think about gender similarities and differences (Henriques, 2017). Demore wrote an internal memo that used some basic evolutionary reasoning to argue that part of the reason there were more male programmers and engineers was the evolved psychological differences between men and women. Many people thought the memo basically justified sexism and it resulted in a significant controversy and ultimately Demore was fired for contributing to an unsafe workplace.

From the vantage point of UTOK, much of the confusion in our society on these issues emerges because both scholars and lay people think about human mental behavior from two different vectors, when the reality is that we need three. The two vectors are the so-called biological and social forces. The biological generally refers to the genetic, physiological, evolutionary, and hormonal "nature" dimension, whereas societal roles, norms, ideals, and expectations for how men and women ought to act form the "nurture" dimension. What is missing in this analysis is a clear understanding of mental architecture that is neither biological (Life-Organism) nor social (i.e., Culture-Person) force. Rather, there is a Mind-Animal dimension of mental behavioral investment tendencies that must be understood. What is an example of a Mind-Animal conception?

As a case in point, consider findings from an article in *American Psychologist* on gender stereotypes. In it, Alice Eagly and her colleagues examined the gender stereotypes of people in the United States from 1946 to 2018. Specifically, Eagly et al. (2020) examined perceptions of men and women on agency, which the authors define as the tendency to "orient toward the self and one's own mastery and goal attainment (e.g., ambitious, assertive, competitive)," and communion, which the authors define as the tendency to orient toward the "other and their well-being (e.g., compassionate, warm, expressive)." The authors note: "Communion prevails in the female stereotype, and agency in the male stereotype." Consistent with the idea that most scholars focus on two vectors of explanation aligned with nature and nurture, Eagly and her colleagues argued that people tend to explain these differences in the genders via a function of either "biology" or "society." They state: "Although some people ascribe such trait essences to biology, others instead ascribe them to socialization and social position in society" (p. 302).

The article reports on perceptions of men and women on the dimensions of agency and communion for over 50 years in the United States. They also included a third construct, competency (i.e., the extent to which men or women were perceived as generally more or less competent or intelligent). They found that, over the years, women were seen as increasingly more competent, such that they are now rated as the more competent and intelligent group. Although this is an interesting finding that warrants reflection and commentary, our primary focus here is on the agency and communion variables. Given the remarkable change in attitudes toward gender in the last 50 years, and the huge societal push to see men and women as having no essential (i.e., nonsocially constructed) differences, it seems to be a basic, straightforward prediction that the differences between the stereotypes about men being more agentic and women being more communal would go down. After all, if, as a society, we are awakening to the idea that the gender identity binary is simply a function of the social construction of reality, then shouldn't our newfound freedom allow us to be unshackled from these primitive notions and allow people to toss off the shell of rules imposed upon them by society?

In contrast to this expectation, the researchers discovered that the perceived differences between men and women, somewhat surprisingly, increased over the years. Women are now seen as even more communal, whereas men generally stayed the same on their perceived agentic advantage. *This means that the perceived distance between the key personality features of the two genders is greater now than 50 years ago.* In the words of the authors: "In sum, U.S. poll data show that it is only in competence that gender equality has come to dominate people's thinking about women and men. For qualities of personality, the past 73 years have produced an accentuated stereotype of women as the more communal sex, with men retaining their agency advantage" (p. 313). The authors engaged in what seemed tortured logic to try to defend their social role view that "gender stereotypes stem from people's direct and indirect observations of women and men in their social roles" (p. 302).

Although social roles obviously play an important part in how people experience their gender and sexuality, from the vantage point of the UTOK, they are not the primary origin or source of the gender differences in relational style. Indeed, to understand them, one simply needs to place the different masculine and feminine investment patterns on the quadrants of the Matrix. As mentioned earlier, the upper left quadrant is the "self over other" quadrant, and it emphasizes the poles of dominance, autonomy (i.e., independent goal attainment), and hostility. The lower right, aka the "other over self," quadrant emphasizes affiliation, dependence (i.e., longing for connection and need for approval), and submissive deference to others. Importantly, this means that the Influence Matrix maps the core features of "agency" and "communion." And, as we have seen, this basic mental architecture existed long before the social construction of reality and the emergence of the Culture-Person plane of existence 50 to 100 thousand years ago. Rather, the Matrix goes back to a time when we were hominid primates rather than cultured persons with social roles framed by propositional justification systems; thus, we are talking about millions of years when it comes to its basic architecture grounding attachment and cooperation and competition in the relational world.

That the Matrix frames agency and communion as constructs is a crucial point in that it highlights how it maps the Mind dimension. Placed in the context of the above discussion, we can see how the nature/biology versus nurture/social role dichotomy is essentially blind to Mind. Moreover, we can use its basic logic of evolution and behavioral investment to understand why, on aggregate, human males tend toward the former and human females the latter. To see why, consider that, long before we were humans, females were giving birth and taking care of their young. In addition, virtually no male primates are involved in parental care. Given this crucial fact about a major difference in relational investment patterns, is it any wonder that the mental architecture of the female hominid would be more relationally oriented? Trivers' (1972) important work on parental investment theory makes explicit the reasons why this is the case. Indeed, it is not surprising at all that we can see a bimodal set of tendencies, such that the female hominid system is more oriented to relational networking, whereas the male is more instrumental and agentic. The analysis stemming from UTOK and the Matrix means that Eagly et al. have the explanatory sequence backwards. Rather than social roles driving the perception of difference, it is the mental architecture that is the primary driver of personality differences and people's perception of them.

The point here is that the Matrix helps explain the origin of gender roles and why they remain potent in modern society, which is that they reflect archetypal patterns of masculine and feminine relational styles. But this analysis really makes a deeper point, which is related to UTOK's mental behaviorism and basic issues of ontology. It is clear from this analysis that our society is painfully confused in its understanding of sex and gender. An obvious reason from the vantage point of UTOK is that we have an unhelpful "biology" versus "sociology" binary—as if these are the only two causal domains that impact human mental behavior. The crucial point from UTOK is that there are mind-animal forces that are neither "life-organism-biological" nor "culture-person-social" in essence. In fact, the analysis shows how psychologists who frame human experience via nature versus nurture are blind to the ontology of Mind.

Conclusion

The broad outline of UTOK's metatheoretical framing of human mental behavior is now in place. Our survey of UTOK's key ideas began with Justification Systems Theory, and how it gives rise to a new Updated Tripartite Model of human consciousness that can assimilate and integrate many different lines of thought. Chapters 12 and 13 summarized how Behavioral Investment Theory provides an explanatory metatheoretical framework for both Mind[1] and Mind[2] that assimilates and integrates many different empirical findings and schools of thought to generate a coherent picture of how minded animals emerged from the plane of Life. This chapter argued why the clarity of the picture is greatly enhanced when it is paired with John Vervaeke's 4P/3R metatheory of cognition. Understanding cognition via recursive relevance realization substantially deepens BIT's neuro-computational control principle and its alignment with active inference and predictive processing models.

Vervaeke's work also provides a direct bridge to the Influence Matrix, which can be considered, in retrospect, to be a map of recursive relevance realization in the relational world. Specifically, the RV–SI line represents how humans track changes in relational value and social influence, which provide the core motivational structure. The blue power line, the red love line, and the green freedom line highlight how humans attend to the processes of competition, cooperation, and involvement, and track subtle changes, which in turn activate emotional signals and impulses to guide the relation toward desired states and away from problems.

The broad dimensions mapped by the Matrix can be framed by the merger of Bowlby's work on attachment theory, which specifies the black and green lines, and Leary's work on the interpersonal circumplex, which specifies the blue and red lines. Drawing on Tomasello's work, which demonstrates how human primates have remarkable capacities for developing a shared attentional and intentional "we" space, the suggestion from the Matrix is that the human relationship system advanced by merging parent–child attachment structures with social processing frames for cooperation and competition into a self–other relational field space that enables humans to participate in a much more intimate, shared, and coordinated participatory relational dance. The result is a picture of the human relationship system that can be framed as a kind of relational recursive relevance realization structure that guides humans in their social exchanges. In the final section of the chapter, we saw how we can use the Influence Matrix map of how humans navigate the self–other world to make sense of the finding of gender differences in agency and communion in a way that affords us clear insights into the origin of gender roles grounded in a proper understanding of the ontology of the mental, relative to biological and social forces.

We have now built a picture of human mental behavior as emerging from the world of pre-mental multi-cellular organisms into the world of animals with brains and complex bodies into the world of mammals with subjective conscious experiences into the world of primates with attachments that track their relational value and social influence on the dimensions of power, love, and freedom. With this architecture in place, we have come full circle and can return to where we started with JUST. The next chapter places Mind^3 and the Culture-Person plane of existence on the organism-animal-mammal-primate stack we have constructed so that we can clarify the ontology of human persons.

References

Ainsworth, M. S., & Bowlby, J. (1991). An ethological approach to personality development. *American Psychologist, 46*(4), 333–341. https://doi.org/10.1037/0003-066X.46.4.333

Delle Fave, A., Brdar, I., Wissing, M. P., Araujo, U., Castro Solano, A., Freire, T., & Soosai-Nathan, L. (2016). Lay definitions of happiness across nations: The primacy of inner harmony and relational connectedness. *Frontiers in Psychology, 7*, 30, 1–23.

Eagly, A. H., Nater, C., Miller, D. I., Kaufmann, M., & Sczesny, S. (2019). Gender Stereotypes Have Changed: A Cross-Temporal Meta-Analysis of U.S. Public Opinion Polls From 1946 to 2018. American Psychologist. http://dx.doi.org/10.1037/amp0000494

Eagly, A. H., Nater, C., Miller, D. I., Kaufmann, M., & Sczesny, S. (2020). Gender stereotypes have changed: A cross-temporal meta-analysis of U.S. public opinion polls from 1946 to 2018. *American Psychologist, 75*(3), 301–315.

Forman, R. K. C. (1990). *The problem of pure consciousness: Mysticism and philosophy.* Oxford University Press.

Henriques, G. R. (2017, August 11). An in-depth analysis of the crisis at Google. [Blog] Theory of Knowledge on Psychology Today. https://www.psychologytoday.com/us/blog/theory-knowledge/201708/in-depth-analysis-the-crisis-google

Leary, M. R., & Baumeister, R. F. (2000). The nature and function of self-esteem: Sociometer theory. In *Advances in experimental social psychology* (Vol. 32, pp. 1–62). Academic Press.

Leary, T (1957). *Interpersonal Diagnosis of Personality*. New York: Ronald.

Neisser, U. (1967). *Cognitive psychology*. Appleton Press.

Pfaff, D. W. (2015). *The altruistic brain: How we are naturally good*. Oxford University Press.

Pylyshyn, Z. (2004). Some puzzling findings in multiple object tracking: I. Tracking without keeping track of object identities. *Visual Cognition, 11*(7), 801–822.

Ryff, C. D. (1989). Happiness is everything, or is it? Explorations on the meaning of psychological well-being. *Journal of Personality and Social Psychology, 57*, 1069–1081. https://doi.org/10.1037/0022-3514.57.6.1069

Safina, C. (2015). *Beyond words: What animals think and feel*. Henry Holt & Company.

Siegel, D. J. (2020). *The developing mind: How relationships and the brain interact to shape who we are* (3rd ed.). Guilford Publications.

Tomasello, M. (2019). *Becoming human: A theory of ontogeny*. Belknap Press.

Trivers, R. L. (1972). Parental investment and sexual selection. In B. Campbell (Ed.), *Sexual selection and the descent of man, 1871–1971* (pp. 136–179). Aldine.

van Wolkenten, M., Brosnan, S. F., & de Waal, F. (2007). Inequity responses of monkeys modified by effort. *PNAS, 104*(47). https://doi.org/10.1073/pnas.0707182104

Vervaeke, J., & Henriques, G. R. (2020, September 4). *Untangling the world knot of consciousness: Grappling with the hard problems of matter and mind*. [Video]. The Cognitive Science Show. https://www.youtube.com/watch?v=bD6Szbf1cHo

Vervaeke, J., Henriques, G. R., & Hovhannisyan, G. (2022, April 2). *Psychepathology and well-being*. [Video]. The Cognitive Science Show. https://www.youtube.com/watch?v=J-eHKAIi1cE

Vervaeke, J., Henriques, G. R., & Mastropietro, C. (2021, April 2). *The elusive "I": The nature and function of the self*. [Video]. The Cognitive Science Show. https://www.youtube.com/watch?v=_Fnmp6UVafM

Vervaeke, J., Henriques, G. R., & Stein, Z. (2021, August 27). *Toward a metapsychology true to transformation* [Video]. The Cognitive Science Show. https://www.youtube.com/watch?v=wZKV8uVe8lQ

Vervaeke, J., Lillicrap, T., & Richards, B. (2009). Relevance realization and the emerging framework in cognitive science. *Journal of Logic and Computation, 22*(1), 79–99. https://doi.org/10.1093/logcom/exp067

CHAPTER 15

Mind^3 and the Culture-Person Plane of Existence

This chapter continues with UTOK's project of synthesizing metatheoretical formulations with descriptive metaphysical frameworks to clearly delineate the relevant ontological referents. We begin by demonstrating how JUST is deeply consonant with several prominent perspectives for defining personhood. We then show that it aligns with how persons behave in everyday life by corresponding commonsense folk psychology with UTOK's metapsychology of the dynamics of investment, influence, and justification (i.e., "JII dynamics"). The analysis is strengthened via research in social psychology that conclusively demonstrates that Mind^3 can be framed as a justifying mind in a social context concerned with social influence and relational value.

We then shift to the Culture-Person plane of existence and the social sciences proper. Just as basic psychological science struggled to define the mental, the social sciences have failed to effectively define culture. Much confusion is clarified when we differentiate between small "c" culture as learned and shared mental behavioral repertoires and the Culture-Person plane of existence as consisting of verbally mediated systems of justification. Via this core distinction, we obtain a much clearer picture regarding the social construction of reality. This allows us to shift our philosophical perspective from the epistemological challenges raised by a social constructionist perspective to see the social construction of reality in the context of a scientifically coherent naturalistic ontology. Put plainly, the thing

G. Henriques, *A New Synthesis for Solving the Problem of Psychology*, Palgrave Studies in the Theory and History of Psychology,
https://doi.org/10.1007/978-3-031-18493-2_15

that is being constructed is the Culture-Person plane of existence via evolving networks of justification systems.

We then align this ontology of the Culture-Person plane of existence with a useful and well-known socio-ecological perspective that affords an effective framing of the structural context of justification across aggregate scales of sociocultural operation. We then briefly explicate how JUST enables us to frame the functional aspects of the domain of justification by delineating a continuum of contexts that range from the con game to the courtroom to the courtyard to the science lab. The chapter ends via a zoomed-out macroscopic perspective that frames the evolution of cultural justification systems over the past 50,000 years. It tracks the shift from oral-indigenous cultures 50,000 years ago to traditional civilizations that appeared approximately 5000 years ago to modern scientific worldviews that emerged approximately 500 years ago to the postmodern sensibility that has grown in the last 50 years. This progression hints at the possibility that a new "metamodern" cultural sensibility is emerging in the near future. The overall conclusion is that UTOK's metapsychology affords us a clearly specifiable metaphysical, ontological, and metatheoretical picture that readily bridges us from Mind^1 and Mind^2 in the animal mental behavioral plane into Mind^3 and the Culture-Person plane of existence.

Toward a Coherent Ontology of Human Persons

Peter Ossorio's *descriptive psychology* is a little known but powerful approach to the field that represents a significant break from mainstream psychology. There are two important ways in which Ossorio's analysis connects to the current work. First, Ossorio began his work by laying out the argument that without the careful "pre-empirical" delineation of psychology's core concepts, the empirical psychological enterprise was doomed to fail. This is what I called the sandcastle problem in Chap. 3. Consequently, Ossorio was deeply concerned with developing a framework that was up to the task of coherently describing the field's core subject matter.

A second alignment pertains to descriptive psychology's central focus, which was on answering the question: "What is a person?" In *The Behavior of Persons*, Ossorio (2006) identified four key concepts that he argued are central in understanding people and their actions: (1) the concept of a person; (2) the concept of behavior; (3) the real world; and (4) language. Ossorio proceeded to develop a powerful "parametric" breakdown of the components that go into the behavior of persons. In it he described how intentional action requires motives for specific states of affairs, knowledge about the action, knowledge about the current and desired states, and procedural know-how for performing the behavior. He also argued that the achievement or outcome of the behavior and its personal and social significance must be considered as part of the act.

Ossorio's (2006) analysis of the concept of a person relates directly to his analysis of the behavior of persons, which refers to the things that persons do. In Ossorio's language: "A person is an individual whose history is, paradigmatically, a history of deliberate action in a dramaturgical pattern." A "dramaturgical pattern" refers to the concept of engaging in action that takes place on a social stage and has social significance over time. Although his technical language is slightly different, Ossorio's analysis regarding the nature of persons and how they behave corresponds directly with the descriptive metaphysics found in UTOK. Deliberative action is a form of behavior in which a person (a) engages in an intentional action, (b) is cognizant of that, and (c) has chosen to do that. The social stage is the stage of influence and justification, upon which deliberative action takes place. Placed in the language of UTOK, Ossorio is pointing to Mind^3 and the idea that persons operate on the Culture-Person plane of existence, and the behavior of persons is behavior that is regulated and justified by self-conscious reflection.

Ossorio argued that these are the ingredients that constitute *persons as such*. Other animals act and have perceptual-motivational-emotional intentions as framed by BIT. In addition, they communicate and have meaningful long-term relations. Indeed, the Influence Matrix highlights some of the key relational process dimensions of attachment (i.e., care and dependency) and social exchange (i.e., cooperation and competition) that humans share with other animals. However, other animals do not act with explicit, verbally mediated self-conscious deliberation on the stage of social justification. Nor do they have developmental practice with asking or answering questions, giving reasons, or collectively engaging in the deliberate legitimizing of actions and claims. It is this process of justification in the matrix of social life that allows a young child—who does not initially have the capacity to act as a full person—to eventually acquire such capacities via socialization and development and thus to become one.

Ossorio's analysis makes a key point about the relationship between the concept of persons and human beings, which is that the two are quite distinct and should not be confused as being equivalent. A human being is an organism and an animal, specifically a kind of great ape. In contrast, properly conceived, a person is an entity with a capacity, namely the capacity to self-consciously reflect and work as a deliberative actor on a social stage. Although the only empirically documented examples of persons are human beings, it is easy to imagine other kinds of entities that have the capacity for personhood. This becomes apparent in science fiction, such as in the worlds shared by *Star Trek* and *Star Wars.* Although they are not human beings, Klingons, Vulcans, and massive worm-like creatures like Jabba the Hutt are all persons in the conceptual sense of the term.

Ossorio's analysis corresponds well with the work of the philosopher Harry Frankfurt. He too emphasized that persons are entities that engaged in second-order recursive self-reflection. He put it as follows (1971):

> Human beings are not alone in having desires or motives, or in making choices. They share these things with the members of certain other species, some of whom even appear to engage in deliberation and to make decisions based upon prior thought. It seems to be peculiarly characteristic of humans, however, that they are able to form what I call second order desires or desires of the second order. (p. 6)

In his analysis of human agency, the psychological theorist Charles Taylor (1985) agreed with Frankfurt that the key concept is a kind of second-order reflective thought, such that one can consider one's beliefs and desires and engage in a recursive looping about what one ought to believe and desire.

More recently, the sociologist Christian Smith (2011) gave a parallel argument regarding what constitutes a person in his excellent work *What Is a Person? Rethinking Humanity, Social Life, and the Moral Good from the Person Up*. Drawing on Roy Bhaskar's critical realism, Smith passionately argued that the social sciences need a clear ontological framework for how human primates become human persons. He identified 30 different capacities of typical persons that he argued set the stage for the emergence of the ontology of personhood. Although some capacities like subjective conscious awareness (i.e., Mind^2) are shared with other animals, many, like abstract reasoning, forming virtues, and making moral judgments, are not. Christian Smith (2011) defined persons as follows (p. 36):

> What then is a person? By person I mean a conscious, reflective, embodied, self-transcending center of subjective experience, durable identity, moral

> commitment, and social communication who—as the efficient cause of his or her own responsible actions and interactions—exercises complex capacities for agency and intersubjectivity in order to develop and sustain his or her own communicable self in loving relationships with other personal selves and the nonpersonal world.

Woven throughout the book is a powerful critique of empirical paradigms in sociology that fail to grapple with the ontology of human persons, which of course parallels the current work's emphasis that mainstream empirical psychology has failed to frame the ontology of the mental. Given our concern regarding the nature of scientific psychology and its relationship with both the natural and social sciences, we should not overlook the fact that the central concern of Ossorio's descriptive psychology overlaps directly with Smith's central concern in sociology. Both are centered on the nature of persons and their behavior.

In making this observation, we can ask: *Do psychology and sociology have precisely the same subject matter?* Thankfully, with its levels and dimensions of behavioral patterns in nature, UTOK directly addresses this issue. Via the ToK System, PTB, and Map of Mind^{1,2,3}, UTOK makes clear that scientific psychology, joined with ethology, picks up from neuroscience on the eighth floor of natural behaviors, and maps the behavior of minded animals. It then joins sociobiology and extends into animal group behavior on the ninth floor. It then moves into human cognitive science on the tenth floor and then human psychology proper (i.e., developmental, personality, and social psychology) at the 11th floor, with its analysis of human mental behavior at the individual level. As the shift in analysis occurs from human individuals to human groups, the baton is passed from human psychology to anthropology into the other social sciences (i.e., sociology, economics, and political science). These disciplines have their focus on human group structure and behavioral patterns, and thus correspond to the 12th floor on the PTB. Via this ontological analysis of behaviors in nature, we can state that the human person is the fundamental unit of the social sciences, and is where human psychology intersects with, and forms the base for, anthropology and sociology.

In addition to clarifying how the scientific disciplines align with various subject matters, UTOK also affords us a clear metatheoretical architecture to specify the causal explanatory dynamics at play in the transformation from primates to persons. Specifically, it enables us to see that humans are primates that follow the principles of behavioral investment and live in intersubjective social-relational fields of influence that are mapped by the process dimensions of the Influence Matrix. In addition, it states that humans are persons who navigate the sociocultural context, framed as large-scale systems of justification. Together, the processes of investment, influence, and justification afford us a system for understanding human mental behavior (i.e., the behavior of human persons). The section that follows aligns these "justification-investment-influence dynamics" with folk psychology.

From Belief-Desire Folk Psychology to JII Dynamics

Folk psychology refers to the process by which people engage in commonsense descriptions of human activity and use everyday language and cultural convention to explain why people do what they do and what should follow from those actions. The most basic, intuitive, and highly useful approach can be framed in terms of "belief-desire" psychology. That is, in everyday situations, people use the concepts of beliefs and desires to explain why they do what they do. Broadly speaking, beliefs refer to what the person consciously knows, sees, and understands, whereas desires refer to what the person wants to happen or wanted to have happen.

Folk psychology is a highly useful frame for giving accounts of human mental behavior. Consider that if we were to observe "Mark" leaving the house to go to the movies, that action would be framed by assuming that Mark *believes* that the movie theater is playing a certain movie at a certain time in a certain place and that seeing the movie is something he *desires* to do. We can be flexible with this frame. For example, we can shift the content of our explanations depending on whether Mark is headed out because he wants to see the movie or he wants to critique the movie for his job or he wants to be with his girlfriend.

In comparison to major psychological paradigms like psychoanalysis and radical behaviorism, UTOK is much more closely aligned with folk psychology and its belief-desire constructs. In fact, unlike these two approaches, with its theories of investment, influence, and justification, UTOK provides a set of constructs that readily transposes with folk psychology. For example, if Mark is going to the movies, both commonsense psychology and UTOK frame this as a form of invested work effort toward a desired outcome. In addition, terms like power, love, and freedom are common in folk psychological exchange. UTOK also translates verbally mediated beliefs into systems of justification. This is a slight reframe, but nonetheless one that is not far removed from everyday language. It just requires the adjustment that justification is both a kind of verbal act and a description for the propositional networks that coordinate action. The summary point is that UTOK's core metatheories create a "JII dynamic" framework that can recast beliefs, desires, and actions into a much more theoretically coherent picture, while at the same time maintaining the fluidity and flexibility of folk psychology.

To see how this can be done, we can start by empathizing with Mark's first-person subjective experience of being (i.e., his $Mind^2$). BIT's P – M => E learning control theory frames Mark's embodied perspective and includes his felt perceptions (P) of where he is relative to the movie theater, his motives (M) to approach the movie theater, and his energized motions (E) in that direction. We can also employ the Map of $Mind^{1,2,3}$ to posit that Mark has a private narrative for why he is doing what he is doing ($Mind^{3a}$), and we can relate that egoic narrative functionally to the way Mark explicitly shares what he is doing with others in the social environment ($Mind^{3b}$). For example, if Mark wanted to be with his girlfriend and she asked him if he wanted to see the movie, he might say "yes" in the $Mind^{3b}$ space to make her feel good, even if he would not go to see the movie on his own. That would be an everyday example of private-to-public filtering via the Rogerian filter.

In addition to framing Mark's experiences, UTOK's theory of human mental behavior situates these mental events within the domain of $Mind^1$ and the neurocognitive processes of behavioral investment. For example, it is likely that Mark will follow the path of least effort when walking out to his car from his house. That is, he will intuitively calculate the shortest distance relative to the effort necessary to get there. This is because his system of behavioral investment tends to intuitively calculate the shortest path relative to obstacles and risk. Folk psychology simply assumes these processes, but it does not provide a credible causal explanatory framework for them. Indeed, folk psychology struggles with concepts like unconscious mental processes. In contrast, with its six core principles, BIT affords us scientifically grounded reasons why humans, as primates, tend to value safety, territory, food, sex, and higher social status over situations that involve the loss or absence of these things.

In addition to going "down" into neuroscience and biology, and back into our evolutionary history, a JII dynamic view also allows us to go "up" into the sociocultural context. UTOK's metaphysics allows us to see clearly that both viewers like Mark and movies like *Star Wars* exist in a sociocultural context. If Mark were employed as a movie critic, there would exist an entire set of cultural justifications and societal institutional structures that legitimize and support this role. Indeed, as the shift in the last two decades away from formal news organizations makes clear, society can change dramatically. This brings us to the processes of justification, which operate in micro-relational contexts between people, but also function to structure macro-level societal operations and regulate patterns of human behavioral investment. At the macroscopic level of analysis, if no one buys newspapers and there is an endless array of free opinions about movies on the internet, then newspaper companies will not be able to justify paying for Mark's salary, and he will be let go.

In addition to moving "down" to see patterns of investment grounded in animal psychology and biology and "up" to see the context of justification, we can move "horizontally" and examine the matrix of social influence, and the intersubjective dynamics therein. In the current example, social influence processes frame the way Mark was headed to the movies, either via an assignment from his boss or a desire to be with his girlfriend. Switching to the microscopic social level of interpersonal interaction, if Mark gets a call and his editor asks if he has completed the write-up, there is a shared, if implicit, justification that Mark needs to do what his boss wants. Likewise, shifting the relational context to Mark being on a date, if he explains to his girlfriend, "I saw the first two movies in the series and thought they were great," or says, "I am just happy to be with you, we can see whatever you want," both are "justifications" that frame and legitimize Mark's investment patterns.

Considering the latter, we can analyze the processes by which Mark and his girlfriend decided to go to the movies (e.g., did he ask her, did she ask him, was there any tension in the process, etc.). The power, love, and freedom dimensions on the Matrix show how the dynamics and processes of social influence involve elements such as competition, cooperation, and shifts in relational engagements toward greater dependency and involvement or toward autonomy and separation. When social influence is framed as a resource, it maps on to the RV–SI (Relational Value–Social Influence) line in the Matrix. The harder Mark had to work to get her to go with him the more questions that are raised about the extent to which she values being with him. Of course, if she breaks up with him, that is a powerful indication of a loss of social influence and relational value.

The point here is that UTOK not only aligns with folk psychology, but it goes deeper to address the core structural and functional dynamics at play. In making this point, it is crucial to keep in mind that UTOK was born out of the problem of psychotherapy, which is tasked with operating in the real world and helping real people deal with maladaptive patterns and work toward more optimal living. In the essay *Maggie's Story: The Many Reasons Why Not*, I provide a clinical case example of how UTOK blends core principles of psychotherapy, such as the focus on relationship quality and outcome, with key insights from cognitive behavioral, emotion-focused, psychodynamic, and interpersonal theories to help a young woman struggling with depression, anxiety, and suicidal ideation (Henriques, 2017). In the section that follows, a more personal account is given to show how human mental behaviors can be framed via JII dynamics.

A Lived Example: JII Dynamics in Real Relations

It was late spring last year when my wife, Andee, entered our bedroom after coming home from work and found me emptying my dresser and my clothes all over the bed. "What are you doing?" she asked, in a tone that was simultaneously somewhat curious and carried more than a hint of annoyance. "I got tired of the clutter and am getting rid of the stuff I do not wear anymore," I replied, with a degree of defensiveness that rose to match her edge. "You remember the dishes?" she asked. I paused for a second or two, searching my memory. Then, upon locating the needed image, I said, with a slight sigh and turning away, "Yes, I remember." Another brief pause as she gave me a look I knew well. Then, in a neutral, slightly distancing tone, I said, "I just got pulled into this. I will get to them." She responded with a quick and sharp, "Fine," and left the room.

This everyday exchange between two well-acquainted individuals highlights several important features regarding the relational world (i.e., our Mind^{2b} space), social influence and relational value, and how people justify their actions, both publicly and privately. To see these dynamics clearly, let me set the stage to provide the micro-relational context for our justifications and actions. It was a Friday afternoon, and my normal routine was to work and write until about noon and then go for a walk and proceed to tidy up the house. This routine had become a bit of a ritual, and it carried some notable elements of care and investment, which Andee appreciated. Another important piece of information that contextualizes the exchange is what happened between us earlier that morning. Just before leaving for work, Andee started putting the dishes in the dishwasher, as it is a pet peeve of hers if they pile up. I came into the kitchen and told her not to worry about it, and that I would be happy to take care of it.

With this contextual backdrop, we can more clearly see the recursive relational dynamics that were being played out in our exchange. The Mind^{2b} space that emerged between us in the bedroom starts with Andee coming home from work and finding the dishes in the exact same place she left them. She experienced the mess in notable contrast to the rewarding feeling of care a clean sink would have given her. This discrepancy resulted in a jolt of irritation tied to an underlying relational sense of being either neglected or ignored. Using the map of self–other processes provided by the Influence Matrix, we can see that this experience activated an upper right quadrant self-over-other response, meaning that she felt disrespected by my neglect and then was charged with anger toward me. Seeing the mess in the bedroom added to that attitude, and thus prompted her to question me with an edge.

According to JUST, a central feature of human language is that persons can ask for and be pressed to give accounts for our actions. Doing so, especially with an edge, is often experienced as a form of dominance. In the language of JII dynamics, we can say that Andee's requiring me to give an account for my actions, coupled with her tone, quickly created a "power dynamic" between us (i.e., the blue line on the Matrix). Put differently, because she felt disrespected seeing the dishes left in the sink, she entered the room with a "self-over-other" framing that my relational recursive relevance realization system quickly picked up on. Because I can be absent-minded, I had forgotten about the dishes. As such, I had no expectation that she would be annoyed with me, and thus I was caught off guard and reacted with some defensiveness.

As her comment about the dishes sparked recognition of my oversight, I quickly found my experiential theater of awareness (i.e., my Mind^{2a}) flooded by a mixture of emotions and images, which were quickly accompanied by Mind^{3a} egoic narratives and defensive justifications. We can use the learning control P – M => E formulation of perceptions, motivations, and emotions coupled with recursive relevance realization to see why. First, she caught me off guard, meaning I was surprised and my predictive processing about what was happening needed conscious reflection. This was the pause I engaged in to process what was occurring. Upon scanning my episodic memory and finding the relevant exchange of the promise about the dishes and the fact I had forgotten, I was then flooded by awareness of the appropriate relational frame for what was transpiring in the Mind^{2b} space between us.

As this awareness emerged, my relational system was processing the emotional valence relative to my motivated state of being. Her valence was clearly negative. My desired state is for her to value me, so experiencing her as being annoyed with me created a discrepancy that energized a response. Interestingly, and directly consistent with the self–other quadrant structure of the Matrix, two different and opposed feeling states and impulses emerged. One set was "other-oriented" and comprised regret and guilt. As soon as the dishes flashed in my mind, I knew instantaneously that we both would have been better off if I had not gotten off track and had remembered and cleaned the kitchen. As this set of images and associated feelings ran through my mind, it pulled forth a justification narrative that had me apologizing and saying that I meant to kick off our Friday evening on a more positive note. This set of motives and emotional responses is represented by the lower right quadrant of the Matrix, which is a relational frame marked by dependency, submission, and affiliation, and feelings of guilt, shame, and love. If I had been made aware of the dishes initially and the task was done with care, I likely would have expressed this apologetic sentiment without much hesitation.

As has been noted, the Matrix also includes a set of "self-oriented" motives and feelings in the upper left quadrant. Andee's initial and unexpected edge primed me to be defensive, and this defensive irritation activated another set of thoughts organized by the emotion of anger. This stream of images and justifications took the form of a private narrative along the lines of the following: "Here I am, a full professor, working hard on a book laying out a new vision for psychology. I have taken much of the afternoon doing three loads of laundry, and I am now tidying my bedroom and organizing my clothes. And my wife is pissed at me because I also failed to do the dishes? This is ridiculous." Notice how this narrative activates a completely different set of self–other justifications having to do with power and place (i.e., blue line dynamics). In this social-relational space, my actions are legitimized, and it is her criticisms that are misplaced.

The contrast between the two narrative streams is quite striking. One emphasized my error and regret, which oriented me to defer to her interests and apologize and move toward her via guilt and affiliation. In contrast, the set of self-defense relational frames oriented me to dismiss and defy her claims and assert my power. The divergent streams highlight the complicated dialectic between other-oriented and self-oriented strategies for social influence and relational value. The example also highlights the role and connection between social motivations and emotions and various justification narratives. When tinged with guilt, I became oriented toward how I overlooked an easy act of love that would have been well received. In that space, I want to make amends and apologize. However, when feeling angry and defensive, power and autonomy themes that involve resentment about being controlled and already giving too much gave rise to an opposing set of justifications.

All of this happened quickly, and the whole exchange probably took less than 15 seconds. Nonetheless, we can apply the JII dynamic lens from UTOK and deconstruct the recursive relational processes with a high level of resolution. My "yes" to her query about the dishes really meant, "now I recall." The guilty-affiliative-defer line of thought was quickly tangled up in my mind with the angry-power-defy line of thinking, but neither dominated, and what emerged was a compromise formation in the form of inaction. The "I just got pulled into this" functioned to diminish my responsibility by emphasizing forces in the environment. It conveys the idea that I did not intend to not do the dishes. The neutral, concluding, "I will get to it" signaled my intent to do what I said and to resolve the issue. But because I was defensive, I did not offer an explicit apology.

The tenseness lingered in the air for a while. I went down and did the dishes. I took several deep breaths, and I was able to see my desired outcome for the evening. I then mentalized her experience, seeing the world through her point of view. In a short time, I went upstairs, apologized for not doing the dishes and explained why I reacted with some defensiveness. She reminded me that it was a "pet peeve," said it had been a long day, and appreciated the work I had done around the house. In so doing, we managed to restore our relational equilibrium for the rest of the day.

The Social and Political Implications of Mind[3] as the Justifying Mind

The above analysis shows how we can use JII dynamics to track human mental behavior in a relational context that aligns with folk psychology and deepens it via connections to scientific psychology and findings from psychotherapy to give a rich and nuanced account of how such processes unfold. The UTOK metapsychology posits that we can also use JII dynamics to move up the human aggregate scale into social psychology and political processes. We can see this by turning to the work of Jonathan Haidt, a social psychologist who has developed powerful analyses of the political polarization that has gripped the United States in the past two decades. Consistent with the JII dynamic formulation, Haidt's (2012) excellent book *The Righteous Mind: Why Good People are Divided by Politics and Religion* characterized the human mind via the metaphor of an elephant and a rider. Directly paralleling JUST's formulation that the human ego functions more as a motivated lawyer than a disinterested scientist, Haidt reviewed the literature on reason-giving and concluded that humans generate attributions and explanations more like politicians seeking re-election than analytic reasoners.

Not only does Haidt's analysis of the rider align directly with JUST, his analysis of the elephant (i.e., the underlying animalistic motives and values that frame primary process cognition) aligns strongly with BIT and the Matrix. He reviewed how decades of research in social psychology have shown that: (1) humans are obsessively concerned about their reputations, even though they are often unaware of how much that concern influences them; (2) when humans engage in conscious reflection to generate reasons for their beliefs and actions, the relationship can be framed as that of a press secretary that justifies the position of the president; (3) humans often lie and bias facts and then deceive themselves about the extent of the data manipulation; (4) reason-giving can result in almost any conclusion and is heavily biased toward belief-narrative construction that accords with the primary processes; and (5) people tend to be "groupish" rather than "selfish" in moral and political matters and engage in reasoning that often functions to demonstrate loyalty to the team they are on.

Haidt advanced these findings via analyses that frame how humans engage in political, social, and moral reasoning. *The Righteous Mind* identified five "moral tastes" that he argued drove much of the discourse regarding social and political values and debates. They were: (1) care/harm; (2) fairness/cheating; (3) loyalty/betrayal; (4) authority/subversion; and (5) sanctity/degradation. In later analyses, he collaborated with colleagues and added a sixth moral taste of liberty/oppression (Iyer et al., 2012). The primary political concerns that he claimed were associated with each are: (1) protect and care for children; (2) benefit from reciprocity; (3) form cohesive ingroup relations; (4) forge beneficial relations in hierarchies; (5) avoid contaminants from impure outside forces; and (6) create the freedom from control and obligation.

From a UTOK perspective, we can see that the social motives and moral tastes that Haidt highlights can be framed via the lens of the Matrix. The first two moral tastes directly align with positive "red line" concerns, pertaining to affiliating with kin and reciprocating with friends and others, and creating a context that affords and stabilizes those processes. The third taste bridges from the red line of affiliation with an ingroup to a more general process of social influence, which is found in belonging and group identification and the pressure to be a good contributor (i.e., to have and receive social influence as a valued member of the ingroup). The fourth moral taste is clearly a blue line dynamic, in that it relates to structuring power relations to achieve functional stability within a hierarchy. This is what respect for authority and dominance is about. The fifth taste is a negative red line motive, meaning it pertains to allowing for contempt and hostility against things that threaten the sanctity of the group (i.e., the group comes together to defend against the world of the outgroup). Finally, the sixth taste is directly aligned with the green line and levels of freedom relative to obligation and control by the governing authority.

It should be noted that the Matrix would slightly shift Haidt's analyses by pointing first to the black line. That is, much like the more general relationship between taste and eating (i.e., eating is ultimately about calories and nutrients, which tastes signal), the Matrix posits that we should first identify the general factor governing social exchange, which is the need for social influence and relational value. Akin to calories relative to taste, the RV–SI line provides the core backbone that situates the self–other process dimensions of power, love, and freedom. In addition, with its broadband differentiation of the self- and other-oriented quadrants, the Matrix also provides a readily available frame for understanding why political processes are so often divided into two parties. Specifically, and directly consistent with the section on the origin of gender roles analyzed in the previous chapter, we can see that the party of "care" and "fairness" is archetypally feminine and concerned with governance structures that emphasize equality and the redistribution of wealth and power toward those in need. This emphasis exists in dialectical tension with a party of independence and an appreciation of and loyalty to authority in a competitive hierarchical structure, which is more archetypally masculine.

A primary motivation for Haidt's analysis is to foster a better understanding of the political divide in the United States. More specifically, Haidt was coming from the left side of the political continuum and was concerned with the increasing political polarization in the country and the academy. His research showed that a major difference between the political orientations was found in the extent to which they valued the various moral tastes and emphasized their role in governance. Specifically, he found that those oriented on the left side of the spectrum primarily valued the first two moral tastes in political contexts, whereas those on the right embraced all the moral tastes more equally. He attempted to use that frame to help those on the left be more empathetic to the logic and values that energized the conservative position. In addition, he emphasized the justifying, self-righteous set point of the rider (i.e., Mind^3) in playing a key role in driving the two political poles toward a destructive polarization. This social psychological analysis thus moves us from individual human psychology on the 11th floor on the PTB into the social sciences proper and 12th floor analyses, which we elaborate on in more detail in the next section.

Defining Culture and Developing an Ontological Perspective on the Social Construction of Reality

The jump from social psychology into the social sciences proper brings us to one of the most important but also most nebulous concepts in science and the humanities, the definition of culture. In the nineteenth century, E. B. Tylor (1871) famously characterized culture as "that complex whole which includes knowledge, belief, art, law, morals, custom, and any other capabilities and habits acquired by man as a member of society." In the twentieth century, social scientists of all stripes attempted to define the term with precision. In their well-known review of these attempts, which included more than 160 definitions, Kroeber and Kluckhohn defined the term in 1952 as follows (p. 357):

> Culture consists of patterns, explicit and implicit, of and for behavior acquired and transmitted by symbols, constituting the distinctive achievement of human groups, including their embodiments in artefacts; the essential core of culture consists of traditional (i.e. historically derived and selected) ideas and especially their attached values; culture systems may, on the one hand, be considered as products of action, on the other as conditioning elements of further action.

Although Kroeber and Kluckhohn did the field an important service by delineating the difficulties and diversity of approaches associated with defining culture, they failed to generate a universally agreed upon conception of the term. More recently, Gustav Jahoda (2012) reviewed the literature and concluded the following (pp. 299–300):

> More than half a century after Kroeber and Kluckhohn, and a literature that could easily fill a sizeable library, the most striking feature of these definitions is their diversity. While some are based on classic formulations, others are newly invented. Moreover, many of them are logically incompatible with each other. Here are a couple of examples: 1. the supposed location of culture is variously said to be (a) only in the mind or (b) both in the mind and in the material world created by humans; (c) external only (without specifying where). 2. (a) culture is treated as a 'variable' by tough-minded advocates of measurement, while (b) others maintain that such a position entails a misconception of what constitutes culture. Some writers explicitly state that their own definition is the correct one and others are wrong....
>
> In sum, the concept of 'culture' is probably indispensable, yet there is no way of escaping Alfred Lang's (1997, p. 389) conclusion 'that attempts at defining culture in a definite way are futile'. So what, if anything, can be done? My answer would be that much of the time it is quite practicable and defensible simply to use the term without seeking to define it. However, if either for a theoretical or empirical reason clarification is essential, then the author should explain the specific manner in which she employs the term 'culture' in that particular context.

Modern cultural anthropology has largely followed this advice. Many texts provide a general summary of the concept, but back away from attempting to clarify the ambiguities. Consider, for example, Mark Sutton's (2021) recent definition of culture in *A Concise Introduction to Cultural Anthropology* (p. 26):

> Simply put, culture is learned and shared behavior in humans. All animals, and even some plants, also have learned and shared behaviors but here we only consider humans. Humans have only minimal instincts, such as self-preservation, reproduction, and being maternal. Thus, virtually all of human behavior is learned: what one likes, how one thinks, what language one speaks, one's beliefs, one's values, one's biases, what is good to eat, how one views the world, and so on. All of these traits are socially transmitted, learned from other members of one's society. Everything one experiences is filtered through the lens of their cultural background. It forms the basis for the generation of appropriate behavior, which is defined by the society. In

essence, then, culture is integrated into all social systems and forms one's view of reality.

As is the case for empirical psychology's concept of "behavior and mental processes," this definition is deeply problematic. Thankfully, the descriptive metaphysics afforded by the ToK System provide us a vision logic to both understand why scientists have had such difficulty defining culture and to see what the solution is. The problem is very similar to the problem of psychology, and the key is to understand how human culture is both simultaneously continuous with and discontinuous with the cultural practices of other animals. And to do that, we need to divide culture into two different domains corresponding to the Mind-Animal and Culture-Person planes of existence.

First, there are learned and shared mental behaviors, which we can call patterns of behavioral investments or behavioral repertoires, which would involve procedural and participatory knowledge capacities. For example, learning how to tie one's shoes is a procedural knowledge skill. Developing these procedural skills and sharing them among a group is a key feature of culture. However, such learned activities are properly construed as taking place on the Mind dimension. It is thus not surprising that other animals have exhibited such patterns. To give just one example, an extensive review of behavior patterns in chimpanzee communities found "39 different behavior patterns, including tool usage, grooming and courtship behaviors, [that] are customary or habitual in some communities but are absent in others where ecological explanations have been discounted" (Whiten et al., 1999, p. 682). These are important patterns that play a key role in the lives of many social animals, and they should be considered phenomena that take place at the ninth level of behavioral complexification on the PTB.

We can call these examples of learned, shared procedural behaviors small "c" culture. They are apparent in many animals and play a major role in human culture. However, as mapped by the ToK System, there is a whole separate dimension that makes up the capital "C" Culture-Person plane of existence. As we have seen, a qualitative shift in complex adaptive behavior is made via language, propositional knowledge, recursive self-conscious reflection, and the emergence of large-scale systems of justification. The problem of justification gave rise to reason-giving and thus reasoning, which afforded much greater cognitive flexibility across expansive timespans. Moreover, the capacity to generate, store, and transmit what Leigh Shaffer called "recipe knowledge" across populations and generations meant that an explosion of technological advances could be shared and maintained (Shaffer & Shaffer, 2007). In short, JUST and the ToK System help us see the nature of the Culture-Person plane and why it generated an accumulation of cultural knowledge that is so different from the small "c" cultural repertoires of animals. This distinction between Culture and culture becomes clearer when we shift into how scholars have analyzed and interpreted the social construction of knowledge.

The Social Construction of the Culture-Person Plane of Existence

In their classic text, Berger and Luckmann (1967) lay out a powerful analysis of the sociology of human knowledge. In it we can see that they are concerned not with learned and shared behavior repertoires per se, as might be found in various groups of chimpanzees. This is because for Berger and Luckmann social knowledge in human groups *presupposes* language and propositional knowing. Chimpanzees do not construct propositional knowledge systems, but operate in the world via perspectival, procedural, and participatory knowledge. In contrast, human persons develop propositional networks that provide explicit semantic representations of is and ought. Through the vision logic of UTOK, we can see clearly that the social construction of knowledge refers to the shared, interlocking network of justifications that human groups generate to make coordinated sense of the world and legitimize action. It is the Culture-Person plane of existence, coupled to the evolution of technology, that makes human culture so different.

As Tylor (1871) pointed out in his analysis of culture as "that complex whole," this domain of complexification includes human knowledge about the world. The fact that human propositional knowledge can be thought of as residing inside of human culture creates some significant philosophical conundrums that have divided and confused both modernist and postmodern thinkers alike. In *Social Constructionism*, Vivien Burr (2015) provides an overview of what social constructionism is and how it gives rise to an alternative for psychology grounded in a postmodern approach to epistemology. She noted that rather than consisting of a clearly defined set of claims, axioms, or propositions, the perspectives that constitute social constructionism are best framed as a kind of "family resemblance" that tends to cluster around a set of key assumptions.

Burr (2015) began by highlighting how social constructionism adopts a critical epistemological stance toward the conventional "taken for granted" assumptions of the mainstream worldview. In contrast to the positivist or modernist assumption that one can observe and interpret the world simply as it is, social constructionist viewpoints assume that metaphysical constructs about the world obtain their validity from groups of people that work together to develop an agreement about what is real or useful, rather than any knower-independent facts that are simply available for observation. She used the example of pop versus classical music to highlight that what might appear to a naïve cultural participant to be a category in the world in fact emerges from the norms, shared agreements, and unique historical trajectories of particular people at a particular time. From this example she claimed that the same analysis can be applied to domains like human personality, sex and gender, and the concept of what it means to be healthy. Such is the primary focus of many scholars working in the social constructionist tradition.

Burr (2015) then shifted into some of the core affirmative claims made by the social constructionist perspective. She began by emphasizing how human knowledge resides in a historical context and is constituted by that process (p. 30):

> Not only are [such knowledge frames and claims] specific to particular cultures and periods of history, they are products of that culture and history, dependent upon the particular social and economic arrangements prevailing in that culture at that time. The particular forms of knowledge that abound in any culture are therefore artefacts of it, and we should not assume that our ways of understanding are necessarily any better, in terms of being any nearer the truth, than other ways.

She then elaborated on the claim that human knowledge is shaped and sustained by social processes, and that knowledge is intimately tied to social action. She then challenged the historical assumption that language is simply a vehicle of thought and argued that most social constructionists consider language to be constitutive of thought. The focus is not on how words serve as arbitrary signifiers, but rather how sentences, paragraphs, arguments, and texts construe and ultimately frame and come to constitute how people think about what is real and good.

As is hopefully apparent, UTOK's metapsychology readily embraces the basic insights from the social constructionist perspective. However, it recasts the perspective in a radical new light. Via JUST and the descriptive metaphysics given by the ToK System, UTOK gives an ontological perspective on the social construction of reality. That is, the key insight of social constructionism as framed by the postmodern sensibility can be framed as stating that language-based propositional thought emerges to create an intersubjectively shared cultural context that frames human knowledge and action. As Burr notes, the focus of many social constructionists has been on the epistemological and critical implications of this point. However, via JUST and the ToK System, UTOK explicitly frames the ontological emergence of the human capacity to construct propositional knowledge systems.

The Socio-ecological and Developmental Context of Justification

The vision logic of UTOK's Culture-Person plane enables us to see the socio-ecological context in which humans are socialized as they develop and learn to operate as persons. Bridging this to the work of Uri Bronfenbrenner (1979) affords greater clarity on how this works. He developed a structural map of the socio-ecological context that aligns closely with UTOK's on human mental behavior. He identified three major levels of aggregate structure that can be useful in parsing how the Culture-Person plane operates across different scales. We have already encountered these frames, as I wove them into the account of human mental behavior via a JII dynamic lens. That I was able to seamlessly incorporate this perspective demonstrates how Bronfenbrenner's structural analysis of the sociocultural context has already been adopted by many who analyze the sociocultural influences on human mental behavior (see, e.g., McGoldrick et al., 2015).

The largest aggregate sociocultural scale is the macroscopic level of norms, values, religious ideologies, and other cultural codes that shape the broad context. In the previous example, this is the context that gives rise to whether movie critics are valuable and the roles that the genders take in going on a date. Because the macro-level cultural context is so pervasive and enveloping, it can be difficult to see from within it. However, when one steps into a different macro-level context, its relevance and assumptions will become much more apparent. For example, several years ago I traveled to Costa Rica for a couple of weeks. In the context of a conversation with a Costa Rican, I identified as being an American. The response I received was "In the Western Hemisphere, we are all from America, and its North and South continents. You are from the United States." In my macroscopic cultural bubble, America and the United States are essentially synonymous. However, from the vantage point of someone in Costa Rica, that is reflective of an insensitive nationalistic bias that is perceived by many in Latin America to be a prominent feature of people from the United States.

One level down is the meso level, which refers primarily to the community and status dimensions pertaining to issues of class. Bronfenbrenner (1979, p. 25) technically defines the mesosystem as comprising "the interrelations among two or more settings in which the developing person actively participates (such as, for a child, the relations among home, school, and neighborhood peer group; for an adult, among family, work, and social life)." To put this in plainer language, we can say that whereas the macroscopic level refers to the sociocultural context of, say, a city, the meso level refers to where in the city the person lives and goes to work or school. Arguably the most salient and general variables that specify the dynamics that operate at the meso level of analysis are those that go into one's socioeconomic status (i.e., levels of wealth, education, and the nature of one's occupation).

Whereas the meso level refers to which part of the city one lives in, the microscopic level refers to the specific house one lives in and the classroom or specific office in which one participates in school or work. That is, it refers to the actual relationships and the exchanges between the participants. In the previous section on JII dynamics, this is the domain of the relationship between Mark and his girlfriend or me and my wife. In sum, human mental behavior resides in socio-ecological spheres ranging from the broadest macro-level context where customs, values, roles, and norms function as the large-scale justification systems that coordinate the population, to the more intermediate range of community-level influences, such as local cultural tones and socioeconomic status, to finally the micro-level relational environment consisting of the individual's family and friends.

In addition to these three levels, Bronfenbrenner added two more elements. One was the "exosystem." He defined the exosystem as "settings that do not involve the developing person as an active participant, but in which events occur that affect, or are affected by, what happens in the setting containing the developing person." Here we will use UTOK's descriptive metaphysics to slightly redefine Bronfenbrenner's concept of the exosystem and to refer to the material cultural artifacts that structure how human cultures are physically instantiated in the world. That is, the exosystem refers to the technologies that groups generate to structure their lives and control the flow of resources. As such, the exosystem consists of the buildings, roads, tools, and other objects that constitute the physical aspects of human society. This allows us to specify how UTOK frames human societies as consisting of assemblages that are constituted by: (a) the small "c" patterns of behavioral investment that are learned and shared; (b) the capital C systems of justification that coordinate the behaviors of persons and can be structurally framed by the macro, meso, and micro levels of analysis; (c) the material cultural artifacts that make up the exosystem; and (d) the biophysical ecologies in which the societies are anchored.

Finally, there is the time and history context across the levels of analysis, which Bronfenbrenner called the chronosystem. He conceived of the chronosystem as the major life transitions, environmental events, and historical events that occur during development. This can be framed in terms of both the events that influence a specific individual (e.g., a person being deeply influenced by the moon landing or WWII) and in terms of the historical evolution of the culture itself. In the final section of this chapter, we use UTOK's framework to zoom out as far as possible and engage in a chronosystem analysis of the Culture-Person plane of existence across the long arc of its history.

A Brief Functional Taxonomy for Framing the Context of Justification

Whereas Bronfenbrenner's work is useful for specifying the basic structure of the Culture-Person plane across scales and time, JUST enables us to analyze the sociocultural context of justification via a more explicitly functional lens. The section on the JII dynamics of human behavior made clear how UTOK can track the relational processes between people such that we can place how they demonstrate functional awareness and responsivity in a clear theoretical framework. However, JUST also allows us to see the larger sociocultural context via a functional perspective. To see how, it is useful to recount an exchange between John Vervaeke and Guy Sengstock, as they were exploring the implications of JUST and what it means to be a person (Henriques et al., 2020).

In discussing how important the social context of justification was for development and socialization as persons, Vervaeke and Sengstock identified two central sociocultural contextual frames that had different functional dynamics in framing the processes of justification. One context they considered was the "courtroom." This is the functional context whereby one is being challenged by another or a social group to defend one's actions or else face sanctions. The example I offered of the exchange with my wife can be framed as a micro-relational courtroom dynamic. Indeed, the fact that our social lives can quickly turn into the functional context of the courtroom is crucial to the logic of JUST. JUST posits that the courtroom dynamics of social influence is one of the primary reasons that we have psychodynamic defense mechanisms. Here we can interpret "defense" as referring to the need to be able to justify our actions on the social stage and defend our position or status in the social influence matrix of relational exchange.

The conversation between Vervaeke and Sengstock was part of a larger project that involves developing an ecology of practices that foster *dialogos* (Vervaeke & Henriques, 2020). Dialogos refers to a form of deliberate, reflective dialogue that cultivates deep interpersonal relating for existential insight. According to Vervaeke, dialogos can be framed as a call to return to the great Socratic tradition of philosophy as the practice of dialogical engagement to foster well-being, self-realization, integrity, and the exploration and cultivation of wisdom. Given the goal of dialogos (which itself can be framed as attempting to create a new sociocultural context of justification), Vervaeke and Sengstock shifted the metaphor of the context of justification from "the courtroom to the courtyard." The courtyard here can refer to the social commons, the process by which groups of individuals come together in a cooperative endeavor to cultivate a shared narrative for the group and its interests, understandings, and activities. In this functional context, communication is not framed to delineate status or dole out sanctions, but to generate a collective understanding. In short, it is where people come together to explore and generate shared systems of justification, as well as little cultural practices.

We can extend this analysis and "bookend" it by identifying two other functional contexts of justification. The first is the "con game." Here the functional context is one of deception and deliberate efforts at lying and misdirection for the purposes of manipulation. The broader grammar of justification here is the lie, whereby conscious deceit is used to give the appearance of truth in the service of some other gain. The other context which moves to the opposite side of the continuum is the scientific research lab. The idealized goal of the scientific research lab is objective truth claims. That is, the goal of science is to develop experiments based on quantitative measurement and the manipulation of variables that yields clear conclusions. Theoretically, such experiments are done "blind" in the sense that the goal is to factor out, as much as possible, the biases of the experimenters and yield results that generate deductive or inductive conclusions that can be derived from logic and evidence as truth statements about reality that exist independently from the preconceived notions and desires of the populace.

The result of adding these two functional contexts is a continuum that can be helpful in framing the contextual dynamics of justification that drive the social operating system. By bookending the courtroom and courtyard with the con game and the research lab, we can lay out these four different contexts of justification in the continuum depicted in Fig. 15.1.

It is helpful to recall that the problem of justification is framed by three different aspects. There is the problem of accuracy in determining what is the case, along with the problems of personal and social interests. We can see this functional continuum in relation to these three aspects because each describes a different relationship between the personal and social motives and the truth. In the case of the con game, the perceived reality

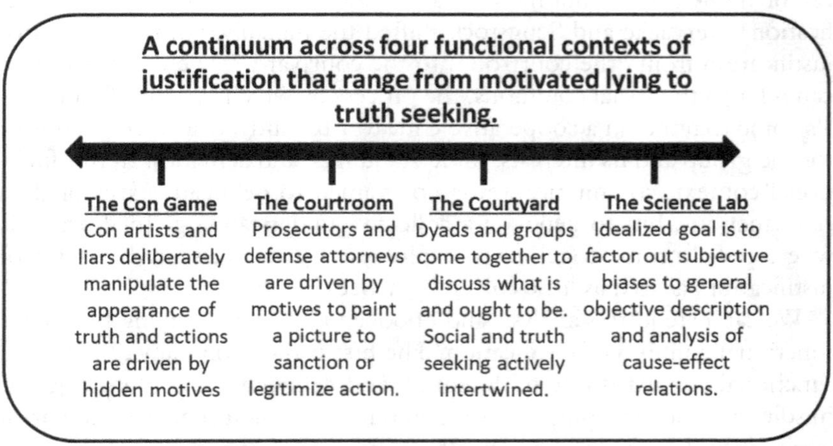

Fig. 15.1 A continuum across the context of justifications from lying to truth seeking

of the victim is deliberately manipulated and distorted based on hidden motives, so that the social context is completely at odds with the personal goals and there is a negative valence for accuracy. In the courtroom, we see that the personal and social are again opposed, but there is a shared understanding of the context and a desire for accuracy. In the courtyard, social and personal motives are mingled with pragmatic truths to coordinate human activity. In the lab, in theory at least, the personal and social motives and desires are attempted to be factored out to generate truth claims about how the world is or truly works. I should be clear that this continuum is not offered as a complete taxonomy or analysis of the functional context of justification. Rather it is simply an example that illustrates how JUST is situated to frame both the structure and the functional organization of the Culture-Person plane of existence.

A Brief History of the Culture-Person Plane of Existence in Four Phases

The vision logic of the ToK System enables us to clearly see the nature of the Culture-Person plane of existence and that it accumulates complexification over time. Given that it is placed on a time axis, we can think about the phases by which it has developed. To accomplish this, we can sync up UTOK with the perspective given by Lene Rachel Andersen (2019) in *Metamodernity: Meaning and Hope in a Complex World*. In this work, the author traces the evolution of the Culture-Person plane of existence across four sensibilities or cultural codes: (1) Premodern Oral/Indigenous; (2) Premodern Traditional; (3) Modern; and (4) Postmodern. As will become clear, Andersen's tracking of these cultural sensibilities can be framed, structurally and functionally, as large-scale systems of justification. Moreover, a review of these four sensibilities sets the stage for potentially understanding an emerging metamodern sensibility that could arise in the twenty-first century. In laying this out, it is important to note that this is not framed either as a hierarchy of progress or something that represents an inevitable shift toward complexification. Rather it represents a zoomed-out view of four different cultural sensibilities that have heuristic value in understanding the sensibilities that are shaping our current justification systems.

The Oral/Indigenous Cultural Code and Sensibility

The archeological record clearly shows that by 50,000 years ago a remarkable transition in human Culture and self-consciousness was well under way. In *Origins of the Modern Mind*, Merlin Donald (1991) made a compelling argument that the emergence of this pattern was associated with a transformation in human communication, such that our hominid ancestors went from a fractured "mimetic system" into an open language system with syntactical rules that enabled the emergence of abstract reasoning, symbolic thought, and the ability to pose questions. This latter stage is what Donald refers to as "mythic" knowledge, which can be defined as the linguistically mediated, intersubjective, sociocultural construction of knowledge. Framed by JUST, it is the question-answer tipping point that arises with propositional language that generates the complexity building feedback loop that results in a shared mythos in the form of narrative justification systems that coordinate people.

Many aspects that separate us from the other great apes, such as language, music, animism, shamanism, art, and the creation of complex tools, are shared across all human indigenous cultures. During this period, all justification systems were orally communicated. At the communal level, stories would be shared, and the identity of the group and its members were embodied in practices such as rituals, body ornaments, song, and dance. In terms of cosmology and worldviews, hunter-gatherers tend to view nature as animated by spirits, and see humans as an integral part of nature. In addition, indigenous societies are generally nomadic and do not amass more artifacts and tools than can be transported. Importantly, this means that the members do not differ much in material wealth.

In *Sand Talk: How Indigenous Thinking Can Save The World*, Tyson Yunkaporta (2019) shared the key features of his Australian aboriginal oral tradition and attempted to apply some of the principles to modern society. The book is structured around "yarns," which are meaningful conversations about life and the state of the world that enable lessons to be drawn and patterns of living to be extracted and woven together to create a tapestry of understanding. He also utilized the Aboriginal dual-person language frame that he translates as "us-two," which enables him to form a kind of kinship pair with the reader. This explicit second-person viewpoint highlights how conversation allows for a shared intersubjectivity and joint attention between human minds that is central to the oral traditions. It is a style that many argue is relatively underdeveloped in our current way of life that we would do well to rediscover.

Yunkaporta emphasized how relationships, a deep appreciation for context, a frame of embeddedness in relations, nature, and across time, as well as an emphasis on cooperation and sharing rather than competition and acquisition define the key features of the oral-indigenous sensibility. In an interview laying out his perspective in *The Alternative* (2020, July 18), Yunkaporta narrated how society would be different if the key elements of the oral tradition were woven back into modern lifestyles:

> For a start there would be a lot less angry people yelling opinions into a void of cyberspace. People would have time to connect and think deeply, grounded in profound relation to the land and adapting to changing environments in cooperative communities... Children would be embedded in community and work spaces rather than locked up during the daylight hours. There would be no struggle for work/life balance, because these things would not be separate domains... Economies would be recession-proof and welfare would be unnecessary because families and communities would be structured to support every person as needed.

Of course, modern worldviews have shifted quite dramatically from those adopted by indigenous cultures. However, Yunkaporta makes a compelling argument that is crucial for modern peoples to realize the wisdom in the oral-indigenous sensibility. To see the world via this lens, consider one's family and one's personal and intimate relations, and the way of being such face-to-face and day-to-day interactions cultivate. Consider how embedded and familiar one is in those contexts. Now imagine that one is also embedded in nature and develops a tradition of thinking this way writ large.

As Darcia Narvaez and her colleagues (2019) explain in *Indigenous Sustainable Wisdom: First-Nation Know-How for Global Flourishing*, "the ancient world view considers the cosmos to be 'unified, sacred, and moral.' Communities who live in this worldview are connected to the lifeways of this particular landscape, and they promote flourishing of the local biocommunity, and the view that the human species is just one member of that community; one member among many members" (p. 4). In her work, Andersen (2019) also emphasizes the connectedness and embeddedness of the oral indigenous sensibility and encourages us to reflect on ways in which this can be reclaimed and enhanced in our lives going forward.

The Traditionalist Cultural Code and Sensibility

Approximately 10,000 years ago, we see the beginnings of a significant shift in human mental behavior and, especially, its societal products. In the Middle East and gradually elsewhere, over a period of several thousands of years, many human groups transitioned from hunter-gatherer and horticultural lifestyles into a fully agrarian way of life. One account for why this transition might have occurred was offered by Aslan (2017) in *God: A Human History*. He puts forth the idea that it was humanity's search for meaning and the idea of God (or gods) that drove the construction of temples. As these temples grew in scope and complexity, they required longer stays at the site, and it was this need that resulted in a shift from a more nomadic and transient existence to one that was much more stationary. One such example is the archeological site at Göbekli Tepe. Such temples required the effortful alignment of investment practices, technology, and meaning-making over long stretches of time in one place. Although other interpretations are plausible, his arguments nevertheless highlight how the functional dynamics associated with the problem of justification can drive the evolution of the Culture-Person plane and the technologies associated with it.

Most historians identify Sumer, which emerged in Mesopotamia between 5500 and 4000 BCE, as the first great human civilization (Bertman, 2005). Other civilizations appeared around the same time in Ancient China, the Indus Valley, and Ancient Egypt, such that by 5000 years ago the Bronze Age had fully emerged. The consequences for human living and lifestyles are hard to overstate, as life in the city is remarkably different than the nomadic world of the hunter-gatherer. In contrast to being fundamentally embedded in nature, city living means a profound separation from nature. More and more activities take place in the technological bubble of material culture. Relationships also change radically. Instead of being on personal terms with virtually everyone, city dwellers are constantly engaged in exchanges with strangers. Moreover, civilization meant a fundamentally different mode of production and control, such that humans began accumulating resources and wealth. The combination of the loss of personal bonds and the dramatic increases in material wealth and technologies such as money resulted in a massive shift from the largely shared power into hugely differentiated hierarchies.

Living with strangers, massive power differentials, specialized work forces, and complicated trading practices required new ways of regulating the increasingly complex social reality, and thus shifts were required in the basic structure of the Culture-Person plane of existence. Oral systems of justification are not sufficient to coordinate such social and institutional arrangements. Rather, a different kind of justification system or code was needed, and formal or traditional systems of justification were birthed. These involved much more systematic institutional codes, laws, and borders that could then be enforced by governmental authorities like priests, kings, or military officials. Consider, for example, the code of Hammurabi (Richardson, 2004). He was the sixth king of Babylon who developed a code of law that was carved on a massive seven-and-a-half-foot stone in the shape of an index finger that is dated to 1754 BCE. It is one of the oldest writings that can be decoded. Consisting of 281 laws, it includes the famous "eye for an eye and tooth for a tooth" aphorism and deals with matters of punishment, trade regulations, wages, liabilities, and other forms of law and social contract.

These ancient societies exist only as a distant echo today. A major reason is that, as Eric Cline (2014) narrates in his book *1177 BC: The Year Civilization Collapsed*, the Bronze Age civilizations all experienced a dramatic decline in a short period of time. Specifically, within a 50-year period from 1200 BC to 1150 BCE, nine major civilizations interconnected by trade, cooperation, and competition all failed. There were likely many reasons that intersected to create a perfect storm, including earthquakes, drought, invading nomads, and loss of crucial resources. Whereas the Bronze Age civilizations have largely drifted into history, the echoes from the next wave of major civilizations continue to shape our sensibilities to the present day. The philosopher Karl Jaspers (1953) coined the term Axial Age to describe a crucial historical period that followed the Bronze Age and lasted from about the eighth to the third century BCE. He argued that it was a pivotal age because new ways of thinking in religion and philosophy emerged in Persia, India, China, and the Greco-Roman worlds. Such systems likely relate to the widespread appearance of writing and elite literacy and begin to emerge at various places in both the East and West during this period. Mathematics also emerged, which played a crucial role in the transition from a purely social epistemology to a more formal-analytic knowledge.

Especially relevant for those situated in the cultural tradition that Richard Tarnas (1993) called the "Western Mind" is the emergence of the trio of Greek Philosophers of Socrates (470–399 BCE), Plato (427–347 BCE), and Aristotle (384–322 BCE), whose views on knowledge would change the world. Building off the work of the Pythagoreans and others, Socrates gives rise to a method of inquiry that we can recognize today as a "formal" approach to epistemology. Via systematic analysis and reflective questioning, with mathematics perhaps representing the logical ideal, Socrates has the insight that much social epistemology is potentially vacuous.

Some modern philosophers have (re)framed this problem in terms of the problem of "bullshit." In "Bullshit as a Problem of Social Epistemology," Wakeham (2017) explores the function it serves in social groups, such as how groups of humans are inclined to construct realities based on their shared interests and goals. His point was that bullshit works to serve crucial social functions, such as persuading other people, enhancing one's position, and getting people to discount other viewpoints at odds with one's interests. This point is an important and humbling one to digest and be reminded of as it results in the following considerations: *If each person develops in the context of a socially constructed reality, how can we ever acquire knowledge that is separate from the contingencies that have shaped the group's knowledge systems in which we are embedded? More bluntly, how can we even think about social knowledge as being different from socially constructed bullshit?*

From a perspective informed by JUST, this analysis has relevance for effectively framing the insights of Socrates. The basic argument is that the insights of the Greek philosophers resulted in a radical shift in how humans thought about knowledge. Prior to the Greeks, knowledge in the West was generally framed in terms of local traditions, religious accounts of the world, and pragmatic or procedural know-how (i.e., knowing how to build a sword or pyramid). That is, there was not a systematic way to question and analyze the explicit relationship between subjective experiences of the world, human discussions about why things happened, and acting in the world with demonstrable skills, from a systematic analysis of what was true. However, the early Greeks changed that and turned their eye toward more abstract and deeper analytic problems in justification. In other words, the Greeks had the vision that philosophy could save humanity from living based on bullshit.

Plato and Aristotle took up the mantle to develop formal-analytical philosophies that could withstand Socratic-like philosophical criticism. In Plato, we get a constructive philosophy that attempts to spell out a proper theory of knowledge (i.e., justified true belief), idealized forms of governance, aesthetics, and a conception of ontology grounded in absolute forms. In Aristotle, we get the shift to a more materially grounded ontology, out of which forms emerge. We also get a great taxonomy from Aristotle across a multiplicity of domains, from logic to morality to biology. For the West, the Platonic and Aristotelean metaphysical systems represent the beginning of formal, refined academic knowledge, as separate from commonsense social epistemology. Of course, the dominant force in the justification systems of Western Europe in the millennia prior to modernity was the Judeo-Christian worldview. It is difficult to overstate how much every aspect of life was infused with Christian theology. Scholarship, the calendar, politics, art, and the entire frame of social life were shaped by the Bible and the Church. Together, the Greek and Christian sensibilities framed the context out of which natural philosophy and a modernist mindset would emerge.

The Modernist Cultural Code and Sensibility

Although Aristotle and others encouraged systematic investigation, modern empirical natural science as it emerged from Christian natural philosophy nevertheless gave rise to a new form of justification. Whereas the Greeks emphasized metaphysics and the logical coherence of their philosophical systems, the pioneers of modern science emphasized that empirical evidence must be gathered and analyzed and found to be in support of one's conjectures for the idea to be deemed scientifically justifiable. By the sixteenth century, European intellectuals had divided the natural and supernatural domains and determined that the former could be investigated systematically. This set the stage for the invention of the empirical method and a move in the conception of truth from coherence and logic (i.e., the primary focus of the ancient philosophers) to a focus on the correspondence theory of truth grounded in systematic observation tied to measurement and experimentation.

Via the work of scientists like Galileo, Descartes, Newton, Maxwell, Darwin, and Einstein, the power of modern science in revealing universal truths shined. The shift from the earth-centric to the heliocentric view of the solar system is emblematic of the shift from traditional coherence models of epistemology to empirically grounded correspondence models. Prior to this shift, true accounts of reality were considered in how the shared framed of reference conformed to observations and practical utility. It was obvious to people that the sun rose in the east and traveled across the sky and set in the west. And yet, despite how obviously true that was to human subjective senses shared by the group, the heliocentric view blew up that conception and placed subjective and socially constructed pragmatic sense-making in doubt. Instead, a deeper trust was placed in scientific epistemology.

The modernist mindset that emerged out of the Enlightenment can be framed by several additional elements over and above the growth and power of natural scientific knowledge. First, there were massive changes in the social order. For example, there were major shifts from feudal structures into capital and labor relations, which dramatically accelerated during the Industrial Revolution. There were also significant shifts in the political relations between nation-states and individuals, and liberty, democracy, and individuality became dominant themes. We also see the emergence of new political philosophies enacted in the French and American Revolutions.

Engineering advances were closely intertwined with the exploding knowledge of matter in motion, and the Industrial Revolution was launched, embraced by class liberal values of freedom and progress, and put into motion by a massively powerful capital labor engine and the invisible hand of economic markets that would literally transform much of Mother Earth into human material culture. The modern capitalist engine would result in us becoming, from the vantage point of other species, a ruthless consumptive virus that would spread into every corner of the globe.

Especially relevant to our concerns was the central role that the emerging worldview had in shaping the direction and character of cultural evolution. As Ken Wilber (2007) notes, prior to modernity there was a much tighter and more coherent and aligned relationship between theology, political structure, philosophy, morality, and science. Indeed, science was not considered separate from philosophy, but simply an extension of it into the natural world. However, the success of modern science, coupled as it was to technology via observation, measurement, and experimentation, fundamentally changed the structure of the pre-modern worldview and how humans lived on planet Earth. The matter-in-motion worldview was remarkably successful in its domain, but also emerged in a way that was inconsistent with the available understanding of the rest of the world. As such, it largely split off from the other ways of knowing, such that science emerged as a separate discipline from philosophy and the other human knowledge systems.

The Postmodern Cultural Code and Sensibility

Although modernism remains a dominant force in society, many have argued that the modernist sensibility did not deliver a fully adequate picture of either human knowledge or humanity and our place in the cosmos. The senseless destruction associated with WWI, the subsequent emergence of Nazism and the horrors of WWII, and the Soviet Union gulags resulted in a powerful challenge to the idea that humanity was on a linear path toward progress ensured by a pristine rational vision of the truth. As the brutal oppression via racist categories became seared into human consciousness by the revelations of the Holocaust, movements for universal human rights, civil rights, and feminism were sparked across the West. Central to these movements was the notion that White heterosexual Christian men had dominated other social categories and colonized the world in unjust ways that needed to be corrected.

In addition to these social justice movements, developments occurred in philosophy that supported a postmodern critique of modernist visions of knowledge and truth. One of the major developments in the twentieth century was that philosophy largely gave up on the dream to develop a workable, comprehensive, and consilient picture of the whole of human knowledge. The failure of synthetic philosophy to succeed in this regard is made clear in Will Durant's *The Story of Philosophy*. Durant narrates the biographies and systems of the great philosophers, from Ancient Greece to the early twentieth century. The story is of one knowledge system after another being developed and critiqued and found wanting and replaced by another. When the book was written in the mid-1920s, Durant feared that the dream of synthetic philosophy was coming to an end, and explicitly called for its revival.

Although his book was a popular success, his fear was well placed. Philosophy's consensus in the twentieth century was that there was no systematic way to solve the problem of creating a big picture synthesis of human knowledge. This is evident in several aspects of the discipline. First, the discipline splits into analytic, continental, and pragmatic approaches that are so different in nature and focus that they almost represent different disciplines. Second, none of the approaches were keen on synthetic philosophy. Third, the philosophy of language, often called the linguistic turn, emphasized context, history, and the sociocultural grammar that shaped human understanding. The argument that there is an inevitable fusion of truth systems with social power was consolidated by influential thinkers like Michel Foucault, Jacques Derrida, and Paul Feyerabend. This conception helped set the stage to blur the boundaries between objective scientific knowledge and social or normative systems of justification that were fully dependent on cultural context, history, and power relations. In 1979, Lyotard captured the essence of the postmodern sensibility as being focused on demolishing grand narratives. Given the fact that Nazism and Stalin's totalitarian communism had soaked the world in blood, such critiques carried much emotional force.

As this brief review suggests, the postmodern sensibility provides a potentially powerful critique of modernist knowledge systems. Most notably, the postmodern view highlights the crucial role of language, hermeneutic processes, and power in generating what groups of people believe to be true. It also highlights the fact that modernist knowledge systems fail to effectively account for the complicated role of social epistemology in influencing the supposedly objective truth claims of science. However, despite offering a powerful critique, postmodernism has long been known by most sophisticated thinkers to be a temporary phase in the evolution of human consciousness and culture.

This fact becomes clear if we take postmodernism's core insight to its logical extreme. The claim that it is universally true and always will be true that "there are no universal truths" is an unworkable contradiction. Moreover, we can see the problems associated with postmodernism, such that it can be considered a good description of how human knowledge systems are currently structured. The failure of philosophy to generate a synthetic picture of the relationship between human knowers and the ontic reality, the proliferation of ever-increasing domains of specialized knowledge systems in science and elsewhere, and the critiques of the post-structuralists and continental philosophers all played a role in the fragmented pluralism that currently characterizes human knowledge.

An Emerging Metamodern Cultural Code and Sensibility

From the vantage point of the UTOK, the current situation is deeply problematic. We are creating dangerous technologies, entering into a new digital world, and damaging the planet, and face debt/financial issues, climate issues, international conflict patterns, and resource degradation and depletion. Additionally, we are also facing profound meaning and mental health crises. And we may not have long to right the ship before we sail into an irreversible state of global civilization collapse. Shifts are needed in how we organize our knowledge and structure our universities. This can be done if we properly diagnose the nature of the Enlightenment Gap and its potential resolution.

The metamodern sensibility is one that emphasizes a holistic vision that is about including and transcending that which came before. In *Metamodernity*, Andersen (2019) lays out a vision for how such changes might be framed. She embraces the idea that the time is right for a metamodern way of thinking to emerge, and offers a clear justification for it as follows (p. 7):

> Metamodernity is an alternative to both modernity and postmodernism, a cultural code that presents itself as an opportunity if we work deliberately towards it. It is a vision, an option and a possible future scenario. As a cultural code, metamodernity contains both indigenous, premodern, modern, and postmodern cultural elements and thus provides social norms and a moral fabric for intimacy, spirituality, individuality, and complex thinking. It has the potential to protect our cultures and cultural heritage as the economy, the internet and exponential technologies are going global and disrupting our current modes of societal organization and governance. Indigenous culture can provide a connection to nature that we have lost and the circularity that we need to solve a host of problems. Premodern culture provided strong existential frameworks through what we today call religion. Modern culture emancipated humans and gave us science, human rights, democracy, prosperity, and safety. Postmodern analysis has the capability to expose hidden power structures and take an outside perspective on our own culture. We need it all. We also need to choose right; otherwise, we risk having to live with the worst of everything.

Conclusion

This book began laying out the details of UTOK in Chap. 5 by showing how JUST functions as a metatheoretical framework for human self-consciousness and the evolution of Culture. From there we followed the trail into the ToK System, which, along with the PTB, gave us a new map of Big History that allowed us to effectively define the levels and dimensions of behavioral complexification from quarks to Culture and the major domains of scientific inquiry that map them. We then bridged the metaphysical and ontological analyses of behavior, mental behavior, neurocognition, consciousness, and the self in the relational world to the Map of Mind^{1,2,3} and the metatheoretical formulations given by BIT and the

Influence Matrix, allowing us to trace the evolution of organisms into animals into mammals and finally primates.

This chapter continued with the core theme of this book, which is the project of synthesizing metatheoretical formulations with descriptive metaphysical frameworks to clearly delineate the ontological referents. It also brings us full circle by returning to JUST as a metatheory and shows why, when coupled to the rest of UTOK's conceptual architecture, it can effectively frame the ontology of Mind^3, human persons, and the Culture-Person plane of existence. The conclusion is that we now have an adequate grip on the multiplicity of dynamics that are necessary for understanding human mental behavior and how it emerged from the animal mental behaviors that preceded it. This optimal grip provides the grounds for a new vision that solves the ontological problem of psychology. Moreover, this solution enables us to fill in the Enlightenment Gap and achieve a coherent, integrative pluralistic synthetic philosophy that might be up to the task of guiding humanity toward wisdom in the twenty-first century.

References

Andersen, L. R. (2019). *Metamodernity: Hope and change in a complex world.* Nordic Bildung.

Aslan, R. (2017). *God: A human history.* Random House.

Berger, P., & Luckmann, T. (1967). *The social construction of reality: A treatise in the sociology of knowledge.* Doubleday.

Bertman, S. (2005). *Handbook to life in ancient Mesopotamia.* Oxford University Press.

Bronfenbrenner, U. (1979). *The ecology of human development.* Harvard University Press.

Burr, V. (2015). *Social constructionism.* Routledge.

Cline, E. (2014). *1177: The year civilization collapsed.* Princeton University Press.

Donald, M. (1991). *Origin of the modern mind: Three stages in the evolution of culture and cognition.* Harvard University Press.

Frankfurt, H. G. (1971). Freedom of the will and the concept of a person. *The Journal of Philosophy, 68*(1), 5–20. https://doi.org/10.2307/2024717

Haidt, J. (2012). *The righteous mind: Why good people are divided by politics and religion.* Vintage.

Henriques, G. (2017, June 28). *Maggie's story: The many reasons why not.* [Blog] Theory of Knowledge on Psychology Today. https://www.psychologytoday.com/us/blog/theory-knowledge/201706/maggies-story-the-many-reasons-why-not

Henriques, G., Vervaeke, J., & Sengstock, G. (2020, August). From the con game to the research lab: Identifying four different contexts of justification. [Blog] Psychology Today, Theory of Knowledge. https://www.psychologytoday.com/us/blog/theory-knowledge/202008/the-con-game-the-research-lab

Iyer, R., Koleva, S., Graham, J., Ditto, P., & Haidt, J. (2012). Understanding libertarian morality: The psychological dispositions of self-identified libertarians. *PLoS One, 7*(8). https://doi.org/10.1371/journal.pone.0042366

Jahoda, G. (2012). Critical reflections on some recent definitions of "culture". *Culture & Psychology, 18*(3), 289–303. https://doi.org/10.1177/1354067X12446229

Jaspers, K. (1953). *The origin and goal of history*. Yale University Press.

Kroeber, A. L., & Kluckhohn, C. (1952). Culture: A critical review of concepts and definitions. *Papers. Peabody Museum of Archaeology & Ethnology, Harvard University 47.*

Lang, A. (1997). Thinking rich as well as simple: Boesch's cultural psychology in semioticperspective. *Culture & Psychology, 3*, 383–394.

McGoldrick, M., Preto, N. G., & Carter, B. A. (2015). *Expanding the family life cycle: Individual, family, and social perspectives* (5th ed.). Pearson.

Narvaez, D., Jacobs, D. T., Halton, E., Collier, B. S., & Enderle, G. (2019). *Indigenous sustainable wisdom: First-Nation know-how for global flourishing*. Peter Lang Inc.

Ossorio, P. G. (2006). *The behavior of persons*. Descriptive Psychology Press.

Richardson, M. R. (2004). *Hammurabi's laws: Text, translation, and glossary*. T. & T. Clark.

Shaffer, V. A., & Shaffer, L. S. (2007). The use of recipe knowledge in medicine. *The International Journal of Technology, Knowledge, and Society, 2*(7), 73–80. https://doi.org/10.18848/1832-3669/CGP/v02i07/55668

Smith, C. (2011). *What is a person? Rethinking humanity, social life, and the moral good from the person up*. University of Chicago Press.

Sutton, M. Q. (2021). *A concise introduction to cultural anthropology*. Routledge.

Tarnas, R. (1993). *The passion of the Western mind: Understanding the ideas that have shaped our world view*. Ballantine Books.

Taylor, C. (1985). The concept of a person. *Philosophical Papers*. Vol. 1. Cambridge University Press.

The Alternative. (2020, July 18). *A startling new book from Australia—"Sand Talk: How Indigenous Thinking Can Save The World"—Gives lessons in complexity*. Retrieved October 25, 2021, from https://www.thealternative.org.uk/dailyalternative/2020/7/19/sand-talk

Tylor, E. B. (1871). *Primitive culture: Researches into the development of mythology, philosophy, religion, art, and custom* (Vol. 1). J. Murray.

Vervaeke, J., & Henriques, G. R. (2020, September 4). *Untangling the world knot of consciousness: Grappling with the hard problems of matter and mind.* [Video]. The Cognitive Science Show. https://www.youtube.com/watch?v=bD6Szbf1cHo

Wakeham, J. C. (2017). Bullshit as a problem of social epistemology. *Sociological Theory, 35*(1), 15–38. https://doi.org/10.1177/0735275117692835

Whiten, A., Goodall, J., McGrew, W. C., Nishida, T., Reynolds, V., Sugiyama, Y., Tutin, C. E., Wrangham, R. W., & Boesch, C. (1999). Cultures in chimpanzees. *Nature, 399*(637), 682–685. https://doi.org/10.1038/21415

Wilber, K. (2007). *Integral spirituality: A startling new role for religion in the modern and postmodern world.* Shambhala Publications.

Yunkaporta, T. (2019). *Sand talk: How indigenous thinking can save the world.* Text Publishing Company.

PART VI

Conclusion

CHAPTER 16

A New Vision of Mind and Psychology that Transcends the Enlightenment Gap

The Unified Theory of Knowledge provides a new framework for human knowing and being, one that seeks to include and transcend both modernist and postmodernist ways of looking at the world. Introduced at the end of the last chapter, the term metamodern can refer to: (1) a cultural phase; (2) a developmental stage of society; (3) a stage of personal development; (4) an abstracted meta-meme; (5) a philosophical paradigm; and (6) a sociopolitical movement (Henriques & Görtz, 2020). Despite the multiplicity of specific meanings, it is nevertheless also the case that there is a straightforward meaning of the term. Metamodern refers to the values, mindset, and cultural code that comes after postmodernism. Moreover, it is situated and structured to do so in a way that includes, synthesizes, and transcends the tensions between the modernist and postmodern positions on the nature of knowledge.

The differences between the modernist and postmodern sensibilities are at the heart of the intellectual disputes that characterize the science versus social and subjective knowledge aspects of the Enlightenment Gap. They can be framed in terms of the following question: *What is science and what is its relationship to reality?* The straightforward answer is that science functions as a method and institution grounded in empirical epistemology that enables greater confidence in the correspondence between one's models and maps and the territory that is reality. As we saw in Chap. 3, this aligns with how Gauch characterizes science. It is true in so far as the

G. Henriques, *A New Synthesis for Solving the Problem of Psychology*, Palgrave Studies in the Theory and History of Psychology, https://doi.org/10.1007/978-3-031-18493-2_16

epistemological and institutional aspects of science are currently at the center of what science means generally. However, this standard answer does not address questions about ontology, nor does it speak to the idea that natural science operates from a particular worldview, such as the naturalism E. O. Wilson outlined in *Consilience* (1999).

It is the nature of this confusion regarding ontology that results in individuals with different sensibilities seeing different aspects of science operating in different ways. That is, most modernists and postmodernists agree that science is a method of knowing that yields some truths about the world. The real debate between these two perspectives is found in: (a) the extent to which science can achieve a coherent worldview; (b) the relationship and differences in the epistemological validity between scientific knowledge and knowledge obtained either subjectively or in the social context; and (c) the authority science has in making truth claims in relationship to other forms of human knowing. Advocates for modernist science tend to argue that science can generate the outline of a worldview that carries implications for other worldviews, that it reveals truths that are generally independent of social context, and that it should be granted authority in making truth claims over other kinds of epistemological systems. Postmodern critics emphasize that there is no generally agreed upon scientific worldview and that any worldview that is espoused will be predicated on a priori assumptions because all knowledge is framed by context, and, consequently, scientific knowledge systems should not be granted any special authority as grand narratives.

Like Bhaskar's critical realism, UTOK embraces both points and frames science as a specific kind of justification system with the potential to generate a transcendent realist ontology. However, unlike Bhaskar, UTOK explicitly fills in the missing gaps in our knowledge systems regarding matter and mind to achieve clarity and coherence. Specifically, it homes in on the problem of psychology as residing at the epicenter of our epistemological and ontological confusions. And then, via the ToK System extended into the PTB, it shows how scientific knowledge, from physics to biology to psychology to the social sciences, can be organized and represented as corresponding to the ontic reality. This macro-level view of a scientific onto-epistemology sets the stage for a novel approach to the philosophical debates between modernity and postmodernity because it brings a zoomed-out view to the whole. It helps clarify the kind of knowledge science generates and it organizes the basic scientific insights together across the stratified dimensions in nature. In addition to framing the core of science as being about the behaviors in the natural world stratified across levels and dimensions of complexification, science can also be effectively framed as a kind of justification system. Effectively holding both truths coherently is what bridging the modernist versus postmodern divide is about.

Inverting Wittgenstein and Finding a Coherent Picture of Psychology

The scope of the vision afforded by UTOK is such that the implications are hard to overstate. One way to think about how UTOK is positioned to revolutionize our thinking about knowledge and psychology is to view the two primary insights that launched UTOK (i.e., JUST and the ToK System) in relationship to the developments in the thought of one of the most influential philosophers of the twentieth century, Ludwig Wittgenstein. Early in his career, Wittgenstein argued that analytic knowledge was much like a geometric picture of reality, described via logical relations. Strongly supported by leading philosophers of the time (e.g., Bertrand Russell), his "picture theory of language" emphasized how truth claims were claims about factual states of affairs and nothing more. It played an important role in the development of an approach to analytical philosophy known as logical positivism. Logic positivists conceived of the truth about the world as propositional statements that were akin to a pristine mathematical description of empirically derived factual relations. Everything else was deemed subjective "nonsense," meaning that opinion and preference was of little interest to analytic truth seekers. Put differently, the logical positivists essentially tried to eliminate qualitative impressions and reduce everything to logic and data to generate a purely factual correspondence theory of knowledge.

Although it was an influential line of thought from the 1920s through the 1940s, logical positivism ran into some serious challenges, such that by the 1980s few philosophers were logical positivists. An important reason for the shift was that Wittgenstein himself came to change his mind regarding the ultimate nature of human knowledge. Simplifying what is a rather complex argument, Wittgenstein came to see language—and human life forms more generally—as being inevitably framed by history and context. Specifically, he argued that words and truth claims were more like tools that humans used to navigate the world rather than analytic terms that generated a pristine correspondence to reality. This meant that human knowledge systems could be thought of as being akin to what he called "language games" that were framed by the needs and goals of the players.

This conception of knowledge is different from the logical positivist view. It has more of a pragmatic and social constructionist bent, meaning that human knowledge is more inevitably tied to the context and utility which are framed by the tasks and goals of the human knowers. Nevertheless, the concept of language games is not meant to confer a flippant or unserious meaning. The idea is that language usage operates via ways of living and frames of understanding, much like games operate via a shared set of rules. Thus, quantum mechanics—one of the most serious and analytic knowledge systems ever constructed—can be considered a language game, in that it has a set of concepts and categories that individuals are employing toward some end (i.e., making sense of what the fundamental particles of matter are and how they behave at the quantum level).

To better understand the shift in our knowledge that UTOK affords, it is helpful to contrast Wittgenstein's evolution in thought from his picture theory of reality into a vision of human knowledge as language games with the development of JUST and how it gave rise to the ToK System. The two lines of thought have many interesting parallels and potential interrelations, but the point I want to make here is that they are developmentally inverted. That is, consistent with the tensions between modern and postmodern philosophy, both lines of thinking deal with vision logic and picture theories that attempt to orient toward truth relations and include insights about the social construction of knowledge as contextualized pragmatic systems that evolve. Wittgenstein initially set out to achieve a philosophy of language and conception of truth that extracted the subjective and social dimensions and left behind the unassailable analytic, geometric correspondence between statements of fact and states of affairs in the world. However, over time, he came to see just how deep the social contextual dimension to human knowing was and this resulted in the shift into the idea of language games and a philosophical focus on the pragmatics of language.

The Unified Theory develops via an inverted sequence. That is, my conception of justification systems overlaps significantly with Wittgenstein's language games. Justification systems are akin to the rules and grammar of the game, and justifications and investments in the context of social influence are how individuals play the game and live their lives. And justification systems are formed by, feedback on, and evolve like forms of life.

However, JUST is different in that it functions to place the emergence of language games in a clear, macroscopic evolutionary context. This meant the stage was set for me to shift from that view to a depiction of knowledge that framed human justification processes in the larger context of cosmic emergence. And out popped the Tree of Knowledge System as a theory of reality mapped by the methods, taxonomies, and theories of science. The consequence was a picture theory of knowledge that included human language games.

Although Wittgenstein was not really a postmodern thinker, his work was tremendously influential in the poststructuralist/postmodern movement. My hope is that UTOK can be part of the shift toward a new cultural and philosophical sensibility. According to UTOK, the most obvious shortcoming of the modernist vision of knowledge is seen in the failure to solve the problem of psychology and related ontological and epistemological problems of clarifying the relationships between matter and mind and social and scientific systems of justification. With UTOK we are finally able to bridge the Enlightenment Gap and solve psychology's subject matter conundrums, specified by the BM^3 problem. And because UTOK emerged first via JUST, it contextualizes human knowledge as systems of justification, much as the postmodern sensibility does. However, rather than stopping at the epistemological implications of this insight, UTOK places human justification processes in the context of a naturalistic ontology. As such, like Bhaskar's critical realism, it can synthesize the modernist assertions regarding science with the postmodern critique and orient us toward a metamodern, coherent integrative pluralistic sensibility.

Framing UTOK's Psychology as Providing a Coherent Language Game for the Scientific Study of Mental Behaviors

In 1949, Wittgenstein completed his two-volume *Remarks on the Philosophy of Psychology*, which was published several years after his death. Given his deep attention to obtaining clarity about language and the meaning of our concepts, it should not come as a surprise that Wittgenstein found psychological science to be conceptually confused and in need of some foundational re-examinations. He wrote:

> the confusion and barrenness of psychology is not to be explained by calling it a 'young science'; its state is not comparable with that of physics, for instance, in its beginnings.... For in psychology there are experimental

> methods and conceptual confusion. The existence of the experimental method makes us think we have the means of solving the problems which trouble us; though problem and method pass one another by. (Wittgenstein, 1949)

The scientific psychology that emerges out of UTOK agrees with Wittgenstein on several key points. The first key insight is that our forms of life and language are contextual, and thus meanings per se are not to be discovered via analysis of foundational essences that exist beneath the context. Second, description is essential and, given the intersubjective highway of justification that connects human minds, care must be taken in making any facile distinction between our inner and outer worlds. For example, we should certainly reject mainstream psychology's methodological behaviorist approach that sloppily defines the outer world in terms of behavior and the inner world in terms of mental processes without diving into the complexity of the assertion. Rather, Wittgenstein helps us see that this is simply a way of talking and a form of life. Moreover, it is a particular and somewhat convoluted language game that fails to recognize itself as such.

In contrast, UTOK's psychology starts self-consciously as a system of justification. That is, it self-identifies as a particular language game that exists as a particular form in relationship to specific sociocultural contexts of justification. It is explicitly placed in the context of modern empirical natural science, the European Enlightenment, and the conceptual and philosophical problems that have long beset scientific psychology, framed here as the BM^3 problem. Moreover, in a manner that is consistent with Wittgenstein's astute reflections, UTOK attends first and foremost to careful descriptions that emerge out of specific language games. For example, as noted in the previous section, UTOK identifies science as a particular kind of epistemological process that relies on the concept of behavior, which generates a grammar of entities, fields, and change, and attaches that to an observer that is systematically placed in an exterior empirical perspective who proceeds to engage in pattern analysis via experimentation and quantification. Similarly, UTOK transforms Wittgenstein's forms of life that characterize folk psychology into patterns of justification, investment, and influence. As such, UTOK is, first and foremost, a descriptive psychology. Specifically, it is a psychology based on a clear descriptive metaphysical system that can specify the right relations between ontology and epistemology, and crisply delineate the ontology of the mental and of human persons.

In addition to focusing on a clear description of the flow of our lives, UTOK embraces the importance of coherence. A coherent system is one that is intelligible, comprehensive, and internally consistent. The problem of psychology renders mainstream psychology incoherent, which is a damning indictment of the field. In addition, given psychology's place in the academy, it ripples through our current knowledge systems writ large, a point we have highlighted via the Enlightenment Gap. In contrast, UTOK uses the ToK System and PTB to develop a coherent naturalistic ontology that affords us the capacity to see the continuity and discontinuity between material objects, living organisms, mental animals, and cultured persons. It then gives the Map of Mind[1,2,3] to specify the domains of the mental and their epistemological and ontological interrelations. With them, we can now clearly see two of the great conceptual errors of mainstream empirical psychology, which are that (a) mental behaviors are a viable class of things in the world, in contrast to behaviors and mental processes, which creates a false dualism based on scientific epistemology; and (b) there are different kinds of mental processes that have different ontological referents and are available via different epistemological frames.

Without a doubt the most glaring conceptual problem that spilled forth from the Enlightenment was the infamous mind-body problem. The convoluted state of the philosophy of mind combined with the problem of psychology, coupled with the flawed grammar of folk psychology, provides powerful evidence that the basic conceptual architecture for understanding the proper relation between the physical and mental worlds is broken. Indeed, as Cahoone properly analyzed it, the problem can be seen in the framing of mind *versus* body. The bipolar split that emerged between matter and mind arose as a function of both the epistemology and ontology of modern empirical natural science. Its exterior epistemological position, tied to its physical reductionism, generated a powerful language game that factored out the unique, subjective idiographic knower. It also left no real place for epistemic processes. Rather, it simply gave a metaphysical flatland of physics, such that cognition, consciousness, and the specific knower were metaphysically unspecifiable.

Although scholars generally attempt to tackle the mind-body problem via either the philosophy of mind or the science of consciousness, the best place to see it is to look at the problem of psychology. The problem of psychology shows how deep and broad the issue really is. We are not simply talking about an explanatory gap between neurobiological mechanisms and the subjective conscious experience of being. We are talking about the meaning of the terms body, behavior, brain, mind, cognition, consciousness, and the self, and the absence of a broad descriptive metaphysical system that can coherently define them and their interrelations. We have seen how the BM^3 problem emerged in psychology and why it could not be solved with the traditional frames of understanding.

By framing the Mind as the set of mental behaviors operating at the third dimension of complexification, UTOK disentangles and coherently reassembles the mentalist versus behaviorist divide. For example, it affords an obvious bridging point between Skinner's radical behaviorism and mainstream psychology's weak neurocognitive functionalist account of mental processes. To see how mental behavior does this, consider Skinner's (1987, p. 784) characterization of behavior and mind in his essay "Whatever Happened to Psychology as the Science of Behavior?":

> Cognitive psychologists like to say that 'the mind is what the brain does,' but surely the rest of the body plays a part. The mind is what the body does. It is what the person does. In other words it is behavior, and that is what behaviorists have been saying for more than half a century.

Skinner is engaging in a bit of a sleight of hand here. It is much more accurate to say that for more than half a century, behaviorists argued that the mental *had* to refer to an underlying cause of behavior that was inconsistent with the physical world, which is a very different claim. But his rhetoric is valuable from a UTOK perspective because it shows that the key concept is mental behavior, which can be defined as the activity of animals doing the things they do in the world. If behaviorists had embraced the concept of mindedness and the idea that mental is an adjective for the kind of behavior psychologists were interested in, the entire history of behaviorism and psychology more generally would likely have been different.

Of course, this is not what happened, and confusion abounds between the concept of mental as a description of a kind of behavior and as a causal explanation for behavior. And the situation has not been resolved, and mainstream empirical psychology is largely guilty of Skinner's primary critique of the mental being framed as some unobservable cause of behavior. Not surprisingly, we are now seeing similar debates being played out between traditional neurocognitive approaches and those who advocate for a more enactivist 4E cognitive science perspective. Namely, enactivist approaches want to define "the mind" in terms of complex adaptive (mental) behaviors framed by the agent–arena relationship as a whole, and this contrasts with traditional cognitive neuroscientists, who generally define it as the information instantiated within and processed by the nervous system (Gallagher, 2017).

This book has laid out a new language game that affords us the needed terms and corresponding vision logic to map the terrain and transcend the debate and its potential confusions. Mind, the set of mental behaviors, refers to the complex, functional activities of animals with brains and complex active bodies. This is the set of minded behaviors that enactivist 4E cognitive scientists are attempting to explain. Modern 4E cognitive science overlaps more with Skinner's concept of operant behavior, only it is framed via the view from within the animal and does not eschew information processing broadly defined, as Skinner did. In contrast, traditional cognitive neuroscientists generally frame their understanding of the mind as the domain of Mind^{1a}. With its PTB and Map of Mind^{1,2,3}, UTOK maps the proper description of "behavior and mental processes."

If we follow David Chalmers' distinction between the hard and easy problem of consciousness, we can say that the cognitive revolution allowed us to develop a relatively "easy" understanding of mental behavior. That is, it showed how information processing could give rise to intelligent behavior that demonstrates functional awareness and responsivity. However, the hard problem of Mind^2 was largely ignored by traditional cognitive approaches, and it has only been seriously considered in the last 20 years. Here again, UTOK brings conceptual clarity to the issues. It allows us to frame the hard problem in both ontological and epistemological terms, which, as we have seen, is a distinction that the behaviorists overlooked and was never fixed. The problem of Mind^2 is hard both in terms of it specifying the ontological mechanisms that engender subjective conscious experience and the epistemological problem that arises with the emergence of modern science and its grounding in an exterior empirical position that frames the world in terms of observable and quantifiable behaviors. Given that, for many, the mental is largely synonymous with Mind^2, UTOK makes it easy to see why the science of psychology is fraught with language game complications. Because Mind^2 is not available to be seen via the exterior behavioral lens of science, this means that a central aspect of the subject matter is potentially defined out of consideration simply by the rules of the game.

Thankfully, UTOK places the subjective, idiographic qualitative human knower back into the equation. It does this on two accounts. First, it gives us the necessary vocabulary between neurocognitive activity and subjective conscious experience. Second, it clarifies the nature of Mind^3 and why it is a different domain. This enables us to see Descartes' primary meaning of mind and how and why humans are so discontinuous in some ways with other animals. The domain of Mind[3] finds many of the major conundrums found in the relationship between the physical and mental worlds, such as recursive self-consciousness and the problem of free will. Via JUST, UTOK explicates clearly how the human mind's big bang exploded into being, giving rise to the evolution of the Culture-Person plane of existence. Moreover, via the ToK System and the PTB, UTOK explains how science is a particular kind of justification system that is focused on the exterior, generalizable, quantitative behavioral patterns in the world and why this language game is different than the idiographic, particular, unique, qualitative life of the subject. Thus, UTOK gives a map of science that also has a clear placeholder for the unique subject.

UTOK's New Vision for Psychology

Mainstream empirical psychology is built on a conceptual house of sand. Consequently, the field will continue to produce much information, but will fail to generate a deep, cumulative ontological picture of the mental. The metapsychology provided by UTOK shifts the focus of scientific psychology from empirical investigations via operationalized constructs to the field's deep metaphysical, ontological, and metatheoretical structure. As such, the mental behaviorist approach to psychological science grounded in UTOK exists in sharp contrast to the methodological approach that Psychology currently embraces. Given the important differences, one might be tempted to conclude that I would advocate for a wholesale rejection of methodological behaviorism to be replaced with a mental behaviorism. However, my view is more nuanced and synergistic.

My perspective for where scientific psychology needs to go can be found in the metaphysical–empirical continuum I delineated in Chap. 4. We need a Psychology that is aware of and committed to working from both ends of the spectrum. This would allow us to clap with both hands and achieve a truly viable and functional scientific psychology. Of course, because mainstream psychology is grossly over-committed to the empirical side of the continuum, there is much work that needs to be done to build an infrastructure to support the development of the metaphysical, ontological, and metatheoretical architecture. The good news is that: (a) there is already some groundwork in this area; (b) fewer people are needed in theory than are needed for empirical investigations; and (c) UTOK has already done much of the necessary work.

Consider, for example, that there have already been calls for a clearly defined theoretical psychology (Slife & Williams, 1997), and there is a division of APA devoted to theoretical and philosophical psychology. Unfortunately, at least from the vantage point of UTOK, the field of theoretical psychology is currently dominated by a postmodern, critical sensibility. However, it is possible that the field could add perspectives and shift toward a more constructive approach to building new and better theories, metatheories, and big picture metaphysical frameworks. The good news is that there is a small infrastructure in place that could be built upon.

What would a synergy between mental and methodological behaviorism look like? First, the mainstream institution would need to explicitly acknowledge its past denial and fully accept and recognize the fact that the problem of psychology exists and threatens the status of the science. That is, the field needs to acknowledge upfront that the ontological domain of inquiry has not been consensually agreed upon and that the terms behavior, mind, cognition, consciousness, and the self have different meanings to different traditions. This means that the science is pre-paradigmatic in the Kuhnian sense. From this explicit awakening, serious work can be undertaken by other theorists, researchers, and scientists to critique the current proposal, compare it and contrast it with the few other proposals that have been generated for a more general and unified approach to the field (e.g., Staats' psychological behaviorism), or generate novel solutions. The point here is that the field needs to accept the BM^3 problem or its equivalent framing and actively search for genuine solutions. Unfortunately, because this kind of scholarship is more theoretical and philosophical, it cuts against the institutional grain. But that is exactly part of UTOK's metapsychological message. We need to coherently fill in the gaps between science, psychology, and philosophy.

It is worth noting here that I have worked on developing a methodology for helping psychology clap with both hands. In ways that parallel methodological behaviorism, the method works via UTOK's structure to generate what can be called an "assimilative integrative hypothesis" for psychological constructs. The idea here is that the metaphysical, ontological, and metatheoretical architecture given by UTOK should be able to assimilate and integrate key ideas and key findings about major constructs and coherently organize them to generate a descriptive picture that can be theoretically framed. This assimilative integrative method was first applied to the concept of depression, and demonstrated that, seen through the Unified Theory's conceptual architecture, depression can be defined as a state of mental behavioral shutdown. The assimilative integrative method has also successfully been applied to the concepts of depression, well-being, psychological mindfulness, borderline personality disorder, character functioning, psychological assessment, neurotic tendencies, intimate partner violence, and dreams. As the compendium of psychological constructs that can be coherently assimilated and integrated grows, the landscape of the field takes shape, and the viability of UTOK's mental behaviorism grows.

Another avenue for helping the field grow into a more coherent, mature discipline is found in the bridge between psychological science and psychotherapy. As the 2022 President of the Society for the Exploration of Psychotherapy Integration, my theme was *Toward a Common Core of Psychotherapy: The Problems, Mechanisms, Processes and Principles at the Center of Psychotherapy.* My presidential address (Henriques, 2022) showed how the Unified Theory provides the conceptual framework for bridging psychological science with the major schools of thought in psychotherapy. As noted in Chap. 4, delineating how the Unified Theory gives rise to a unified approach to psychotherapy requires a separate book. But the point is that the linkages here are broad and deep and the proper location of UTOK is as a metapsychology that bridges science with humanistic living.

Perhaps the fastest, most effective way to advance a mental behavioral approach would be to develop an introductory psychology textbook grounded in UTOK's metapsychology. Such a textbook would not foreground UTOK's technical arguments, but instead would move to capitalize on students' intuitive knowledge of the world and of folk psychology to help them see that there is a coherent domain of scientific inquiry that can be framed as the mental and from there deconstruct their folk understanding and replace it in terms of JII dynamics. What follows is a possible introduction to such a text.

Psychology as the Science of Mind

This textbook adopts a new approach to psychological science. Most textbooks on psychology focus on how the methods of science can be applied to understand behavior and mental processes. In such frames, behavior is defined as what scientists can see or measure and mental processes are inferred based on the results from research. The approach here is different. It is focused on helping you understand the field's subject matter, the domain of the mental. Yes, we will be relying on science and the understanding we will give you is most definitely a scientific picture. But rather than understanding psychology as a field that takes a scientific approach to behavior and mental processes, we want to help you understand exactly what mental behavioral processes are and how to understand them, and learn how they operate in the world.

Although it may be surprising to you, psychology has not been able to take this approach in the past because there was no agreed upon way to coherently define psychology's subject matter. As such, many different schools of thought, like behaviorism, cognitivism, humanism, and psychodynamic frameworks, would be shared, and students would be told that the key to psychology is that it is a science, meaning that it uses the methods of science to ask and answer questions about behavior and mental processes. Thankfully, recent developments in theoretical psychology have helped us overcome this difficulty, so we can now put the puzzle pieces together and see the whole. And that means we can give new students of psychology a much more coherent view of the field's subject matter.

A substantial reason for the advance is that there was a shift in how psychologists frame the relationship between mind and behavior. To help you understand this, we will need to teach you a bit about how to think about these concepts. To start, we define psychology as the science of Mind. Mind is capitalized here because it is a new word, and we need to help you get clear about what it means. We can start by noting that there is one aspect of Mind that refers to something you have intimate experience with, which is your own personal consciousness. To be clear about this, let me share with you what happened when I woke up this morning. For me, the buzzing sound of my alarm jolted my senses, and I slowly emerged from a cloud of darkness. The darkness had not been just external but was internal as well. That is, as I slept, my conscious awareness basically went offline, and time passed for me with little or no awareness. Likewise, although my body regularly shifted positions, I had no awareness of that either. I can recall a brief period in which I was roused and glanced at the clock—3:38 am—after which I rolled over and was again unconscious. I also had vague recollections of my consciousness flickering on and off with strange dreams, the images of which I can now only dimly recall.

As my senses became more fully aroused, there was a strong sense of continuity with the "me" that went to bed. I then got out of bed, brushed my teeth, chatted some with my children, and then sat down to work on my email and write my book. Unlike my dreams, these actions would have been readily available to other people in that they could be filmed and analyzed. As I obtained a cup of coffee, I proceeded to pet my dog Benji, and then fed the fish in my tank. Benji's tail wagged as he nudged up against me. And my fish swam to the top of the tank in anticipation as soon as the light came on. It is obvious that Benji and the fish, and even the housefly flying around my head, act very differently in the world than other organisms like cells or plants. This brings us to a key point. Animals like flies, fish, and dogs have brains and complex active bodies and they operate very differently in the world than organisms or inanimate objects. Indeed, we can say that they operate on a different complex adaptive plane. We can call this plane of existence Mind. And we say that creatures that exhibit it are "minded".

Animal mental behaviors, or the behavior of minded animals, are all around you. For example, if you go for a walk in the woods, you will see the squirrels, birds, and bees acting very differently than the trees, mushrooms, and flowers. The trees, mushrooms, and flowers are behaving as living organisms. The rocks and streams are behaving as inanimate entities. In contrast, the squirrels, birds, and bees are behaving as minded animals. The actions of these creatures make up what we call Mind. More specifically, Mind refers to the set of mental behaviors of minded creatures. A minded creature is a creature with a complex, active, segmented body, and a brain that functions as a centralized control center that coordinates the animal's movements as a whole. Indeed, the brain can be thought of as the organ of mental behavior. And just as physicists study the behaviors of material entities, and biologists study the behavior of living entities, psychological scientists study the behavior of minded animals. These creatures include houseflies, fish, dogs, and people. This framing of Mind as having the property of mindedness is a new framing that helps us solve many of psychology's old problems with definitions. It is important to know that our society has traditionally not viewed the mind in this way. Rather, our society has tended to view the mind more along the lines of the inner world of thoughts and feelings, or in terms of higher thought processes like reason or self-reflective awareness.

The divide between the seemingly outer world of behaviors and the inner world of thoughts and feelings has been a central issue for most approaches to scientific psychology. In fact, the traditional academic approach in psychology has been to split the field's subject matter into "behavior," which supposedly can be seen from the outside, and "mental processes" that are presumably unobservable or can only be seen from within. By focusing on the concept of mental behavior and mindedness, we take a different approach. Fortunately, it gives rise to new and more coherent ways to put our inner and outer worlds together.

To see how it achieves this, we can start by dividing the world of mental behaviors into three broad categories, called Mind^{1}, Mind^{2}, and Mind^{3}. As we will see, Mind^{1} and Mind^{3} each has two subsets, so the result is a map that divides mental processes up into five different categories. The first domain of mental behavior is the overt activity that can be observed from the exterior and filmed with a camera. We can call this the domain of Mind^{1b}, and most simply it can be framed as what the animal or person can be seen to be doing from an exterior point of view. Using the methods of science, psychologists can see how animals operate on the environment and how the effects of their actions shape their future behavioral investments. The second domain refers to the information inside the nervous system that is being processed as the animal acts. This is not as easy to observe directly, but we can see that the central nervous system is a hierarchically arranged input-output system that plays a key role in coordinating animal activity. These neurocognitive processes inside the nervous system are the domain of Mind^{1a}. Together, the overt activity and neurocognitive processes make up the broader domain of Mind^{1}.

If we return to my description of my morning activities, many Mind^{1a} neurocognitive processes were operative as I slept, but my subjective conscious experience of being went offline. This highlights that there is a difference between the nonconscious mental activity and my subjective conscious experience of being. We can call this Mind^{2}. The domain of Mind^{2} is special because, unlike my overt actions, or my neurocognitive activity, it can only be seen directly from the inside. That is, there is no camera that can allow us to see the domain of Mind^{2}. This fact of Mind^{2} is important, especially from our vantage point as scientists. Science is a general system of knowing that is based on observing the patterns in the world from an objective exterior position (i.e., stuff that can be filmed with a camera). This makes Mind^{2} difficult to study from a scientific perspective. However, as we will see, it is not impossible. For example, our scientific knowledge of Mind^{2} has advanced to the point where there is a high degree of confidence that my dog Benji has Mind^{2} experiences. However, there remains much that we do not know about it. For example, there is much debate about whether animals like fish and houseflies have Mind^{2} experiences. Part of the reason for this confusion is that scientists still do not fully understand when and how exactly Mind^{1} gives rise to Mind^{2}.

When we consider humans, there is yet another major domain of mental processes, which we are communicating through right now. We can call this Mind^{3}. The domain of Mind^{3} includes propositional language, explicit self-conscious reflection, and the dynamic dialogue that takes place between people as they try to justify their actions and beliefs on the social stage. As with Mind^{1}, we can divide Mind^{3} processes into those that take place within the individual, as when someone engages in private speech, and those that take place between people. We can call the former Mind^{3a} (i.e., private) and the latter Mind^{3b} (i.e., public). Crucially, Mind^{3} is associated with a whole new plane of existence. That is, just as the Mind plane emerged out of Life, the Culture plane of human persons emerged out of the Mind plane of minded animals.

This brief analysis of the domains of Mind raises several important points. First, it shows that the concept of mental processes refers to several different kinds of things, and so we need a new vocabulary to get clear about the things in the world we are talking about. This is one of the features of the mental behavioral approach to psychology we are taking. The second point it makes pertains to human psychology and the nature of human mental behavior. Specifically, the above analyses highlight how animal mental behavior can be framed as Mind^{1a} and Mind^{1b} and extends into Mind^{2} for many animals. In contrast, Mind^{3} is the domain of humans. This means that humans have an entirely new set of mental processes (i.e., the domain of Mind^{3}) that can be thought of as residing on top of the animal mental processes (i.e., domains of Mind^{1} and Mind^{2}). This gives rise to a key feature of human mental behavior, which can be stated in the form of a fact: Mentally, human beings are both primates and persons. More specifically, human beings are primates that are socialized to become human persons. Understanding both our animal mental behavior patterns as the domains of Mind^{1} and Mind^{2} and the processes by which we learn to operate on the Culture-Person plane of existence and how that relates to the domain of Mind^{3} is central to achieving a comprehensive picture of who and what we are and how we came to be.

Mapping the Institution of Psychology to the Domains of Mind

In addition to being concerned with how to define psychology's subject matter, the mental behavioral approach is simultaneously concerned with how to define the institution of psychology. The reason is that we need a close and coherent correspondence between the subject matter and the way the institution is defined. The fact that there is a qualitative jump when we shift from animal mental behavior into human mental behavior has important implications for how we organize and define the institution of Psychology. Specifically, the difference between Mind^3 from Mind^1 and Mind^2 means that we need to divide the field of scientific psychology accordingly. Thus, the science of psychology is divided into "basic psychology," which is the science of animal mental behavior, and "human psychology," which is the science of human mental behavior at the individual and small group level.

We need to note that in addition to these two branches of scientific psychology, there is a third branch of the field. This is the domain of professional psychology, and it is the domain that many think about when they imagine a psychologist. Whereas human psychology is tasked with scientifically describing and explaining human mental behavior, professional psychology is tasked with reducing psychological suffering and enhancing human well-being. Figure 16.1 provides a "Map of Psychology" that depicts how the institution can be effectively divided into these three great branches.

Basic psychology rests on biology and the neurosciences. It consists of fields such as behavioral science, the (basic) cognitive (or computational) neurosciences, and psychophysics. It overlaps with ethology and sociobiology, as these are fields that scientifically study animal mental behavioral patterns in the world. Human psychology consists of the science of human mental behavior at the individual and small group levels of analysis. This includes human cognitive science, human development, personality, and social psychology, as well as human psychology that grounds itself first in the cultural context. Human psychology resides at the base of the core social science disciplines. That is, it bridges into anthropology and then sociology, economics, and political science.

The professional practice of psychology is a different discipline because it has a fundamentally different goal structure. Unlike scientists, who are tasked with describing and explaining how the world works, professional psychologists are trained to generate change in the world. That is, their justification for doing what they do is to reduce psychological suffering and enhance psychological well-being. Formally designated by the APA as "health service psychology," this branch of the discipline consists of an amalgamation of clinical, counseling, and school psychology. These are the licensed psychological doctors who are formally trained to assess and treat psychological disorders. As suggested by the circle in the middle, the

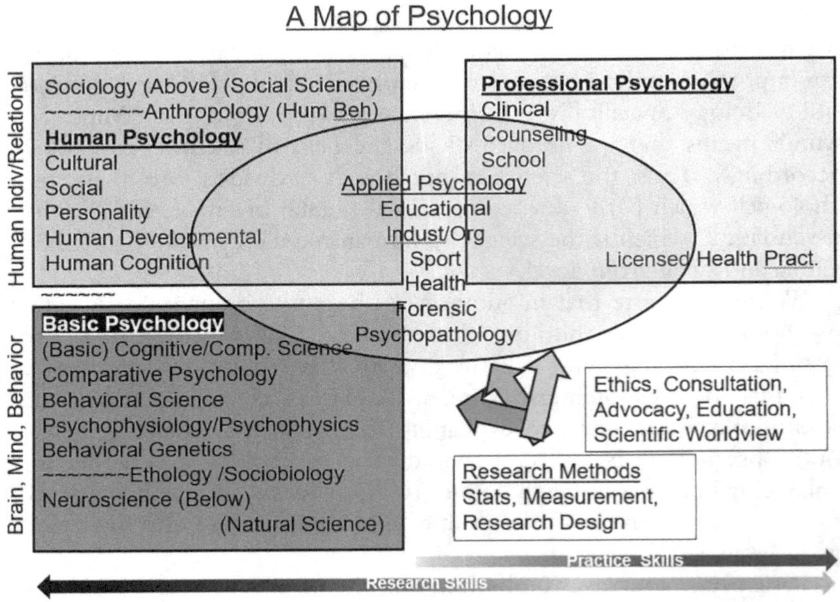

Fig. 16.1 A map of the institution of Psychology grounded in UTOK's metapsychology

line between understanding something scientifically and developing techniques for improving the world is not a sharp one but is more of a continuum. The applied psychological sciences of sport, educational, industrial organizational, abnormal, and forensic psychology reside between the basic descriptive-explanatory sciences and the profession. As such, they are placed accordingly on the map. Finally, all psychologists are committed in some ways to the methods and findings of science, as well as to humanistic ethics.

We share this because it is important to understand the relationship between the institution of Psychology and the subject matter of psychology. The approach we take focuses on developing a coherent relationship between the two. That is, we want to help you understand what psychology refers to in the world (i.e., the science of mental behavioral processes, defined by the domains of Mind[1,2,3]) and we want to help you see how that relates to the way the institution of Psychology should be organized (i.e., divided into basic, human, and professional branches). Finally, there is one additional point we need to make about our terms and that is the relationship between the science of psychology's focus on mental behavior and your particular psyche.

The Science of Psychology and Your Particular Psyche

The term psychology has its roots back in Aristotle's term for the psyche or soul. Because of the associations many make between the soul and supernatural forms of existence, the term psyche is rarely used in scientific psychology. We, however, think the psyche is a very important concept for psychology, even though it is not a traditional scientific concept. Here, the psyche refers to the specific, unique, subjective first-person experience of being in the world from the vantage point of that particular person. Framed by the domain of Mind[2], it can only be experienced by that person, and one can never directly observe another's psyche.

The reason that the psyche, as defined here, is not a scientific concept has to do with the rules of science and how science functions as a system of knowledge. As a system of knowledge, science is based on mapping behaviors based on generalizable, lawful, natural processes that can be observed via a systematic third-person empirical data gathering and experimental method. In contrast, the psyche consists of instances of experiential awareness that are unique, qualitative, and particular. Thus, scientific psychology is about understanding the general patterns and causes of animal and human mental behavior, whereas the psyche is your real experience of being as a unique particular individual in the world. Traditionally, modern science has, in many ways, tried to develop a worldview that essentially eliminates the psyche. The current framework for scientific psychology suggests this is a mistake. Rather, we need to understand that science is a particular kind of justification system that plays by particular rules. It is remarkably useful for understanding how general causal forces give rise to the behavior patterns we see in the world. However, it does not include everything, and what is needed is a way to coherently place science in relation to subjective knowing. Framing the psyche as your unique particular experience of the real world from the inside and synthesizing that with scientific psychology's frame for mental behavior viewed from the outside provides us with a coherent picture of the whole.

By the time you are finished this book, you will have access to a whole new language for understanding the mind and Mind, as well as psychology and Psychology. It will teach you about what science is in general and why psychology is the science of mental behavior. You will then learn how to think clearly and coherently about cognition, consciousness, and the self. You will be able to place human mental behavior in a grand vision of the cosmos and understand how we emerged from simple inanimate entities like molecules into single-celled organisms into simple animals into mammals and primates and finally human persons. You will be shown the key insights from each of the major historical schools of thought and be told how each offers a set of partial truths that can be woven together into a coherent whole. And you will be able to relate what science tells us about the general, behavioral state of the world from a third-person view to the unique perspective of your psyche. By joining the science of psychology with the unique perspectives of our psyches, we can achieve a more coherent and holistic view of the cosmos and our place in it.

Conclusion

Although this book has been an academic work focused on an academic problem, what is ultimately at stake here is of far greater importance than a shift inside academic psychology from methodological behaviorism to mental behaviorism. Rather, what is at stake is how we, humanity writ large, frame our scientific knowledge and our natures and our place in the cosmos. Put bluntly, humanity needs to wake up to the situation we find ourselves in. It is a kairotic "time between worlds" (Stein, 2019) that requires us to make a fundamental shift in how we make sense of ourselves and our modes of being.

The Enlightenment shined a great light of understanding on the material and living worlds. It also generated new ideas for governance and markets, and we have seen an explosion of technological advances in the last 300 years. Unfortunately, although we have achieved much progress on many fronts, there have been serious problems that have been accelerating in magnitude. Our monumental impact on nature has resulted in us shifting into a new geological epoch, the Anthropocene, which is defined by the changes that our institutions and industries have brought to the planet's ecological systems. The world of the digital is opening the potential for a new complex adaptive plane of existence, the likes of which we have never seen. And as has been painfully highlighted by the Covid-19 pandemic, the world's enormous, stacked interdependencies mean that system-wide crashes and changes can ripple through the entire system, making it potentially fragile and vulnerable to large-scale collapse.

In addition to these natural and technological challenges, we are facing crises of meaning and mental health. We are overloaded with information, bullshit is ubiquitous, our collective mental health is deteriorating, especially in our youth. Moreover, the state of our knowledge is a chaotic, fragmented pluralism. The Unified Theory provides a clear diagnosis of the epicenter of the problem within our academic knowledge systems. The Enlightenment produced a conceptual gap between matter and mind and failed to generate a coherent philosophical system for understanding scientific knowledge relative to social and subjective knowledge. The evidence for this is readily apparent and found in the convoluted state of the philosophy of mind, the problem of psychology, and the confusion and disarray observed between modernist and postmodernist sensibilities.

The Unified Theory offers a new vision forward. The ToK System extended into the PTB provides a new map of Big History that properly delineates the levels and dimensions of behavior in nature and how they align with the various domains of science. This map enables us to see much more clearly the ontology of the mental as a class of minded behavioral patterns in nature. Our conception of mental processes is substantively enhanced by the Map of Mind^{1,2,3}, and its capacity to differentiate epistemological vantage points and the ontological referents of overt and covert neurocognitive behavior, subjective conscious experience, and private and public self-conscious reflection via processes of justification. When coupled to the ToK System's planes of existence, we can now clarify the nature of Mind^{3} and the ontology of human persons. In other words, by carving nature at its joints from Energy Information to Material Objects to Living Organisms to Minded Animals to Cultured Persons UTOK finally allows us the proper metaphysical description of the world and our place in it.

Of course, the effective description of the key domains in the landscape of knowledge is only part of the problem. The Unified Theory advances the ball further by weaving together these descriptive taxonomies with metatheoretical formulations that afford dynamic causal explanatory frameworks. BIT and the Matrix, especially when joined with Vervaeke's metatheory of cognition, clarify the emergence of the mental behaviors that constitute the Mind plane of existence. It frames the function of the mind as a predictive, neuro-information processing system that engages in recursive relevance realization to carve out paths of investment in the agent arena environment. Such processes are complexified and enriched as the relational environment becomes increasingly complicated, fluid, and participatory. And via JUST we obtain a much clearer frame for the evolution of human consciousness and Culture and clarity about the ontology of human persons. JUST also enables us to see the evolution of justification systems that give rise to science, and the ToK System effectively maps the relationship between the ontic reality and scientific maps of that terrain. Ultimately, UTOK affords us the opportunity to transform from the current chaotic fragmented pluralist state of our knowledge into a coherent integrative pluralism that can both include and transcend the modern and postmodern sensibilities and launch us into a new intellectual era of Enlightenment focused on unifying knowledge and orienting toward wisdom.

References

Gallagher, S. (2017). *Enactivist interventions: Rethinking the mind.* Oxford University Press.

Henriques, G. R. (2022, April). Toward a common core: The problems, mechanisms, processes, and principles at the center of psychotherapy. Presidential Address given at the 38th Annual Society for the Exploration of Psychotherapy Integration Conference (online).

Henriques, G. R., & Görtz, D. (2020, April). *What is metamodernism?* [Blog] Theory of Knowledge on Psychology Today. https://www.psychologytoday.com/us/blog/theory-knowledge/202004/what-is-metamodernism

Skinner, B. F. (1987). Whatever happened to the science of behavior? *American Psychologist, 42*(8), 780–786. https://doi.org/10.1037/0003-066X.42.8.780

Slife, B. D., & Williams, R. N. (1997). Toward a theoretical psychology: Should a subdiscipline be formally recognized? *American Psychologist, 52*(2), 117–129. https://doi.org/10.1037/0003-066X.52.2.117

Stein, Z. (2019). *Education in a time between worlds: Essays on the future of schools, technology, and society.* Bright Alliance.

Wilson, E. O. (1999). *Consilience: The unity of knowledge.* Vintage Books.

Wittgenstein. (1949). *Remarks on the philosophy of psychology.* University of Chicago Press.)

Glossary of Terms for a New Synthesis for Solving the Problem of Psychology

Architecture of the human mind A diagram that maps the layers of neuro-information processing that have evolved from reacting to learning to thinking to talking and combines these layers with modern models of memory to provide a gestalt for how neurocognitive information is instantiated within and flows through the human cognitive system.

Attentional filter Refers to how top-down focus influences what is brought on to the stage of subjective conscious experience, as well as how bottom-up signals compete for attentional resources.

Behavior Change in entity–field relations. It is the central concept in modern natural science, in that science can be defined as the systematic analysis of the behavior of entities in nature. The Tree of Knowledge (ToK) System posits that different behaviors take place on different planes of existence, such that the behavior of inanimate objects takes place on the Matter plane, the behavior of living organisms on the Life plane, the behavior of minded animals on the Mind plane, and the behavior of self-conscious persons on the Culture plane.

Behavior Investment Theory The "Life-to-Mind" joint point on the Tree of Knowledge System. It is a metatheoretical formulation that frames the evolution of Mind via the positing that the nervous system functions as an investment value system that coordinates animal actions on a cost to benefit ratio on the principles of (1) energy economics; (2)

G. Henriques, *A New Synthesis for Solving the Problem of Psychology*, Palgrave Studies in the Theory and History of Psychology,
https://doi.org/10.1007/978-3-031-18493-2

evolution; (3) behavioral genetics; (4) neuro-computational control; (5) learning and environmental feedback; and (6) developmental stage/life history. It bridges evolutionary biology, ethology, cybernetics, complex adaptive systems, traditional cognitive neuroscience, 4E cognitive science, and behavioral science. In combination with the ToK System's descriptive metaphysics it resolves the mentalist versus behaviorist division via framing animal activity in terms of mental behavior or mindedness.

Big History The interdisciplinary movement launched by David Christian that functions to map our knowledge of the cosmos and our place in it on the dimensions of time (from the Big Bang to the present) and complexity (from subatomic particles to societies). The ToK System provides a new map of Big History that includes the distinction between levels and dimensions of complexification and maps the Mind dimension as different from Life and from human culture.

BM^3 problem Another term for the mind-body problem, it refers to the need for a descriptive metaphysical system that can place behavior, neurocognition, subjective conscious experience of being, and self-conscious justification in proper relation. BM^3 stands for behavior-mind-mind-mind, although the "b" can also reference the physical body, the living body, and the brain.

CALM-MO Refers to the fourth key idea in the Unified Approach to psychotherapy. An integrative approach to psychological mindfulness, it is an acronym that stands for developing a Meta-cognitive Observer as one's modus operandi (the M. O.) that is Curious, Accepting, Loving-compassionate and Motivated toward valued states of being in the short and long term.

Character Adaptation Systems Theory (CAST) Refers to the first key idea in the Unified Approach to psychotherapy. It maps the processes of human adaptation via three contexts (i.e., biophysiological, learning and developmental, and sociocultural) and five systems of character adaptation (habit, experiential, relational, defensive, and justification).

Cognition Refers to the information processing that leads to functional effects. Neurocognition refers to the information instantiated within and processed by the nervous system and corresponds to the domain of $Mind^{1a}$ on the Map of $Mind^{1,2,3}$.

Consciousness Refers most broadly to functional awareness and responsivity; however, the more narrow and precise definition refers to one's subjective experience of being in the world. This corresponds to the domain of $Mind^2$ on the Map of $Mind^{1,2,3}$.

Culture Refers to the large-scale system of justification that coordinates the actions of people and legitimizes sanctions and rewards. It is also the fourth dimension of behavioral complexification on the Tree of Knowledge System, often referred to as the Culture-Person plane of existence. It should be differentiated from small "c" culture, which refers to the learned and shared behavioral investment repertoires, as such processes present in many animals.

Dimensions of existence or complexification Refers to the four dimensions mapped by the Tree of Knowledge System, which are: Matter-Object, Life-Organism, Mind-Animal, and Culture-Person. The latter three dimensions emerge as a function of novel information processing and communication networks (i.e., genes-cells, neurons-animals, language-people) that give rise to new complex adaptive planes.

Double hermeneutic A term drawn from Anthony Giddens' work that refers to the way in which knowledge in the social sciences is different from knowledge in the natural sciences because the justifications that social scientists develop circle back into the Culture-Person plane of existence and influence how humans think about themselves.

Ego Refers to the private self-conscious system on the Updated Tripartite Model of human consciousness that can verbally reflect on one's own mental experience and report on that experience. In the Unified Theory of Knowledge (UTOK), the ego is framed by Justification Systems Theory as the mental organ of justification that functions as a propositional interpreter of the world and justifier of one's actions, thoughts and feelings.

Empirical (and empiricism) Refers to making observations or bringing data in via the senses. It is crucial to differentiate first-person empirical phenomena from the third-person systematic empiricism of modern natural science. The latter refers to developing measurements or systematic observational procedures that are independent of any specific subjective knower, whereas first-person empiricism is wholly dependent on the subjective conscious experience of the individual.

Energy Information Implicate Order The substance or "implicate order" that resides beneath the Matter dimension of complexification, framed by the state of the universe at the Big Bang and analyzed in quantum field theory.

Enlightenment Gap Refers to the joint problem that emerged in the wake of the scientific Enlightenment of placing mind in proper relation to matter (i.e., the mind-body problem) and scientific knowledge in relationship to social and subjective knowledge systems. The downstream consequence of this gap in our knowledge gives rise to the problem of psychology and a chaotic, fragmented pluralistic state of knowledge more generally.

Epistemological gap Refers to the inability to directly observe another sentient being's subjective conscious experience of being in the world. It can also be framed as the gap between first-person and third-personempiricism, whereby the latter is grounded in data that can be observed by any trained observer.

Epistemological portal Refers to the unique portal of perspectival knowing that subjective conscious experience provides each sentient creature. It also corresponds to the epistemic processes that enable a creature to have a Mind[2].

Epistemology (and epistemic) In philosophy it refers to the systematic process of analyzing how we know and what is justifiable. The term epistemic refers to the knowing process, and can refer to the ways creatures like plants, animals, or humans take in information to make predictions about how to act.

Experiential (or primate or core) self In the Updated Tripartite Model, it refers to the felt sense of being, organized by perceptions, motivations, and emotions. The core of the experiential self tracks self-relevant information, especially social information as it pertains to one's place in the social influence matrix.

Freudian filter Refers to the space between the experiential self and the private self-consciousness system or ego in the Updated Tripartite Model. It functions to track potentially anxiety-provoking images or impulses that are unjustifiable and then works to shape them or redirect them via processes like repression and rationalization so that the conscious egoic state remains in justified equilibrium.

Influence Matrix It maps the human relationship system on the four self–other process dimensions of (1) relational value and social influence; (2) power or dominance-submission; (3) love or affiliation-hostility; and (4) freedom or autonomy-dependency. It also identifies common social emotions as being activated in response to change in these dimensions.

Informational Interface Refers to the process by which different systems of information processing interface and interact with each other. Human beings are understood via the Tree of Knowledge System to be nested systems of informational interface at the cell-living-organism, neuro-animal-mental, and linguistic-person-cultural levels and dimensions of analysis, which are mapped further by the Periodic Table of Behavior.

JII dynamics Refers to the complex interplay between justification, behavioral investment, and influence in human mental behavior. It can be used to analyze human discourse and folk psychology by framing the investment, influence, and justification processes via Behavioral Investment Theory, the Influence Matrix, and Justification Systems Theory.

Joint points Refer to the links between the planes of existence. They provide the causal explanatory frameworks for the relation between the domains. The Inflationary Big Bang coupled with quantum field theory provides the joint point between the Energy-Information Implicate Order and the Matter-Object dimension, the modern evolutionary synthesis coupled to cell theory provides the joint point between the Matter-Object and Life-Organism planes, Behavioral Investment Theory is the Life-to-Mind joint point, and Justification Systems Theory is the Mind-to-Culture joint point.

Justification Refers first to conventional usage relating to legitimizing claims and actions, but it expands in UTOK to characterize virtually all propositional statements, each of which can be thought of as a statement of justification that does or does not carry legitimacy. Justifications carry both truth and value claims that can be potentially challenged in the social context of justification that they are offered. In UTOK, justification is seen as both the central driving selective force on the Culture-Person plane of existence and as a central concept in epistemology, where justification is the relationship between belief and the truth. It also has a "strange loop" aspect, as in: this definition justifies my definition of justification, which gives the term a fluid and dynamic meaning.

Justification Hypothesis Refers to a key aspect of Justification Systems Theory, which is the claim that there was a tipping point in the evolution of language, such that the emergence of propositions was closely tied to the emergence of questions, and together they generated a complexity building feedback loop that led both to the evolution of the human ego as the mental organ of justification and to the emergence of the Culture-Person plane of existence.

Justification Systems Theory The metatheoretical joint point of the Mind-to-Culture transition on the ToK System. It connects the Justification Hypothesis with the Updated Tripartite Model of Human Consciousness with the evolution of the Culture-Person plane of existence.

Life When capitalized, it refers to the second dimension of behavioral complexification on the ToK System, also called the Life-Organism plane of existence. It is a novel complex adaptive network of behavioral patterns that emerge as a function of novel information processing and communication networks found in the genetic material and the complex adaptive structure of cells. The biological sciences work to describe and explain the behavior of living organisms.

Map of Mind[1,2,3] A diagram that provides a descriptive metaphysical system that divides the domains of the mental into three broad domains (i.e., neurocognitive/mental behavioral activity, subjective conscious experience, self-conscious justification) based on their ontological referents and epistemological vantage points.

Matter When capitalized, Matter refers to the first dimension of behavioral complexification on the ToK System. It is also referred to as the Matter-Object plane of existence. It is framed as emerging out of the Energy-Information Implicate Order at the Big Bang. The physical sciences (i.e., physics, chemistry, geology, astronomy) work to describe and explain the behavior of material objects.

Mental behavior Refers to the functional awareness and responsivity of animals with brains and complex active bodies as they act on the environment, either directly or via simulations. Mental is an adjective that refers to that which makes animal behavior so different from the behaviors of other living creatures. The set of mental behaviors makes up the Mind-Animal plane of existence.

Mental behaviorism Refers to the approach to scientific psychology that emphasizes the need for a naturalistic ontology of the mental that is clearly specified, which is in contrast to the mainstream approach grounded in methodological behaviorism. Mental behaviorism claims that the mental behavior of animals and humans is the proper subject matter of scientific psychology.

Metamodernism A post-postmodern sensibility that seeks to resolve or transcend the tensions between modernism and its focus on sincerity, reason, and progress, and postmodernism's emphasis on irony and deconstructing the social construction of knowledge. The combination of JUST and the ToK System provides a systemic metaphysics for the ontology of human knowledge that advances a metamodern sensibility by synthesizing the modernist thesis of truth and reason with the postmodern antithesis that knowledge is socially constructed.

Metaphysics Refers to the concepts and categories that are used to generate a worldview. In UTOK, an important distinction is made between pure metaphysics, which are metaphysical claims that have no anchor in empirically observable entities, and descriptive or systematic metaphysics, which refers to metaphysical systems that describe entities in the world and their interrelations. The Tree of Knowledge System is a descriptive metaphysical system for the natural sciences.

Methodological behaviorism The technical term for the current structure of mainstream academic psychology that defines the field as the science of behavior and mental processes, such that behavior is that which is accessible to the methods of science and mental processes are the inferred causes that explain the relations between independent and dependent variables as framed by scientific experiments.

Mind (and the mind and mindedness) When capitalized, Mind refers to the third dimension of behavioral complexification on the Tree of Knowledge. It is often called the Mind-Animal plane of existence and corresponds to the behavior of animals, specifically the sensory-motor looping that generates functional awareness and responsivity in a way that is different from other living organisms. In contrast, "the mind" refers to the information instantiated within and processed by the nervous system. Finally, mindedness refers to the functional property of existing at the Mind dimension.

Mind^1 The mental behavioral activity that differentiates the behavior of animals from the behavior of other organisms. It can be divided into the domain of neurocognition within the nervous system (labeled the domain of Mind^{1a}) and the functional activity of the animal engaged with the environment (labeled the domain of Mind^{1b}).

Mind^2 The mental domain involving the subjective conscious experience of being in the world, which is only available via the first-person point of view. The current work argues that Mind^2 evolved first as valence qualia (i.e., pleasure and pain) and then emerged as a more sophisticated global workspace that represents the self extended across time and, in social animals, in the relational field. Although Mind^2 is largely self-contained inside the epistemological portal, it also is reasonable to argue that people develop implicit intersubjective relational models and modes of being with each other that give rise to the possibility of a domain that could be labeled Mind^{2b}.

Mind^3 The domain of the mental involving self-conscious reflection and justification of one's actions on the social stage. As framed by

Justification Systems Theory, it is present only in humans and is the process by which humans are socialized to participate on the Culture-Person plane of existence. It can be divided into the private narrating domain of the ego (Mind^{3a}) and what is shared publicly with others, the domain of which corresponds to the persona (Mind^{3b}).

Nested Model of Well-Being The third key idea that frames the Unified Approach to psychotherapy. It provides a descriptive metaphysical systemic analysis of the concept of human well-being that divides the construct into (1) Subjective Conscious Experience; (2) Health and Functioning at the psychological and biological levels of analysis; (3) Environment (both material and social); and (4) Values and ideology of the evaluator.

Ontology (and the ontic) Refers to thoughts or theories or beliefs about what is real or what are the things, events, or processes in the world. Scientific ontology refers to the maps of reality that science develops, such as atomic theory and the Periodic Table of the Elements. The ontic refers to reality as it exists independently of human beliefs, although the two concepts are deeply interconnected in that as soon as one makes any claims about the nature of the ontic, such claims will be ontological in nature.

P – M => E formulation The control learning theory formulation that arises out of Behavioral Investment Theory and the bridge between operant conditioning and living control systems. It refers to the process by which perceptions are referenced against motivations, which in turn activate emotions. This sensory-perceptual-motivational-emotional-motor loop can frame how animals are oriented to salient variables and shift their actions based on consequences. It also provides a conceptual bridge between Mind^{1} neurocognitive processes and Mind^{2} subjective conscious experiences.

Periodic Table of Behavior (PTB) Provides a new taxonomy for categorizing natural behavioral patterns mapped by science. It extends the ToK System's four dimensions of existence to include the idea that there are primary units of analysis on each dimension (i.e., atoms, cells, animals, and persons), and that there are parts that make up these units and these units also form groups across aggregate scales. The result is three levels of analysis by four dimensions of existence that give rise to 12 floors that correspond closely to the domains of natural-into-social scientific inquiry.

Person Refers to an entity that has the capacity to self-consciously reflect on one's actions and justify them on the social stage.

Persona Refers to the public self on the Updated Tripartite Model of Human Consciousness. It refers to what is publicly shared with others. According to UTOK, humans generally strive to maintain a justifiable image in the eyes of important others, as a function of the motive for social influence and relational value.

Problem of Psychology Refers to the inability of psychological science to clearly specify the basic ontological referents for its subject matter. There simply is no clear, shared definition of the field, nor is there a clear, shared subject matter for the science, nor is there a general identity of the professional practitioner. The term can be framed as a reformulation of the famous crisis in psychology, to emphasize that the core of the problem is found in the inability of the science to effectively define what constitutes the domain of the mental. With its new map of behavior complexification framed by different planes of existence, the ToK System provides a new metaphysical picture that resolves the problem.

Psyche The individual human's unique particular experience of being in the world. The domain of the specific individual's Mind^2 as framed by the epistemological portal. It is a concept that is outside the general language system of science, because science is framed by the intersubjective analysis of behavior that can be quantified. The UTOK metapsychology provides a way to place the individual's unique particular psyche in relationship to the science of human psychology, defined as human mental behavior.

Psychology In the mainstream, the term refers both to the institution and the science of behavior and mental processes. The current work defines the science of psychology as the science of mental behavioral processes in animals and human persons. It divides the institution of psychology into three great branches of basic psychology (science of animal mental behavior), human psychology (science of human mental behavior), and the profession, which refers to the trained application of knowledge of human well-being toward the greater good.

Recursive relevance realization A metatheory of neurocognitive processes that emphasizes how such systems scan for relevant information that enables them to realize both what is the case and what affordances might be created, while recursively modeling the agent–arena relation across a multi-level information processing hierarchy.

Rogerian filter Refers to how individuals filter private thoughts and deliver them in different public spheres depending on the dynamics of social influence and the public image they are attempting to manage. It can also be described as the "private to public filter" and is placed between the ego and the persona on the Updated Tripartite Model of Human Consciousness.

RV–SI Line It is the central black line on the Influence Matrix and stands for "Relational Value–Social Influence." Relational value refers to the felt sense of being seen, known, and valued by important others. Social influence refers to the extent to which one can instrumentally influence others to act in accordance with one's interests. Both are central to how humans track their place and its affordances and dangers on the social stage.

Sandcastle problem The term for the way psychological science is generated, as it is largely dependent on the operational definitions of the researcher and not connected to a larger, shared map of ontology. The result is findings that are like sandcastles in that they may be fascinating and intricate, but they are not built on a solid foundation and are thus vulnerable to being washed away by the next generation of research with its new operational definitions.

Science A broad term that refers to systematic inquiry. Modern, empirical, natural science emerged in the context of the Enlightenment and is grounded in third-personempiricism, measurement, quantitative analysis, and systematic reasoning.

Self Because all organisms must model both themselves and the environment, there is a broad meaning of the self that pertains to all organisms. The more specific meaning pertains to the manner in which thinking animals project themselves across time and environments, such that the environments change but the self remains quite constant. This generates an internal working model of the agent that tracks what is relevant over time. The self extends significantly as the environment becomes increasingly dominated by relationships with important others. In humans, the self extends further, and can be effectively divided into the primate experiential self, the private narrating ego, and the public persona.

Tree of Knowledge (ToK) System The central framework in UTOK that provides a descriptive metaphysical system that corresponds the planes of existence (Energy, Matter, Life, Mind, and Culture) to the major domains in science (i.e., physical, biological, psychological, and social). A tremendous advantage of the visuospatial gestalt provided by the ToK is that it simultaneously defines extremely broad concepts (e.g., life, mind) and defines how they exist in relationship to one another in a single, coherent knowledge system. The system of interlocking definitions ultimately provides the potential framework for a shared "metaphysical empirical" system from which scientists from all disciplines can work. The ToK System addresses the Enlightenment Gap by providing a clear map for placing the various concepts of "mind" in relation to "matter" and by clarifying science as a unique kind of justification system that emerges out of the Culture-Person plane of existence. The ToK System provides the basic structure for resolving the problem of psychology by crisply defining the Mind-Animal dimension from the Life-Organism dimension from below and the Culture-Person dimension from above.

Unified Approach to psychotherapy An approach to individual psychotherapy that is grounded in the unified theory of psychology and framed by Character Adaptation Systems Theory, the Wheel of Development, the Nested Model of Well-being, and CALM-MO.

Unified Theory of Knowledge (UTOK) A new metapsychology that addresses the Enlightenment Gap and works to solve the problem of psychology and generate a unified approach to psychotherapy. It consists of the unified theory of psychology and the unified approach to psychotherapy, and other ideas that provide a consilient scientific humanistic worldview that can potentially revitalize the human soul and spirit in the twenty-first century. Thus, it functions to unify knowledge and orient toward wisdom.

Unified Theory of Psychology A metatheoretical synthesis of scientific psychology that is achieved via the interlocking network of ideas generated by the Tree of Knowledge System, Justification Systems Theory, Behavioral Investment Theory, and the Influence Matrix, which work together to assimilate and integrate the major findings from empirical psychology and the key insights from the major schools of thought (i.e., evolutionary psychology, cognitive and behavioral neurosciences, personality and social psychology, psychodynamic and humanistic approaches, and social systems).

Updated Tripartite Model of Human Consciousness A diagram that divides human consciousness into the primate experiential self, the ego (private narrator), and the persona (public self). It connects these domains to the three filters (attentional, Freudian, and Rogerian). In addition, it places such processes in the larger context of justification. The dynamic interrelations between the domains and the social context of justification can be mapped by Justification Systems Theory, Behavioral Investment Theory, and the Influence Matrix.

Index

G. Henriques, *A New Synthesis for Solving the Problem of Psychology*, Palgrave Studies in the Theory and History of Psychology, https://doi.org/10.1007/978-3-031-18493-2